Computational and Combinatorial Group Theory and Cryptography

CONTEMPORARY MATHEMATICS

582

Computational and Combinatorial Group Theory and Cryptography

AMS Special Sessions:
Computational Algebra, Groups, and Applications
April 30–May 1, 2011
University of Nevada, Las Vegas, NV
Mathematical Aspects of Cryptography and Cyber Security
September 10–11, 2011
Cornell University, Ithaca, NY

Benjamin Fine
Delaram Kahrobaei
Gerhard Rosenberger
Editors

American Mathematical Society
Providence, Rhode Island

EDITORIAL COMMITTEE
Dennis DeTurck, Managing Editor

Michael Loss   Kailash Misra   Martin J. Strauss

2000 *Mathematics Subject Classification.* Primary 20-XX, 68-XX.

Library of Congress Cataloging-in-Publication Data
Computational and combinatorial group theory and cryptography : AMS Special Session on Computational Algebra, Groups, and Applications, April 30–May 1, 2011, University of Nevada, Las Vegas, Nevada : AMS Special Session on Mathematical Aspects of Cryptography and Cyber Security, September 10–11, 2011, Cornell University, Ithaca, New York / Benjamin Fine, Delaram Kahrobaei, Gerhard Rosenberger, editors.
   p. cm. – (Contemporary mathematics ; volume 582)
   Includes bibliographical references.
   ISBN 978-0-8218-7563-6 (alk. paper)
   1. Group theory–Congresses. 2. Cryptography–Congresses. I. Fine, Benjamin (1948–), editor of compilation. II. Kahrobaei, Delaram (1975–), editor of compilation. III. Rosenberger, Gerhard, editor of compilation. IV. AMS Special Session on Computational Algebra, Groups, and Applications (2011 : University of Nevada). V. AMS Special Session on Mathematical Aspects of Cryptography and Cyber Security (2011 : Cornell University)

QA174.C645 2012
512'.2–dc23

2012023441

   Copying and reprinting. Material in this book may be reproduced by any means for educational and scientific purposes without fee or permission with the exception of reproduction by services that collect fees for delivery of documents and provided that the customary acknowledgment of the source is given. This consent does not extend to other kinds of copying for general distribution, for advertising or promotional purposes, or for resale. Requests for permission for commercial use of material should be addressed to the Acquisitions Department, American Mathematical Society, 201 Charles Street, Providence, Rhode Island 02904-2294, USA. Requests can also be made by e-mail to reprint-permission@ams.org.
   Excluded from these provisions is material in articles for which the author holds copyright. In such cases, requests for permission to use or reprint should be addressed directly to the author(s). (Copyright ownership is indicated in the notice in the lower right-hand corner of the first page of each article.)

© 2012 by the American Mathematical Society. All rights reserved.
The American Mathematical Society retains all rights
except those granted to the United States Government.
Copyright of individual articles may revert to the public domain 28 years
after publication. Contact the AMS for copyright status of individual articles.
Printed in the United States of America.

∞ The paper used in this book is acid-free and falls within the guidelines
established to ensure permanence and durability.
Visit the AMS home page at http://www.ams.org/

10 9 8 7 6 5 4 3 2 1   17 16 15 14 13 12

Contents

Preface	vii
Weyl Gröbner Basis Cryptosystems RASHID ALI AND MARTIN KREUZER	1
A New Look at Finitely Generated Metabelian Groups GILBERT BAUMSLAG, ROMAN MIKHAILOV, AND KENT E. ORR	21
IA-Automorphisms of Groups with Almost Constant Upper Central Series MARIANNA BONANOME, MARGARET H. DEAN, AND MARCOS ZYMAN	39
A Proposed Alternative to the Shamir Secret Sharing Scheme CHI SING CHUM, BENJAMIN FINE, GERHARD ROSENBERGER, AND XIAOWEN ZHANG	47
Improving Latin Square Based Secret Sharing Schemes CHI SING CHUM AND XIAOWEN ZHANG	51
A Hand-Computation Involving Surface Groups, the Reidemeister-Schreier Rewriting Process and Kurosh Subgroup Theorem ANTHONY E. CLEMENT	65
Adjunction of Roots in Exponential A-Groups MARGARET H. DEAN, STEPHEN MAJEWICZ, AND MARCOS ZYMAN	71
Logspace Computations in Coxeter Groups and Graph Groups VOLKER DIEKERT, JONATHAN KAUSCH, AND MARKUS LOHREY	77
Collection by Polynomials in Finite p-groups BETTINA EICK	95
All Finite Generalized Tetrahedron Groups II BENJAMIN FINE, ALEXANDER HULPKE, AND GERHARD ROSENBERGER	105
The Classification of One Relator Limit Groups and the Surface Group Conjecture BENJAMIN FINE AND GERHARD ROSENBERGER	107
Discrimination and Separation in the Metabelian Variety ANTHONY M. GAGLIONE, SEYMOUR LIPSCHUTZ, AND DENNIS SPELLMAN	129

A Secret Sharing Scheme Based on Group Presentations and the Word Problem
  MAGGIE HABEEB, DELARAM KAHROBAEI, AND VLADIMIR
  SHPILRAIN   143

Authenticated Key Agreement with Key Re-Use in the Short Authenticated
Strings Model
  STANISLAW JARECKI AND NITESH SAXENA   151

Publicly Verifiable Secret Sharing Using Non-Abelian Groups
  DELARAM KAHROBAEI AND ELIZABETH VIDAURRE   175

A Note on the Hyperbolicity of Strict Pride Groups
  MATTHIAS NEUMANN-BROSIG   181

An Algorithm to Express Words as a Product of Conjugates of Relators
  ELLEN ZILIAK   187

Preface

This volume consists of contributions by participants and speakers at special sessions at two AMS Conferences. These conferences concerned Computational and Combinatorial Group Theory as well as various aspects of Group-Based Cryptography. One of these sessions was at Cornell University in September of 2011 and the other was the University of Nevada, Las Vegas in May of 2011.

Over the past twenty years Combinatorial and Infinite Group Theory has been energized by three developments: the emergence of Geometric and Asymptotic Group Theory, the development of Algebraic Geometry over Groups leading to the solution of the Tarski problems, and the development of Group-Based Cryptography. These three areas in turn have had an impact on computational algebra and complexity theory. The papers in this volume, both survey and research, exhibit the tremendous vitality that is at the heart of Group Theory in the beginning of the twenty-first century as well as the diversity of interests in the field.

We are grateful to the American Mathematical Society for their help in publication of this volume. In particular, we thank Christine Thivierge for her patience and diligence in assembling this volume.

Bejamin Fine
Delaram Kahrobaei
Gerhard Rosenberger

Weyl Gröbner Basis Cryptosystems

Rashid Ali and Martin Kreuzer

ABSTRACT. The main purpose of this paper is to propose ways of constructing Gröbner basis cryptosystems based on left ideals or two-sided ideals in Weyl algebras. The main advantage of using Weyl algebras over commutative polynomial rings is that Weyl multiplication increases the support of the standard form of the polynomials, slows down the Gröbner basis computation, and changes many coefficients in the ciphertext polynomials. Special care is taken to provide concrete directions for constructing secure instances of Weyl Gröbner basis cryptosystems (WGBC). Their security and possible defences against chosen ciphertext attacks, linear algebra attacks and partial Gröbner basis attacks are analysed. Based on these guidelines, we construct several actual instances of left and two-sided WGBC of which we believe that they are reasonably secure and efficient.

1. Introduction

In 1994, M. Fellows and N. Koblitz introduced in [7] a new type of cryptosystem which became known as Polly Cracker cryptosystem and was based on (commutative) polynomials and the difficulty of computing Gröbner bases for polynomial ideals. Although these cryptosystems could, in principle, encode NP-hard problems, it turned out to be very difficult to construct hard instances. Many methods of attacking Polly Cracker cryptosystems were proposed and carried out successfully. To counter these attacks, it was suggested in [22] and [1] to use Gröbner bases for non-commutative algebraic structures. In this paper we construct specific instances of the general Gröbner basis cryptosystem defined in [1]. We use Gröbner bases for left and two-sided ideals in Weyl algebras. We analyse the resistance of these cryptosystems to some standard attacks and provide computational evidence that secure instances can be built using left and two-sided Gröbner bases in Weyl algebras.

Recall that the Weyl algebra A_n of index n over a field K is the associative algebra $A_n = K\langle x_1, \ldots, x_n, \partial_1, \ldots, \partial_n\rangle$ such that $[x_i, x_j] = [\partial_i, \partial_j] = 0$ and $[\partial_i, x_j] = \delta_{ij}$. For a variety of reasons, it appears beneficial to use a finite base field K. Since Weyl algebras in characteristic p have properties which differ from the well-known case $\mathrm{char}(K) = 0$, we start in Section 2 by recalling them (see

2010 *Mathematics Subject Classification.* Primary 94A60; Secondary 14G15, 16Z05, 68W30.
Key words and phrases. Algebraic cryptography, Groebner basis cryptosystem, Weyl algebra, non-commutative cryptosystem.
The first author was supported by the HEC-DAAD Fellowship Programme.

©2012 American Mathematical Society

Prop. 2.2). One of the main reasons for suggesting the use of Weyl algebras for cryptography is elucidated by Prop. 2.3: every multiplication of Weyl polynomials, even if it is only multiplication by a term, increases the size of the support substantially. Many lower degree terms are added and the coefficients of the lower degree parts change in a subtle way that is hard to predict.

Next we recall the fundamentals of Gröbner basis theory for Weyl algebras. In fact, they were described in [11] for the more general case of algebras of solvable type, and in [16] for the even more general case of G-algebras. For us, the relevant settings are Gröbner bases of left and two-sided ideals, and in these settings most results of commutative Gröbner basis theory hold true after minimal adaptations.

Then, in Section 3, we introduce the class of (left) Weyl Gröbner basis cryptosystems (WGBC). They are a straightforward generalization of the original generalized Polly Cracker systems (see [7] and [12]) and can also be obtained by specializing the very general setting used in [1]. The main contribution of this paper are then specific suggestions for implementing a WGBC such that it resists the standard attacks. These suggestions are given in Def. 3.1 and Remarks 3.2, 3.3, and 3.4. Then we provide two explicit examples (see 3.5 and 3.6) for actual instances of WGBC which we deem to be reasonably secure.

In particular, we examine the resistance of WGBC to various standard attacks in Section 4. By the choices and methods introduced before, we think we can refute a number of attacks: the chosen cyphertext attack, the basic and "intelligent" linear algebra attacks, and the partial Gröbner basis attack. Other attacks to Polly Cracker cryptosystems, such as the differential attack in [10], do not seem to apply directly in the Weyl algebra setting we propose.

In the last section, we also consider WGBC based on Gröbner bases of two-sided ideals of A_n, called two-sided WGBC. They only exist in positive characteristic $p > 0$ and require a secret key contained in the center $C_n = K[x_1^p, \ldots, x_n^p, \partial_1^p, \ldots, \partial_n^p]$. But the public key will not be contained in C_n and may have a Gröbner basis which is difficult to compute. The main advantage of two-sided WGBC is that linear algebra attacks are apparently impossible. We discuss the advantages and disadvantages of two-sided WGBC versus left WGBC and produce an example which we believe to be indicative of the feasibility of two-sided WGBC.

Altogether, we think that WGBC offer plenty of possibilities for constructing secure instances and that they warrant further study. As our Examples show, the fact that we can use comparatively large message spaces means that we can also keep the message expansion under control, i.e. that the transfer rate is quite respectable. Thus well constructed instances of WGBC may offer reasonable security and efficiency.

Unless specifically stated otherwise, we shall use the notation and definitions introduced in [14], [15] and [1]. All computations were performed on a computer with two processors working at 2.4 GHz and having 24 GB of RAM.

It's kind of fun to do the impossible.
(Walt Disney)

2. Gröbner Bases in Weyl Algebras

In the following we let K be a field and $n \geq 1$. Recall that Weyl algebras are defined as follows.

DEFINITION 2.1. Let $\{x_1, \ldots, x_n, \partial_1, \ldots, \partial_n\}$ be a set of indeterminates, and let $K\langle\!\langle x_1, \ldots, x_n, \partial_1, \ldots, \partial_n \rangle\!\rangle$ be the free associative algebra in these indeterminates. Then the **Weyl algebra** of index n over K is the associative K-algebra

$$A_n = K\langle x_1, \ldots, x_n, \partial_1, \ldots, \partial_n \rangle = K\langle\!\langle x_1, \ldots, x_n, \partial_1, \ldots, \partial_n \rangle\!\rangle / I$$

where I is the two-sided ideal generated by the following commutators:
  (1) $[x_i, x_j] = [\partial_i, \partial_j] = 0$ for $i, j \in \{1, \ldots, n\}$,
  (2) $[x_i, \partial_j] = 0$ for $i, j \in \{1, \ldots, n\}$ such that $i \neq j$,
  (3) $[x_i, \partial_i] = -1$ for $i = 1, \ldots, n$.

The elements of A_n will be called **Weyl polynomials**.

For the basic properties of Weyl algebras we refer to standard textbooks such as [6] in the case $\operatorname{char}(K) = 0$ and to the articles [24], [26] and [4] in the case $\operatorname{char}(K) > 0$. Let us collect a few facts which will be useful later.

PROPOSITION 2.2. *Let A_n be the Weyl algebra of index n over K.*

  (a) *If $\operatorname{char}(K) = 0$ then A_n is a simple domain, i.e. it has no left or right zerodivisors and the only two-sided ideals are $\{0\}$ and A_n.*
  (b) *If $\operatorname{char}(K) = p > 0$ then the center of A_n is $C_n = K[x_1^p, \ldots, x_n^p, \partial_1^p, \ldots, \partial_n^p]$. The ring C_n is a commutative polynomial ring in $2n$ indeterminates over K. Moreover, the Weyl algebra A_n is a free C_n-module of rank p^{2n} and an Azumaya algebra of rank p^n over C_n.*
  (c) *The elements of the set $B_n = \{x^\alpha \partial^\beta \mid \alpha, \beta \in \mathbb{N}^n\}$ form a K-vector space basis of A_n. Here we use the multi-index notation $x^\alpha = x_1^{\alpha_1} \cdots x_n^{\alpha_n}$ for $\alpha = (\alpha_1, \ldots, \alpha_n) \in \mathbb{N}^n$, and similarly for ∂^β.*

PROOF. For (a), see [6], Ch. 2, Cor. 1.2 and Thm. 2.1. Claim (b) is shown in [26], Lemma 3. For $\operatorname{char}(K) = 0$, the third claim is shown in [6], Ch. 1, Prop. 2.1, and for $\operatorname{char}(K) > 0$ it follows from (b). □ □

In view of (a) and (b), we will be using mainly left ideals in Weyl algebras, and most of the time the base field will have positive characteristic. Since we plan to perform explicit calculations with Weyl polynomials, we have to bring them into some kind of **standard form**. By part (c) of the proposition, it is natural to write them as K-linear combinations of the elements of the basis B_n. Using the commutator relations of Def. 2.1, this can be done in a straightforward way.

The following result illuminates one of the main reasons why we propose to use Weyl algebras for GB cryptosystems.

PROPOSITION 2.3.   (a) *Let $i \in \{1, \ldots, n\}$, and let $k, \ell \in \mathbb{N}$. Then we have*

$$\partial_i^k x_i^\ell = \sum_{j=0}^{\min\{k,\ell\}} j! \binom{k}{j} \binom{\ell}{j} x_i^{\ell-j} \partial_i^{k-j}$$

  (b) *Assume that $\operatorname{char}(K) = 0$, and let $t = x^\alpha \partial^\beta$ and $t' = x^{\alpha'} \partial^{\beta'}$ be two terms in A_n. Write $\alpha' = (\alpha'_1, \ldots, \alpha'_n)$ and $\beta = (\beta_1, \ldots, \beta_n)$. Then the representation of $t\,t'$ in the basis B_n consists of $\prod_{i=1}^n (\min\{\alpha'_i, \beta_i\} + 1)$ summands.*

PROOF. The proof of (a) proceeds by induction on k. For $k = 0$, the formula is obviously true. To show the induction step, we note that $\partial_i^a x_i^b = 0$ if $a > b$. Thus we may ignore the summation bounds and calculate

$$\begin{aligned}\partial_i^{k+1} x_i^\ell &= \sum_j j! \binom{k}{j}\binom{\ell}{j}[(\ell-j)x_i^{\ell-j-j} + x_i^{\ell-j}\partial]\partial_i^{k-j} \\ &= \sum_j j! \binom{k+1}{j}\frac{k-j+1}{k+1}\binom{\ell}{j} x_i^{\ell-j}\partial_i^{k+1-j} + \\ &\quad + \sum_j (j-1)! \binom{k+1}{j}\frac{j}{k+1}\binom{\ell}{j}\frac{j}{\ell+1-j}(\ell-j+1) x_i^{\ell-j+1}\partial_i^{k+1-j} \\ &= \sum_j j! \binom{k+1}{j}\binom{\ell}{j} x_i^{\ell-j}\partial_i^{k+1-j}\end{aligned}$$

Claim (b) follows immediately from (a). □

If the base field has characteristic $p > 0$, the formula in (b) does not hold exactly, because the numerators of the binomial coefficients in (a) may be divisible by p. However, it is clear that this will usually be true for a small fraction of the coefficients. Thus, if we want to perform concrete computations with Weyl polynomials, part (b) of this propositions means that the supports are going to expand greatly with every multiplication, even if it is only the multiplication by a term.

With this in mind, we now recall the basics of Gröbner basis theory for Weyl algebras. In [11] a Gröbner basis theory for algebras of solvable type was introduced. Weyl algebras are special cases for these algebras (see [11], 1.9.b). We shall use the following terminology.

DEFINITION 2.4. (a) A Weyl polynomial of the form $x^\alpha \partial^\beta$ with $\alpha, \beta \in \mathbb{N}^n$ is called a **standard term**. The set of all standard terms in A_n is denoted by B_n. It is the K-basis of A_n mentioned in Prop. 2.2.c.

(b) The **degree** of a Weyl term $t = x^\alpha y^\beta$ is $\deg(t) = |\alpha| + |\beta|$. (Here we use the notation $|\alpha| = \alpha_1 + \cdots + \alpha_n$ for $\alpha = (\alpha_1, \ldots, \alpha_n) \in \mathbb{N}^n$.)

(c) Let $f = c_1 t_1 + \cdots + c_s t_s$ be a Weyl polynomial in **standard form**, i.e. a Weyl polynomial where we have $c_i \in K \setminus \{0\}$ and $t_i \in B_n$. Then $\mathrm{Supp}(f) = \{t_1, \ldots, t_s\}$ is called the **(standard) support** of f and the number $\deg(f) = \max\{\deg(t_1), \ldots, \deg(t_s)\}$ is called the **degree** of f.

(d) Given a Weyl polynomial $f = c_1 t_1 + \cdots + c_s t_s$ as in (c), let $d = \deg(f)$. Then the Weyl polynomial $\mathrm{DF}(f) = \sum_{\{i | \deg(t_i) = d\}} c_i t_i$ is called the **degree form** of f.

(e) A complete ordering σ on B_n is called a **(Weyl) term ordering** if it has the following properties:

(1) The ordering σ is *multiplicative*, i.e. an inequality $x^\alpha \partial^\beta <_\sigma x^{\alpha'} \partial^{\beta'}$ implies $x^{\alpha + \alpha''} \partial^{\beta + \beta''} <_\sigma x^{\alpha' + \alpha''} \partial^{\beta' + \beta''}$ for all $\alpha, \alpha', \alpha'', \beta, \beta', \beta'' \in \mathbb{N}^n$.

(2) The ordering σ is *well-founded*, i.e. we have $1 <_\sigma t$ for all $t \in B_n \setminus \{1\}$.

(f) Given term ordering σ and a Weyl polynomial $f = c_1 t_1 + \cdots + c_s t_s$ with $c_i \in K \setminus \{0\}$ and $t_i \in B_n$, where $t_1 >_\sigma \cdots >_\sigma t_s$, we call $\mathrm{LT}_\sigma(f) = t_1$ the **leading term** of f and $\mathrm{LC}_\sigma(f) = c_1$ the **leading coefficient** of f.

(g) For two standard terms $t = x^\alpha \partial^\beta$ and $t' = x^{\alpha'} \partial^{\beta'}$ in B_n we say that t **pseudo-divides** t' if $\alpha \leq \alpha'$ and $\beta \leq \beta'$. (These inequalities have to be satisfied componentwise.)

(h) Let I be a left ideal in A_n. A set of Weyl polynomials $G \subseteq I$ is called a **left σ-Gröbner basis** of I if the leading term $\mathrm{LT}_\sigma(f)$ of every element

$f \in I \setminus \{0\}$ is pseudo-divisible by one of the leading terms $\mathrm{LT}_\sigma(g)$ with $g \in G$.

Notice that the condition that a term $t' = x^{\alpha'}\partial^{\beta'}$ is pseudo-divisible by a term $t = x^\alpha \partial^\beta$ means that t' is **left reducible** by t, i.e. there exist $t'' \in B_n$ and $f \in A_n$ such that $t' = t'' \cdot t + f$ and $\deg(f) < \deg(t')$. Thus the condition that G is a left σ-Gröbner basis of a left ideal I is equivalent to the condition that the left rewrite relation $\xrightarrow{G}$ defined by G is confluent (see [11], Section 3).

For two-sided ideals in A_n, the notion of a two-sided Gröbner basis can be defined using left Gröbner bases (see [11], Section 5 and [16], Ch. 2, Sec.3). In the following we denote the left ideal generated by a subset S of A_n by $\langle S \rangle_\lambda$ and the two-sided ideal generated by S is $\langle S \rangle_\tau$.

DEFINITION 2.5. Let σ be a term ordering on B_n, and let I be a two-sided ideal of A_n. A set of polynomials G is called a **two-sided σ-Gröbner basis** of the ideal I if G is a left σ-Gröbner basis and $I = \langle G \rangle_\tau$ is the two-sided ideal generated by G.

Let us collect some useful observations about these notions.

PROPOSITION 2.6. *Let σ be a term ordering on B_n.*

(a) *Given $f, g \in A_n \setminus \{0\}$, write $\mathrm{LT}_\sigma(f) = x^\alpha \partial^\beta$ and $\mathrm{LT}_\sigma(g) = x^{\alpha'}\partial^{\beta'}$ with $\alpha, \alpha', \beta, \beta' \in \mathbb{N}^n$. Then we have*

$$\mathrm{LT}_\sigma(fg) = \mathrm{LT}_\sigma(gf) = x^{\alpha+\alpha'}\partial^{\beta+\beta'}$$

(b) *Let $g \in A_n \setminus \{0\}$ and let $I = \langle g \rangle_\lambda$ be the left principal ideal generated by g. Then $G = \{g\}$ is a left σ-Gröbner basis of I.*
(c) *Let $G \subset A_n \setminus \{0\}$ be a set of Weyl polynomials such that no indeterminate appears in the leading terms of two elements of G, and such that any two elements of G commute. Then G is a left σ-Gröbner basis of the left ideal $\langle G \rangle_\lambda$.*

PROOF. Claim (a) is shown in [11], 1.5.2, claim (b) is an immediate consequence of (a), and claim (c) follows from the *Generalized Product Criterion* (see [16], Ch. 2, Lemma 4.11). □

For Weyl polynomials, there exist natural definitions of S-polynomials, an analogue of the Buchberger algorithm for computing left (and two-sided) σ-Gröbner bases, and a straightforward generalization of the concept of reduced Gröbner bases (see [11], Sect. 2 and 3). The Buchberger algorithm and many of its applications have been implemented in several readily available computer algebra systems. For our experiments, we have relied on the following implementations:

(1) the package `Plural` by V. Levandovskyy (see [17]) for `Singular` (see [9]),
(2) the package `D-modules` by A. Leykin (see [18]) for `Macaulay2` (see [8]),
(3) the computer algebra system `Risa/Asir` (see [21]),
(4) the first author's ApCoCoA package (see [3]).

This concludes our brief overview. Further results about Gröbner bases in Weyl algebras will be recalled as needed.

3. Weyl Gröbner Basis Cryptosystems

In this section we construct instances of the following cryptosystem. It is adapted from the very general setting of [1] to the Weyl algebra case we are considering.

Let K be a field, let $A_n = K[x_1, \ldots, x_n, \partial_1, \ldots, \partial_n]$ be the Weyl algebra of index n over K, let $B_n = \{x^\alpha \partial^\beta \mid \alpha, \beta \in \mathbb{N}^n\}$ be its set of terms, and let σ be a term ordering on B_n. Given a set of Weyl polynomials $G \subset A_n \setminus \{0\}$, the **left rewrite rule** generated by G is denoted by $\xrightarrow{G}$. If G is a left σ-Gröbner basis, every Weyl polynomial f can be reduced using $\xrightarrow{G}$ to its **normal form** $\mathrm{NF}_{\sigma,G}(f)$. (For details, see [11], Sect. 3.) Now we introduce the following cryptosystems.

DEFINITION 3.1. A **Weyl Gröbner basis cryptosystem (WGBC)** consists of the following data.

Secret key: A left σ-Gröbner basis $G = \{g_1, \ldots, g_s\}$ which generates a left ideal $I = \langle G \rangle_\lambda$ in A_n. Moreover, also the set $\mathcal{O}_\sigma(I) = B_n \setminus \{\mathrm{LT}_\sigma(f) \mid f \in I \setminus \{0\}\}$ is secret.

Public key: A set of Weyl polynomials $\{p_1, \ldots, p_r\}$ contained in $I \setminus \{0\}$ and a subset M of $\mathcal{O}_\sigma(I)$.

Message space: The message space is the K-vector subspace $\langle M \rangle_K$ of A_n generated by M.

Ciphertext space: The ciphertext units are elements of A_n in standard form.

Encryption map: Given a plaintext unit $m \in \langle M \rangle_K$, choose Weyl polynomials $\ell_1, \ldots, \ell_r$ and compute the standard form of $c = m + \ell_1 p_1 + \cdots + \ell_r p_r$.

Decryption map: Given a ciphertext unit $c \in A_n$, compute $\mathrm{NF}_{\sigma,G}(c)$. If the result is contained in $\langle M \rangle_K$, return it. Otherwise, return c.

The correctness of this system follows immediately from $\mathrm{NF}_{\sigma,G}(m + \ell_1 p_1 + \cdots + \ell_r p_r) = m$. If an attacker can compute the left σ-Gröbner basis of the ideal $J = \langle p_1, \ldots, p_r \rangle_\lambda$, he can obviously break the system. In this sense the security of the system rests on the difficulty of computing left Gröbner bases in Weyl algebras. To construct a secure instance of a WGBC, we need secure secret and public keys and a secure way to perform the encryption algorithm. In the next two remarks we propose some specific instructions how to proceed with these tasks.

REMARK 3.2 (WGBC Key Generation). In the above setting, to create an instance of a WGBC, perform the following steps.

(1) Choose a set of Weyl polynomials $G = \{g_1, \ldots, g_s\}$ which form a reduced left σ-Gröbner basis of the left ideal $I = \langle G \rangle_\lambda$ they generate.
(2) Let $\mathcal{O}_\sigma(I) = B_n \setminus \mathrm{LT}_\sigma(I)$. Choose a proper subset M of $\mathcal{O}_\sigma(I)$ which is small enough so that it does not leak information about the degrees and the shape of the generators of $\mathcal{O}_\sigma(I)$.
(3) Make sure that every Weyl polynomial $g_i \in G$ contains terms from $\mathcal{O}_\sigma(I) \setminus M$ in its support. If not, modify G and M accordingly.
(4) Choose $h_{ij} \in A_n$ and compute the standard form of $p_i = h_{i1} g_1 + \cdots + h_{is} g_s$ for $i = 1, \ldots, r$. Make sure that following properties hold.
   (a) The degree forms $\mathrm{DF}(h_{ij} g_j)$ of highest degree cancel. The other degree forms $\mathrm{DF}(h_{ij} g_j)$ cancel or their coefficients are changed in p_i by the process of converting the remaining $h_{ik} g_k$ to standard form.

(b) There are sufficiently high powers of $\partial_1, \ldots, \partial_n$ in the terms of the support of h_{ij} such that, after bringing $h_{ij}g_j$ to standard form, no information about $\mathrm{Supp}(g_j)$ is leaked in $\mathrm{Supp}(p_i)$.

(c) There are no gaps in the degrees of the standard terms in $\mathrm{Supp}(p_i)$. In each degree there are sufficiently many terms whose coefficient results from at least two contributions during the process of converting $h_{i1}g_1 + \cdots + h_{is}g_s$ to standard form.

(5) Verify that no Gröbner basis of the left ideal $J = \langle p_1, \ldots, p_r \rangle_\lambda$ can be computed efficiently. In fact, ascertain that not even partial Gröbner bases of this ideal can be computed for large enough degree bounds. (The meaning of "large enough" will be specified in the next remark.)

REMARK 3.3 (Implementation of WGBC Encryption). In the above setting, to implement the encryption function of a WGBC, perform the following steps.

(1) Given a plain text unit $m \in \langle M \rangle_K$, choose $\ell_1, \ldots, \ell_r \in A_n$ and compute the standard form of the cipher text unit $c = \ell_1 p_1 + \cdots + \ell_r p_r + m$.

(2) Make sure that the standard form of $\ell_1 p_1 + \cdots + \ell_r p_r$ contains many terms of $\mathrm{Supp}(m)$ and many other terms of M.

(3) Ascertain that the degree forms $\mathrm{DF}(\ell_i p_i)$ of highest degree cancel in c, and that the other degree forms $\mathrm{DF}(\ell_j p_j)$ cancel or their coefficients are changed in c by the process of converting the remaining $\ell_k p_k$ to standard form.

(4) Choose the terms in $\mathrm{Supp}(\ell_i)$ such that they contain high enough powers of $\partial_1, \ldots, \partial_n$. In particular, verify that the range of the degrees of the terms in the standard form of $\ell_i p_i$ has no wide gaps.

(5) Choose the elements ℓ_i such that the degree of c is high enough such that no partial Gröbner basis of J can be computed up to the degree bound $\deg(c)$. Check that no easily computable partial Gröbner basis H of J satisfies $\mathrm{NR}_{\sigma, H}(c) \in \langle M \rangle_K$. (Here $\mathrm{NR}(\cdots)$ denotes the **normal remainder**, i.e. the result of the Division Algorithm.)

In the next section we shall explain why we believe that these choices render the usual attacks infeasible. But is it possible to satisfy the requirements of these remarks? Our next remark provides some explicit suggestions how this can be done.

REMARK 3.4. Let $K = \mathbb{F}_p$ be a finite field, let $n \geq 2$, let A_n be the Weyl algebra of index n over K, and let σ be a term ordering on B_n.

(1) In order to construct a non-trivial reduced left σ-Gröbner basis in A_n, we can proceed as follows: for $i \in \{1, \ldots, n\}$ (or a subset thereof), let $g_i \in K\langle x_i, \partial_i \rangle$. Then Proposition 2.6.c implies that $G = \{g_1, \ldots, g_n\}$ is the reduced left σ-Gröbner basis of $I = \langle G \rangle_\lambda$. To prevent easy detection of G, we suggest to use polynomials g_i of degree $\deg(g_i) \geq 10$ having at least five terms in their standard support.

(2) If the reduced Gröbner basis G contains sufficiently many elements, use only subsets of G to construct the various polynomials p_i of the public key. Ideally, use disjoint or almost disjoint subsets. In this way, it is certain that the terms appearing in in the standard forms of the polynomials p_i do not leak information about the support of the elements of G or the coefficients h_{jk} that were used.

(3) To satisfy Condition (4a) of Remark 3.2, we can proceed as follows. First choose "random" polynomials h_{ij} such that the other conditions of Remark 3.2 hold. Then try to top-reduce a degree form $\mathrm{DF}(h_{ij}g_j)$ of maximal degree using the polynomials g_k with $k \ne j$. If this is not possible for some of the terms, exchange the corresponding terms in h_{ij} by other terms t such that $t \, \mathrm{LT}_\sigma(g_j)$ is pseudo-divisible by some $\mathrm{LT}_\sigma(g_k)$. Continue in this way until the top degree (and ideally even the top two or three degrees) of the terms appearing in the sum $h_{i1}g_1 + \cdots + h_{is}g_s$ cancels completely.

(4) Similarly, implement the encryption map such that the polynomials ℓ_{ij} are chosen to create syzygies in the top degree part (or parts) of the expansions of the sums $\ell_{i1}p_1 + \cdots + \ell_{ir}p_r$.

Let us execute this prescription in some concrete cases.

EXAMPLE 3.5. Over the base field $\mathbb{F}_{13}$, we consider the Weyl algebra $A_2 = \mathbb{F}_{13}\langle x_1, x_2, \partial_1, \partial_2 \rangle$ of index 2 and the term ordering $\sigma = \mathtt{DegRevLex}$. Then we introduce the following WGBC.

(1) Let $G = \{g_1, g_2\}$ be given by

$$g_1 = 7x_1^7\partial_1^7 + 2x_1^6\partial_1^6 + 3x_1^3 - \partial_1^3 + 4x_1^2\partial_1^2 + x_1^2 - 3x_1\partial_1 - 2\partial_1^2 + 5x_1 - 7\partial_1 + 1$$
$$g_2 = 4x_2^5\partial_2^5 + 3x_2^4\partial_2^4 + 5x_2^4 + \partial_2^4 - 3x_2^3 - 4\partial_2^3 + x_2^2 - x_2\partial_2 + 2\partial_2^2 - 3$$

and let I be the left ideal $I = \langle g_1, g_2 \rangle_\lambda$. By Proposition 2.6.c, the set G is a left σ-Gröbner basis of I.

(2) Let $p_1 = h_{11}g_1 + h_{12}g_2$ and $p_2 = h_{21}g_1 + h_{22}g_2$ where

$$h_{11} = 4x_1^3x_2^{11}\partial_1^3\partial_2^9 + 5x_1^3x_2^{10}\partial_1^3\partial_2^8 + 5x_1 - 3x_2 + 2\partial_1 - 6\partial_2 + 3$$
$$h_{12} = 6x_1^{10}x_2^6\partial_1^{10}\partial_2^4 - 6x_1^9x_2^6\partial_1^9\partial_2^4 - 3x_1 + 4x_2 - 5\partial_1 + 2\partial_2 + 4$$
$$h_{21} = 5x_1^2x_2^{14}\partial_1^6\partial_2^{16} - 4x_1^2x_2^{13}\partial_1^6\partial_2^{15} - 7x_1 + 2x_2 + 4$$
$$h_{22} = x_1^9x_1^9\partial_1^{13}\partial_2^{11} + 7x_1^8x_2^9\partial_1^{12}\partial_2^{11} + 6\partial_1 - 3\partial_2 + 1$$

Then the Weyl polynomial p_1 has degree 36 and its standard form consists of 154 terms. Notice that $\deg(h_{11}) + \deg(g_1) = \deg(h_{12}) + \deg(g_2) = 40$, so that the top four degrees in the expansion of $h_{11}g_1 + h_{12}g_2$ cancel. The Weyl polynomial p_2 has degree 48 and 128 terms in its standard form. Again we note that $\deg(h_{21}) + \deg(g_1) = \deg(h_{22}) + \deg(g_2) = 52 > 48$.

(3) For the message space, we choose the K-vector space generated by $M = \{x^\alpha \partial^\beta \mid |\alpha| \le 11, |\beta| \le 7\}$. There are 13^{2808} different possible plaintext units. Both g_1 and g_2 have terms from $\mathcal{O}_\sigma(I) \setminus M$ in their support.

(4) To encrypt a message m, we choose polynomials ℓ_1, ℓ_2 of sufficiently high degree and form $c = m + \ell_1 p_1 + \ell_2 p_2$. In the case at hand, we should choose $\deg(\ell_1) = \deg(\ell_2) + 12 \ge 55$ and make sure that the top degree parts in the expansion of $\ell_1 p_1 + \ell_2 p_2$ cancel.

For instance, we can choose

$$\ell_1 = -5x_1^{10}x_2^{16}\partial_1^{12}\partial_2^{19} - 2x_1^8x_2^{18}\partial_1^{10}\partial_2^{21} + \text{(lower degree terms)}$$
$$\ell_2 = 4x_1^{11}x_2^{13}\partial_1^9\partial_2^{12} - 6x_1^9x_2^{15}\partial_1^7\partial_2^{14} + \text{(lower degree terms)}$$

The resulting ciphertext has degree 91, whereas $\deg(\ell_1) + \deg(p_1) = \deg(\ell_2) + \deg(p_2) = 93$.

(5) To decipher c, it suffices to compute $m = \mathrm{NF}_{\sigma,G}(c)$. In our case, an efficient implementation of the Division Algorithm recovers m in a few seconds.

In the next section, we shall discuss the resistance of this example with respect to several standard attacks. The reason why we had to go up to rather high degrees is clearly the fact that we used the Weyl algebra of index 2. As soon as we add a few more indeterminates, we gain additional freedom and the attacks of the next section are even more difficult to carry out.

EXAMPLE 3.6. Let $A_3 = \mathbb{F}_7\langle x_1, x_2, x_3, \partial_1, \partial_2, \partial_3 \rangle$, and let $\sigma = \texttt{DegRevLex}$. We construct a WGBC instance as follows.

(1) First we choose polynomials in $\mathbb{F}_7\langle x_i, \partial_i \rangle$ such that the left ideals they generate have non-trivial left σ-Gröbner bases. For instance, let
$$\begin{aligned} f_1 &= x_1^7, & f_2 &= x_1^3 \partial_1^3 + x_1, \\ f_3 &= x_2^7, & f_4 &= x_2^2 \partial_2^2 + x_2 + \partial_2, \\ f_5 &= x_3^7 \partial_3^7, & f_6 &= \partial_3^4 + x_3. \end{aligned}$$

Then the reduced σ-Gröbner basis of $I_1 = \langle f_1, f_2 \rangle_\lambda$ is given by $\{g_1, g_2, g_3\}$ where
$$\begin{aligned} g_1 &= x_1 \partial_1^3 + 3x_1^2 + 3x_1 \partial_1 - x_1 - 3\partial_1 - 3, \\ g_2 &= x_1^3 + 3x_1 \partial_1^2 - 2x_1^2 - 2x_1 \partial_1 - 2x_1 - 3\partial_1 + 2, \\ g_3 &= x_1^2 \partial_1 + 3x_1^2 + x_1 \partial_1 - 3x_1 - 2, \end{aligned}$$

the reduced σ-Gröbner basis of $I_2 = \langle f_3, f_4 \rangle_\lambda$ is given by $\{g_4, g_5, g_6, g_7\}$ where
$$\begin{aligned} g_4 &= \partial_2^4 + 3\partial_2^3 + 2x_2^2 - 2x_2 \partial_2 + \partial_2^2 - x_2 - 3\partial_2 - 3, \\ g_5 &= x_2 \partial_2^2 - \partial_2^3 - 2x_2^2 + 2x_2 \partial_2 + 2\partial_2^2 + 2x_2 - 3, \\ g_6 &= x_2^2 \partial_2 + 3\partial_2^3 + 2x_2^2 - 2x_2 \partial_2 + 3\partial_2^2 - 2x_2 - 2\partial_2 - 3, \\ g_7 &= x_2^3 - \partial_2^3 + x_2^2 - 3x_2 \partial_2 + 3\partial_2^2 - 2x_2 - \partial_2 - 2, \end{aligned}$$

and the reduced σ-Gröbner basis of $I_3 = \langle f_5, f_6 \rangle_\lambda$ is $\{g_8, g_9, g_{10}, g_{11}\}$ where
$$\begin{aligned} g_8 &= x_3^{10} - 2x_3^7 \partial_3^2, & g_9 &= x_3^9 \partial_3 - x_3^8, \\ g_{10} &= x_3^7 \partial_3^3 - 2x_3^9, & g_{11} &= \partial_3^4 + x_3. \end{aligned}$$

Altogether, the Generalized Product Criterion (cf. [16], Ch. 2, Lemma 4.11) implies that the reduced σ-Gröbner basis of $I = \langle f_1, \ldots, f_6 \rangle_\lambda$ is $G = \{g_1, \ldots, g_{11}\}$.

(2) For the message space, we use the linear span of the set
$$\begin{aligned} M = \{ & x_1^{\alpha_1} x_2^{\alpha_2} x_3^{\alpha_3} \partial_1^{\beta_1} \partial_2^{\beta_2} \partial_3^{\beta_3} \mid \alpha_1 \leq 1,\ \alpha_2 + \beta_2 \leq 2,\ \alpha_3 \leq 8, \\ & \beta_1 \leq 2,\ \beta_3 \leq 3,\ \alpha_3 + \beta_3 \leq 7 \} \end{aligned}$$

In this way, the set M has 936 elements, is contained in $\mathcal{O}_\sigma(I) = B_n \setminus \mathrm{LT}_\sigma(I)$, and every Gröbner basis polynomial, except for g_{11}, contains a term from $\mathcal{O}_\sigma(I) \setminus M$ in its standard support.

(3) Next we create the public key. It will consist of polynomials p_i of the form
$$\begin{aligned} p_1 &= h_{11} g_1 + h_{12} g_4 + h_{13} g_8 \\ p_2 &= h_{21} g_2 + h_{22} g_5 + h_{23} g_9 \\ p_3 &= h_{31} g_3 + h_{32} g_6 + h_{33} g_{10} + h_{34} g_{11} \end{aligned}$$

Notice that these public key polynomials involve different elements of the Gröbner basis. Therefore it appears to be impossible to recover G by analyzing the public key. Of course, we still need to enforce Conditions (4) and (5) of Remark 3.2. To prevent the partial Gröbner basis attack (see Section 4.3), we create public polynomials of degrees ≥ 20. For instance, by using

$$\begin{aligned}
h_{11} &= x_1 x_2 x_3^5 \partial_1^3 \partial_2^4 \partial_3^4 - x_2 x_3^5 \partial_1^4 \partial_2^4 \partial_3^4 - 3x_1 x_3^{10} \partial_2^2 \partial_3^5 - x_1 x_3^7 \partial_2^2 \partial_3^7 \\
&\quad - 3x_2 x_3^5 \partial_1^4 \partial_2^3 \partial_3^4 - 2x_2 \partial_2^3 \partial_3 + 3x_2 \partial_2^3 - 2x_3^2 \partial_3 + x_2 \partial_3 - 3\partial_3 + 1 \\
h_{12} &= -x_1^2 x_2 x_3^5 \partial_1^6 \partial_3^4 + x_1 x_2 x_3^5 \partial_1^7 \partial_3^4 + x_1 x_2 x_3^{10} \partial_1 \partial_2 \partial_3^4 - 2x_1 x_2 x_3^7 \partial_1 \partial_2 \partial_3^6 \\
&\quad - x_1^3 \partial_3^4 + \partial_1^3 \partial_2^3 \partial_3 + 3x_3 \partial_3^4 - x_1^3 - 3x_2^3 + 3x_3 \\
h_{13} &= -x_1 x_2 \partial_1 \partial_2^5 \partial_3^4 + 3x_1^2 \partial_1^3 \partial_2^2 \partial_3^5 - 3x_1 x_2 \partial_1 \partial_2^4 \partial_3^4 \\
&\quad + x_3 \partial_3^3 + x_1 \partial_1^2 + x_2 \partial_2^2 - \partial_2^3 + 2x_1 \partial_1 + x_1 \partial_2 - 4\partial_1
\end{aligned}$$

we obtain a Weyl polynomial p_1 of degree 21 having 173 terms in its standard support. Notice that $\deg(h_{11}) + \deg(g_1) = \deg(h_{12}) + \deg(g_4) = \deg(h_{13}) + \deg(g_8) = 22$, so that the top degree in the expansion of p_1 cancels. Moreover, we have reduced p_1 in degree 21 until only one term remains. This will aid us in the construction of secure ciphertexts (see Example 4.7). Next, we use

$$\begin{aligned}
h_{21} &= 4x_3^5 \partial_1^2 \partial_2^4 \partial_3^3 + 3x_1^3 x_3^2 \partial_1 \partial_3^2 + 2x_2 \partial_1 + 4x_1^2 \partial_3 - x_3 - 2\partial_3 \\
h_{22} &= x_1^2 x_2 x_3^9 \partial_1^3 \partial_2^2 \partial_3 + 2x_1^2 x_2^2 x_3^9 \partial_1^3 \partial_3 - 3x_1^2 x_2^2 x_3^9 \partial_2^3 \partial_3 - x_1^2 x_2^2 x_3^9 \partial_3^3 \\
&\quad + x_1 x_2^3 x_3^2 \partial_1^4 \partial_3^3 - 2x_1 \partial_1 \partial_2^2 \partial_3^2 + x_2 x_3 \partial_1^2 + x_1 \partial_1 - 2\partial_2^2 - \partial_3 + 4x_3 \\
h_{23} &= -x_1^2 x_2^2 \partial_1^3 \partial_2^4 + x_1^2 x_2 \partial_1^3 \partial_2^5 - 3x_1^2 x_2^4 \partial_2^4 + 3x_1^2 x_2^4 \partial_2^4 - 3x_1^2 x_2^2 \partial_2^6 \\
&\quad + 3x_1^2 x_2^3 \partial_2^5 - x_1^2 x_2^4 \partial_2^3 + x_1^2 x_2^3 \partial_2^4 + 2x_1^2 x_2^2 \partial_2^5 + x_1^2 x_2^2 \partial_2^4 + x_1^2 x_2^3 \partial_2^2 \\
&\quad + x_1^2 x_2^2 \partial_2^3 + 4x_3^2 \partial_3^3 - 4x_1 x_3 \partial_3^2 + 3\partial_1 \partial_2 \partial_3 + x_1^2 \partial_3 - 2x_2 + \partial_2
\end{aligned}$$

and get a public key polynomial p_2 of degree 20 having 155 terms in its standard support. Since $\deg(h_{22}) + \deg(g_5) = \deg(h_{23}) + \deg(g_9) = 21$, the top degree in the expansion of p_2 cancels. Finally, we use

$$\begin{aligned}
h_{31} &= 3x_1 x_2^2 x_3^8 \partial_1 \partial_2^2 \partial_3^4 + 2x_1 x_3^8 \partial_1 \partial_2^4 \partial_3^4 - x_3^2 x_3^8 \partial_1^2 \partial_2^2 \partial_3^3 - x_1 x_2^2 x_3^8 \partial_1^2 \partial_2 \partial_3^4 \\
&\quad + x_2 x_3^8 \partial_1^2 \partial_2^2 \partial_3^5 + 2x_2^3 x_3^{10} \partial_1^2 \partial_2^2 + 2x_1 x_2^2 x_3^{10} \partial_1^2 \partial_2 \partial_3 - x_2 x_3^9 \partial_1^2 \partial_2^3 \partial_3^2 + x_2^3 x_3^9 \\
&\quad + 3x_2 x_3^9 \partial_2 - 2x_2 x_3^7 \partial_3^3 - 2x_2 x_3^7 \partial_2 \partial_3^3 + x_1 x_2 x_3 \partial_3 - 4x_2 \partial_3^2 + \partial_1^2 \partial_2 \\
&\quad + 2x_1 \partial_3 - 2\partial_3 + 4x_2 - 1 \\
h_{32} &= -3x_1^3 x_3^8 \partial_1^2 \partial_2 \partial_3^4 + 2x_1^3 x_3^8 \partial_1^2 \partial_3^4 - 2x_1^3 x_3^8 \partial_1 \partial_2 \partial_3^4 - 3x_1^2 x_3^8 \partial_1^2 \partial_2 \partial_3^4 \\
&\quad + 3x_1^2 x_3^7 \partial_3^3 + 3x_1 x_3^7 \partial_1 \partial_3^3 + 2x_1 x_3^3 \partial_3^3 + x_3^4 - 2x_1 \partial_2^2 + 4x_2 \partial_3^2 - \partial_3^3 \\
&\quad + x_3^2 - x_2 x_3 + 3\partial_1 \\
h_{33} &= x_1^3 x_2^2 x_3 \partial_1^3 \partial_2 \partial_3 + 3x_1^2 x_2 \partial_1^3 \partial_2^3 \partial_3^2 + x_1^2 x_3^3 x_3 \partial_1^3 \partial_2^2 - 2x_1 x_2^2 \partial_1^2 \partial_3^2 \partial_3 \\
&\quad + 3x_1^2 x_2^3 x_3 \partial_1^2 \partial_2^2 + x_1 x_2^3 x_3 \partial_1^3 \partial_2^2 + 3x_1 x_2^2 x_3^2 \partial_1^2 \partial_2^3 + x_1^2 x_2^2 x_3 \partial_1^3 \partial_2 \partial_3 \\
&\quad + 2x_2 x_3 \partial_1 \partial_2^2 - 3x_1 x_3 \partial_3^3 - x_1^2 x_3 \partial_2^2 + x_1 x_3 \partial_3 - x_3 \partial_2 + 4x_1 \partial_3 - 2x_1 \\
h_{34} &= -x_1^2 x_2 x_3^8 \partial_1^3 \partial_2^2 \partial_3 + 2x_1 x_2^2 x_3^7 \partial_1^2 \partial_2^2 \partial_3 - 3x_1^2 x_2 x_3^7 \partial_1^3 \partial_2^3 \partial_3 \\
&\quad + x_1^3 x_2^2 x_3^8 \partial_1^3 \partial_2 - x_1 x_2 x_3^8 \partial_1^3 \partial_2^2 \partial_3 - 3x_1^2 x_2 x_3^8 \partial_1^2 \partial_2^2 \partial_3 + x_1^3 x_2 x_3^8 \partial_1^2 \partial_2^2 \\
&\quad - x_1 x_2 \partial_1^3 \partial_3 - 2x_1^2 x_2 \partial_2^2 \partial_3 - x_1^3 x_2 x_3^8 \partial_1^3 \partial_2 - 2x_1 \partial_1^3 \partial_3 - 3x_1^2 \partial_2^2 + x_1 \partial_3 \\
&\quad - 3\partial_1^2 + 4x_2 \partial_2 - \partial_1^2 - 3x_3
\end{aligned}$$

to construct p_3. We obtain $\deg(p_3) = 20$, there are 226 terms in the standard support of p_3, and the top degree, namely 21, in the expansion of p_3 cancels.

Let us also check whether the sizes of the homogeneous components of the public key polynomials are reasonably distributed.

deg	1	2	3	4	5	6	7	8	9	10	11	12	13	14	15	16	17	18	19	20	21
p_1	4	7	10	11	15	10	11	7	8	4	5	6	4	1	2	6	9	16	2	15	1
p_2	3	10	14	19	10	7	5	4	8	2	6	3	5	6	10	9	11	10	11	1	
p_3	5	9	19	21	25	17	6	3	5	4	5	9	9	13	14	10	16	18	16	1	

Clearly, there are no gaps in these degrees and plenty of terms are contained in M. They will help to alter the plaintext coefficients during enciphering.

(4) To decipher c, it suffices to compute $m = \mathrm{NF}_{\sigma,G}(c)$. In this case, an efficient implementation of the Division Algorithm recovers m in few seconds.

In the next section we shall examine the security of the WGBC instances proposed in the preceding two examples against various attacks. Finally, we note a suggestion which could possibly be used to make WGBC even more secure.

REMARK 3.7. In a WGBC, both the characteristic $\mathrm{char}(K)$ of the base field and the term ordering σ can be kept secret. This makes it very hard to compute a useful Gröbner basis of the ideal J, and both the linear algebra attack and the chosen ciphertext attack of Section 4 become even harder to mount. For the case of the usual Polly Cracker cryptosystems, this suggestion has been worked out in detail in [**25**].

4. Security of Weyl Gröbner Basis Cryptosystems

As mentioned in the introduction, Polly Cracker cryptosystems have been attacked in a variety of ways. Some basic attack methods are described in [**5**] and [**12**]. Before discussing them, we examine the following weakness of many homomorphic public key cryptosystems.

4.1. Chosen Ciphertext Attacks.
Several ways have been proposed to break Gröbner basis cryptosystems if an attacker has temporary access to the decryption algorithm. Here we describe some methods how to defend against them.

REMARK 4.1. (**Fake Ciphertext Attack**) Suppose that the secret key (i.e. the Gröbner basis) is of the form $G = \{g_1, \ldots, g_s\}$ with $g_i = t_i + h_i$ and $t_i = \mathrm{LT}_\sigma(g_i)$. If the attacker knows or guesses the leading terms $t_1, \ldots, t_s$, he can use his access to the decryption algorithm and enter fake ciphertexts of the form $t_i + \sum_j \ell_{ij} p_j$ for $i = 1, \ldots, s$. As a result, he gets $\mathrm{NF}_{\sigma,G}(t_i) = -h_i$ and then he recombines $g_i = t_i + h_i$. To defend against this attack, the following strategy has been proposed (see [**23**]):

(1) Do not publish the complete set $\mathcal{O}_\sigma(I)$. (This is the complement of the set of leading terms of elements of I.) Publish only a (small) part M of it and use M as a basis for the message space.

(2) Make sure that the *tail* h_i of each Gröbner basis polynomial g_i contains at least one term from $\mathcal{O}_\sigma(I) \setminus M$ in its support. In this way, if the attacker guesses $\mathrm{LT}_\sigma(g_i)$ and tries to decrypt it, countermeasure (3) will make sure that he fails. Moreover, if the attacker tries to decrypt any term t from $\mathcal{O}_\sigma(I) \setminus M$, for instance to find $\mathcal{O}_\sigma(I)$ by trial and error, he fails, too.

(3) If the decryption algorithm computes a normal form which is not contained in $\langle M \rangle_K$, it is clear that an invalid ciphertext was used. Therefore the decryption algorithm does not reveal the normal form, but returns the ciphertext unchanged. In this way, when the attacker decrypts a term outside M, the term is returned and no secret information is revealed.

Notice that we incorporated countermeasures (3) in the very description of the WGBC in Def. 3.1. The other two countermeasures are part of the setup proposed in Remark 3.2.

It has been argued that countermeasure (1) reduces the efficiency of the cryptosystem too much. By restricting M to a proper subset of $\mathcal{O}_\sigma(I)$ we can make the probability for a random polynomial to be a valid cipher text as small as we like. Even if the base field is finite, skipping N terms from $\mathcal{O}_\sigma(I)$ reduces this probability to q^{-N}, where $q = \#K$. Thus the cost for protection against a fake ciphertext attack is rather small.

REMARK 4.2. (**Homomorphic Chosen Ciphertext Attack**) Another way to perform a chosen ciphertext attack was described in [**12**], Ch. 5, Sec. 3, Ex. 11. The attacker exploits the fact that WGBC are *homomorphic*: if c is a valid ciphertext for the plaintext $m \in \langle M \rangle_K$ and c' is a valid ciphertext for the plaintext $m' \in \langle M \rangle_K$ then $c + c'$ is a valid ciphertext for $m + m'$. To decrypt c, the attacker can therefore create c', decrypt it, and recover m from $m + m'$.

To defend against this attack, we can for instance proceed as follows. (A similar method was described in [**19**], Section 4.4.)

(1) Choose a non-homomorphic hash function $h : \langle M \rangle_K \longrightarrow \langle H \rangle_K$ where H is a set of terms in $\mathcal{O}_\sigma(I) \setminus M$ such that the coefficients in $\langle H \rangle_K$ are sufficient to store a hash value.
(2) Given a plaintext $m \in \langle M \setminus H \rangle_K$, compute the WGBC ciphertext c and transmit $c + h(m)$.
(3) When c is decrypted, check whether the result contains a term from $\mathcal{O}_\sigma(I) \setminus (M \cup H)$. If that is the case, return c and stop. Otherwise, write the result as $m + p$ with $m \in \langle M \rangle_K$ and $p \in \langle H \rangle_K$. If $p \neq h(m)$, return c and stop. Otherwise, return m.

Since an adversary cannot create the correct hash value $h(m + m')$ without knowing m, Step (3) will foil his attempt to discover $m + m'$.

4.2. Linear Algebra Attacks. In [**5**] and [**12**] several linear algebra attacks were suggested in order to attack Gröbner basis cryptosystems. Let us first describe a basic version of this attack (called *Moriarty attack* in [**5**]).

REMARK 4.3 (Basic Linear Algebra Attack). Consider the construction of the ciphertext via $c = m + \ell_1 p_1 + \cdots + \ell_r p_r$. The attacker knows the polynomials $p_1, \ldots, p_r$ and c, as well as a set M containing the support of m. He therefore has degree estimates for the coefficient polynomials ℓ_i, namely $\deg(\ell_i) \leq \deg(c) - \deg(p_i)$. Now he writes down the polynomials ℓ_i and the message m using indeterminate coefficients, i.e. he sets $\ell_i = \sum_j a_{ij} t_j$ and $m = \sum_j b_j t_j$ where the first sum ranges over all j such that the terms t_j are all terms of degree $\leq \deg(c) - \deg(p_i)$, and the second sum ranges over all j such that the terms t_j are the elements of M. After bringing $m + \ell_1 p_1 + \cdots + \ell_r p_r$ into standard form and equating its coefficients to those of c, the attacker obtains a linear system of equations for the indeterminates

a_{ij}, b_j which he then attempts to solve. If the linear system has no solution, he increases his degree estimates and tries the same procedure again.

The defence against this kind of attack is clear: one has to make the degree estimates $\deg(c) - \deg(p_i)$ large enough, in order to generate linear systems of equations in too many indeterminates to be solvable in an acceptable amount of time. At this point the first important difference between the standard (commutative) Polly Cracker systems and our WGBC surfaces: by Prop. 2.3, the process of bringing $\ell_i p_i$ into standard form creates a large number of terms. Hence the indeterminates a_{ij} appear in many different linear equations, and the linear equations are not sparse. Except for very small examples, the basic linear algebra attack has no chance of succeeding.

A more refined version of this attack has been presented in [12], Ch. 5, Sect. 6.

REMARK 4.4 (Intelligent Linear Algebra Attack). Consider again the construction of the ciphertext via $c = m + \ell_1 p_1 + \cdots + \ell_r p_r$. Instead of using a dense representation of ℓ_i, we make the following assumption:

If a term $t = x^\alpha \partial^\beta \in B_n$ appears in ℓ_i then there should be a term $t' = x^{\alpha'} \partial^{\beta'}$ in the support of p_i such that the term $t'' = x^{\alpha+\alpha'} \partial^{\beta+\beta'}$ is contained in the support of c.

Then the attacker uses indeterminate coefficients a_{ij} in ℓ_i only for the terms t satisfying this assumption and mounts a linear algebra attack as above.

Since multiplication and addition of commutative polynomials rarely cancel terms completely, this is a serious attack for the usual Polly Cracker system. In the case of the WGBC, we are again protected to a large extent by Prop. 2.3. Namely, by the expansion which takes place during the conversion of $h_{ij} p_j$ to standard form, the support of c becomes rather large and essentially all terms of lower degrees pseudo-divide some term in $\mathrm{Supp}(c)$. Hence even the *intelligent linear algebra attacker* has to introduce a prohibitively large number of unknowns a_{ij}.

Let us illustrate this effect with an extremely simple example.

EXAMPLE 4.5. In the Weyl algebra $A_2 = \mathbb{F}_{101}[x_1, x_2, \partial_1, \partial_2]$ we consider the "public key" polynomials $p_1 = 3x_1^2 \partial_1^3 - x_1^2 + x_1 \partial_1 - 1$ and $p_2 = 2x_2^3 \partial_2^2 + 3\partial_2^3 - 2x_2 \partial_2 + 2$. To encrypt the plaintext $m = 4x_1 + x_2 - \partial_2 - 2$, we use the coefficient polynomials $\ell_1 = -2x_1 x_2^3 \partial_2^2 + 2\partial_2^6 - 3x_1 \partial_2^3 + 2x_1 x_2 \partial_2 - 2x_1$ and $\ell_2 = -3x_1^2 \partial_1^3 \partial_2^3 - 3x_1^3 \partial_1^3 + x_1^2 \partial_2^3 - x_1 \partial_1 \partial_2^3 + \partial_2^3$.

Notice that we started with $\#\mathrm{Supp}(p_i) = 4$ and $\#\mathrm{Supp}(\ell_i) = 5$. The resulting ciphertext $c = m + \ell_1 p_1 + \ell_2 p_2$ has degree 13 and already 47 terms in its standard form. However, the number of terms an *intelligent linear algebra attacker* has to consider for ℓ_1 and ℓ_2 is 352 each. Thus he has to solve a linear system of equations in more than 700 indeterminates. For the reasons explained above, it will be a rather dense linear system.

For more realistic examples, the sizes of the linear systems that have to be solved for the intelligent linear algebra attack quickly become enormous, as our next examples show.

EXAMPLE 4.6. Let us consider the WGBC instance described in Example 3.5, and let us try to encrypt an actual message, e.g. the message

$$m = -6x_2^4 \partial_2^3 + 6\partial_2^6 + 5x_2^4 - \partial_2^4 + 6x_2^3 + x_1^2 + x_2 \partial_2 - 3\partial_1 \partial_2 + 2x_1 - 5$$

Notice that this message involves only 11 of the 2808 terms of M. To encipher m, we use the Weyl polynomials
$$\ell_1 = -5x_1^{10}x_2^{16}\partial_1^{12}\partial_2^{19} - 2x_1^8 x_2^{18}\partial_1^{10}\partial_2^{21} - \partial_1 + 1$$
$$\ell_2 = 4x_1^{11}x_2^{13}\partial_1^9\partial_2^{12} - 6x_1^9 x_2^{15}\partial_1^7\partial_2^{14} + 2\partial_2 + x_2 + 2$$
The resulting ciphertext $c = m + \ell_1 p_1 + \ell_2 p_2$ has degree 91 and 2357 terms in its standard support. Since $\deg(\ell_1) + \deg(p_1) = \deg(\ell_2) + \deg(p_2) = 93$, the top two degrees cancelled in its construction. Only two terms of m appear unchanged in c and one term of m does not appear at all in c.

Of course, both the size of the message and the size of the polynomials p_i have been chosen unrealistically small and cannot be considered safe. Nevertheless, even so the basic linear algebra attack leads to a linear system of size $3.2 \cdot 10^6 \times 0.9 \cdot 10^6$, and the intelligent linear algebra attack results in a linear system of size 345402×75970 having about 43 Mio. non-zero entries. This linear system could not be solved on our computing machine using the interface to the C++ package Linbox in the system ApCoCoA [3]. And even if it could have been solved, the attacker would only have learned that he should have increased his degree estimate, first from $55 = \deg(c) - \deg(p_1)$ to 56, and then to 57. Both increases lead to considerably larger linear systems.

Next we encrypt a message using the WGBC instance of Example 3.6.

EXAMPLE 4.7. In the setting of Example 3.6, consider the Weyl polynomials
$$\begin{aligned}\ell_1 &= -3x_1^5 x_2^4 x_3^4 \partial_1^8 \partial_2^5 \partial_3^3 - 2x_1^5 x_2^2 x_3^4 \partial_1^8 \partial_2^7 \partial_3^3 - x_1^8 x_2^5 x_3^3 \partial_1^2 \partial_2^7 \partial_3^4 + 3x_1^5 x_2^2 x_3^2 \partial_1^7 \partial_2^9 \partial_3^3 \\ &\quad -2x_1^5 x_2 x_3 \partial_1^7 \partial_2^8 \partial_3^3 + 3\partial_2^2 \partial_3 - 3\partial_2 \partial_3 + 2x_1 - 3\partial_1 + 1 \\ \ell_2 &= -3x_1^8 x_2^2 x_3 \partial_1^3 \partial_2^{10} \partial_3^2 + x_1^6 x_2^4 x_3^3 \partial_1^8 \partial_2^5 \partial_3^3 - x_1^5 x_2^2 \partial_1^{11} \partial_2^4 \partial_3^6 - 3x_1^5 x_2^2 \partial_1^{11} \partial_2^6 \partial_3^6 \\ &\quad +2x_1^6 x_2^4 x_3^2 \partial_1^8 \partial_2^5 \partial_3^4 + 3x_1^7 x_2^2 x_3 \partial_1^3 \partial_2^5 + x_3 \partial_1^3 \partial_2^3 - x_1 x_3 \partial_1 - \partial_1^2 \partial_3 - x_2^2 \\ &\quad +x_3 \partial_1 + \partial_1 \partial_3 - 2\partial_2 + x_3 - 1 \\ \ell_3 &= x_1^7 x_2^6 \partial_1^6 \partial_2^7 \partial_3^4 - x_1^7 x_2^6 x_3 \partial_1^5 \partial_2^9 \partial_3 + 3x_1^5 x_2^5 x_3^4 \partial_1^9 \partial_2^6 \partial_3 + 3x_1 x_2^7 \partial_2^7 + 2x_2^5 \partial_2^3 \partial_3 \\ &\quad +x_1 x_2^2 \partial_2^3 \partial_3^2 - x_1^2 \partial_2 + 2x_2 \partial_2 + x_1 \partial_3 + 2x_1 - 3\partial_1 - 3\partial_3 + 3\end{aligned}$$
They yield an element $c' = \ell_1 p_1 + \ell_2 p_2 + \ell_3 p_3$ of degree 49 whose standard form consists of 15649 terms. Of these, 311 terms are contained in M, so that we can use c' to encrypt more than 300 coefficients simultaneously. So, despite the size of c', message expansion is still acceptable. In fact, a more careful setup would have allowed us to use a much larger part of the 936 terms of M, or even of the 1960 terms of $\mathcal{O}_\sigma(I + \langle \partial_1^4 \rangle_\lambda)$.

An intelligent linear algebra attack leads to a linear system of equations of size $7.2 \cdot 10^6 \times 0.8 \cdot 10^6$ having more than $575 \cdot 10^6$ nonzero entries. Therefore it is safe to say that the intelligent linear algebra attack fails for this instance of the WGBC. It should be noted that one should actually also add terms of intermediate degrees to the elements ℓ_i to avoid that the support of c leaks information about the elements ℓ_i. Although this makes the system somewhat less efficient, it clearly makes linear algebra attacks even more hopeless.

4.3. The Partial Gröbner Basis Attack. In [5], the following attack was proposed for the standard Polly Cracker cryptosystem. (It is jokingly called the *Fantômas attack* there.)

REMARK 4.8 (Partial Gröbner Basis Attack). Suppose that an attacker of a WGBC knows the public polynomials $p_1, \ldots, p_r$ and a certain ciphertext c. Then he performs the following algorithm.

(1) Estimate the maximal degree $d = \max\{\deg(q_i g_i)\}$ of the summands in a representation $c = \mathrm{NF}_{\sigma,G}(c) + \sum_{i=1}^{s} q_i g_i$ for which $\mathrm{LT}_\sigma(q_i g_i) \leq_\sigma \mathrm{LT}_\sigma(c)$ holds. (Such a representation exists because the Gröbner basis G satisfies conditions (A_i) of [14], Sect. 2.1.)
(2) Run the Buchberger algorithm on $\{p_1, \ldots, p_r\}$ to compute a Gröbner basis of $J = \langle p_1, \ldots, p_r \rangle$, but skip all operations involving polynomials of degree larger than d. (We shall say that the attacker computes a *d-truncated partial Gröbner basis*.) Let the result be a set H.
(3) Using the Division Algorithm, compute the *normal remainder* $\mathrm{NR}_{\sigma,H}(c)$. If its support is contained in M, return this normal remainder. Otherwise, increase d by one and repeat steps (2) and (3).

This attack is based on a number of arguments favouring the attacker. For instance, in the representation of most polynomials c only comparatively few Gröbner basis elements g_i are involved, and their degrees are relatively low. Moreover, since the perceived difficulty of computing Gröbner bases of (commutative) polynomial ideals is mainly an *output complexity*, it is frequently not so difficult to generate enough Gröbner basis elements for the desired representation of c to exist.

In the construction of a WGBC, we have explicitly requested that the designer checks that partial Gröbner bases of J are hard to compute. But is this realistic? How difficult is it to accomplish? Again we note that in Weyl algebras we have Prop. 2.3 at our disposal. Even if one starts with randomly chosen sparse Weyl polynomials and plans to truncate the computation at a low degree, the expansion of the supports happening at each step of Buchberger's algorithm quickly increases the size of the computation and slows it down enormously. Of course, our polynomials $p_1, \ldots, p_r$ are not entirely random, since they are contained in a larger ideal which has a simple Gröbner basis, namely G. But we have not been able to use this fact to the benefit of the attacker, and in all cases that we tried, the predicted expansion of the supports happened indeed. Most of the time, Conditions (5) of Remark 3.2 and of Remark 3.3 were satisfied automatically.

For instance, in Example 3.5 the partial Gröbner basis attack fails as follows.

EXAMPLE 4.9. Let $J = \langle p_1, p_2 \rangle_\lambda$ be the left ideal generated by the public polynomials of Example 3.5. Using the most efficient implementation of Buchberger's algorithm and a degree bound of 60, we were able to compute the corresponding partial left Gröbner basis H of J. It took about one hour, consists of $\#H = 108$ Weyl polynomials and needs about 183 MB of memory to be stored. This partial Gröbner basis is not even sufficient to decrypt the ciphertext of Example 4.6. Instead, the computation of the normal remainder $\mathrm{NR}_{\sigma,H}(c)$ takes 5 hours and results in a polynomial having 284745 terms in its standard support.

Increasing the degree bound to 65 resulted in a failure to find the partial Gröbner basis, after 150 hours of computation and the consumption of 3.5 GB of memory. Since the degree bound necessary to ensure the success of the partial Gröbner basis attack in this example would be 93, it appears to be out of reach using current technology.

Similarly, for Example 3.6, partial Gröbner bases can be found (with difficulty) approximately up to degree 35, whereas an actual ciphertext such as the one in Example 4.7 would require a degree bound close to 50.

5. Two-Sided Weyl Gröbner Basis Cryptosystems

In characteristic zero, Weyl algebras have only the trivial two-sided ideals. However, in characteristic $p > 0$, this is no more the case and we can try to use these two-sided ideals to construct Gröbner basis cryptosystems. The following proposition tells us where to look for them.

PROPOSITION 5.1. *Let $A_n = K\langle x_1, \ldots, x_n, \partial_1, \ldots, \partial_n \rangle$ be a Weyl algebra of index n over field K of characteristic $p > 0$, and let $I \subseteq A_n$ be a two-sided ideal. Then the reduced Gröbner basis of I with respect to any term ordering is contained in the center $C_n = K[x_1^p, \ldots, x_n^p, \partial_1^p, \ldots, \partial_n^p]$ of A_n.*

PROOF. By Proposition 2.2.b, the Weyl algebra A_n is an Azumaya algebra over its center C_n. Therefore it follows from [20], Ch. 13, Prop. 7.9, that every two-sided ideal I in A_n is generated by elements in $I \cap C_n$. If we start from a system of generators of I contained in C_n and compute a reduced Gröbner basis, we never leave this polynomial ring. Consequently, also the reduced Gröbner bases of I are contained in C_n. □

Therefore we now restrict ourselves to secret keys contained in the center. The definition of two-sided Weyl Gröbner basis cryptosystems is essentially analogous to the definition of left WGBC. For the convenience of the reader, we spell it out explicitly.

Let K be a field of characteristic $p > 0$, let $A_n = K\langle x_1, \ldots, x_n, \partial_1, \ldots, \partial_n \rangle$ be the Weyl algebra of index n over K, let $C_n = K[x_1^p, \ldots, x_n^p, \partial_1^p, \ldots, \partial_n^p]$ be the center of A_n, let B_n be the set of standard terms in A_n, and let σ be a term ordering on B_n.

DEFINITION 5.2. A **two-sided Weyl Gröbner basis cryptosystem** consists of the following data.

Secret key: A two-sided σ-Gröbner basis $G = \{g_1, \ldots, g_s\}$ contained in the center C_n which generates a two-sided ideal $I = \langle G \rangle_\tau$ in A_n. Moreover, also the set $\mathcal{O}_\sigma(I) = B_n \setminus \{\mathrm{LT}_\sigma(f) \mid f \in I \setminus \{0\}\}$ is secret.

Public key: A set of Weyl polynomials $\{p_1, \ldots, p_r\}$ contained in $I \setminus \{0\}$ and a subset M of $\mathcal{O}_\sigma(I)$.

Message space: The message space is the K-vector subspace $\langle M \rangle_K$ of A_n generated by M.

Ciphertext space: The ciphertext units are elements of A_n in standard form.

Encryption map: Given a plaintext $m \in \langle M \rangle_K$, choose Weyl polynomials $\ell_1, \ldots, \ell_q, \varrho_1, \ldots, \varrho_q$ and indices $j_1, \ldots, j_q \in \{1, \ldots, r\}$ and compute the standard form of $c = m + \ell_1 p_{j_1} \varrho_1 + \cdots + \ell_q p_{j_q} \varrho_q$.

Decryption map: Given a ciphertext unit $c \in A_n$, compute $\mathrm{NF}_{\sigma, G}(c)$. If the result is contained in $\langle M \rangle_K$, return it. Otherwise, return c.

Much of what has been noted about left WGBC remains true for two-sided WGBC. In the following we restrict ourselves to pointing out important differences and to providing some actual examples. The obvious advantage of two-sided WGBC encryption is that it offers additional freedom to hide the plaintext and the coefficient Weyl polynomials ℓ_i, ϱ_j. This comes at the price of a somewhat higher message expansion, as the standard form of c will tend to have a larger support. Clearly, the construction of a concrete instance has to balance these factors.

It is clear that an attacker who manages to compute a two-sided Gröbner basis of the public ideal $J = \langle p_1, \ldots, p_r \rangle_\tau$ can break the two-sided WGBC. Moreover, we note that the ideal J is contained in the ideal I which is generated by elements of C_n, and that in the construction of the public polynomials p_i it suffices to take combinations of the form $p_i = h_{i1}g_1 + \cdots + h_{is}g_s$ because we have $\langle G \rangle_\tau = \langle G \rangle_\lambda$. For key generation and encryption, we can use the obvious analogies to the suggestions of Remarks 3.2 and 3.3. For decryption, we have the following advantage.

REMARK 5.3. For a two-sided WGBC, the decryption process is usually faster than for a left WGBC with similar parameters. Since the two-sided Gröbner basis G is contained in C_n, the reduction steps carried out during the computation of $\mathrm{NF}_{\sigma,G}(c)$ do not suffer from the expansion of the standard support described in Proposition 2.3.

As for the resistance of two-sided WGBC with respect to the attacks in Section 4, we have another advantage.

REMARK 5.4. For a two-sided WGBC, to set up a linear algebra attack we have to use indeterminate coefficients for the polynomials ℓ_i, ϱ_i in the representation $c = \ell_1 p_{j_1} \varrho_1 + \cdots + \ell_q p_{j_q} \varrho_q$ of the ciphertext. Bringing the right-hand side into standard form results in a system of *quadratic equations* for these coefficients. In this sense, there does not exist a *linear algebra attack* for a two-sided WGBC.

Furthermore, we note that each public polynomial p_k can be used several times in the construction of the ciphertext, and the attacker does know how often it has been used. Thus he has to make guesses for the numbers q and $\#\{i \in \{1, \ldots, q\} \mid j_i = k\}$ which increases the number of cases to consider and the number of indeterminate coefficients to a point where the attack becomes completely impractical.

Since we can defend against chosen ciphertext attacks as in the left WGBC case, it remains to look at the partial Gröbner basis attack. As we shall see, it suffices to compute a partial Gröbner basis of $J = \langle p_1, \ldots, p_r \rangle_\tau$ up to the highest manageable degree bound and then use coefficient polynomials ℓ_i, ϱ_i in the encryption process which generate ciphertexts of significantly higher degrees. Let us conclude this section with an example indicating that instances of two-sided WGBC can be constructed that are reasonable efficient and secure.

EXAMPLE 5.5. In the Weyl algebra $A_3 = \mathbb{F}_2 \langle x_1, x_2, x_3, \partial_1, \partial_2, \partial_3 \rangle$, we use the term ordering $\sigma = \mathtt{DegRevLex}$ and consider the polynomials

$$\begin{aligned}
f_1 &= x_1^6 x_2^4 + x_1^4 x_2^2 + x_1^2 + 1 & f_2 &= x_2^6 + x_2^4 x_3^2 + x_2^2 + 1 \\
f_3 &= \partial_1^6 \partial_2^4 + \partial_1^4 \partial_2^2 + \partial_1^2 + 1 & f_4 &= \partial_3^8 + x_1^2 \partial_2^2 \partial_3^2 + 1 \\
f_5 &= x_2^2 x_3^{10} + x_3^6 + x_1^2 x_3^2 + x_3^2 + 1
\end{aligned}$$

The reduced σ-Gröbner basis of $\langle f_1, \ldots, f_5 \rangle_\tau$ is the set $G = \{g_1, \ldots, g_{10}\}$, where

$g_1 = x_2^6 + x_2^4 x_3^2 + x_2^2 + 1$
$g_2 = \partial_3^8 + x_1^2 \partial_2^2 \partial_3^2 + 1$
$g_3 = \partial_1^6 \partial_2^4 + \partial_1^4 \partial_2^2 + \partial_1^2 + 1$
$g_4 = x_1^6 x_2^2 + x_1^2 x_2^4 + x_1^2 x_2^2 x_3^2 + x_1^4 + x_2^4 + x_2^2 x_3^2 + x_1^2 + 1$
$g_5 = x_2^2 x_3^{10} + x_3^6 + x_1^2 x_3^2 + x_3^2 + 1$
$g_6 = x_1^4 x_2^4 + x_1^4 x_2^2 x_3^2 + x_1^6 + x_1^2 x_2^4 + x_1^2 x_2^2 x_3^2 + x_1^4 + x_1^2 x_2^2 + x_2^4 + x_1^2 x_3^2 + x_2^2 x_3^2$
$\qquad + x_1^2 + x_2^2 + x_3^2 + 1$
$g_7 = x_1^8 + x_1^2 x_2^4 x_3^2 + x_1^2 x_2^2 x_3^4 + x_2^4 x_3^2 + x_2^2 x_3^4 + x_2^4 + x_1^2 x_3^2 + x_2^2 x_3^2 + x_1^2 + x_2^2$
$g_8 = x_2^4 x_3^6 + x_2^2 x_3^8 + x_3^{10} + x_1^2 x_2^4 x_3^2 + x_1^2 x_2^2 x_3^4 + x_2^4 x_3^2 + x_2^2 x_3^4 + x_3^6 + x_2^4 + x_1^2 x_3^2$
$\qquad + x_2^2 x_3^2 + x_3^2 + 1$
$g_9 = x_3^{14} + x_1^2 x_3^{10} + x_1^6 x_3^2 + x_1^2 x_2^4 x_3^2 + x_1^2 x_2^2 x_3^4 + x_2^4 x_3^2 + x_3^8 + x_1^2 x_2^4 + x_1^2 x_2^2 x_3^2$
$\qquad + x_2^4 + x_1^2 x_3^2 + x_2^2 x_3^2 + x_3^4 + x_1^2 + x_2^2 + x_3^2 + 1$
$g_{10} = x_1^4 x_3^{10} + x_1^6 x_3^6 + x_1^2 x_3^{10} + x_1^2 x_2^4 x_3^4 + x_1^2 x_3^8 + x_3^{10} + x_1^6 x_3^2 + x_1^4 x_2^2 x_3^2 + x_1^4 x_3^4$
$\qquad + x_2^4 x_3^4 + x_3^8 + x_1^6 + x_2^4 x_3^2 + x_1^2 x_3^4 + x_2^2 x_3^4 + x_1^2 x_2^2 + x_3^4 + x_2^2 + x_3^2$

Next we construct the public key $\{p_1, p_2, p_3\}$ by using combinations of the form $p_1 = h_{11} g_2 + h_{12} g_4 + h_{13} g_6$ and $p_2 = h_{21} g_3 + h_{22} g_5 + h_{23} g_7$ as well as $p_3 = h_{31} g_2 + h_{32} g_3 + h_{33} g_9 + h_{34} g_{10}$. In fact, if we choose

$h_{11} = x_1^6 x_2^4 x_3^4 \partial_1^5 \partial_2^5 + x_1^6 x_2^2 x_3^6 \partial_1^5 \partial_2^5 + x_1^8 x_3^4 \partial_1^5 \partial_2^5 + \partial_2^2 + \partial_1 \partial_3 + \partial_1 + 1$
$h_{12} = x_2^2 x_3^6 \partial_1^5 \partial_2^5 \partial_3^6 + x_3^8 \partial_1^5 \partial_2^5 \partial_3^6 + \partial_1^3 + \partial_2 \partial_3 + \partial_2 + \partial_3$
$h_{13} = x_1^2 x_3^4 \partial_1^5 \partial_2^5 \partial_3^8 + x_1^2 x_3^6 \partial_1^5 \partial_2^5 \partial_3^6 + x_3^6 \partial_1^5 \partial_2^5 \partial_3^6 + x_3^4 \partial_1^5 \partial_2^5 \partial_3^8 + x_1^4 x_3^4 \partial_1^5 \partial_2^7 \partial_3^2$
$\qquad + x_1 \partial_1 + \partial_3^2 + \partial_3 + 1$

we get a polynomial p_1 of degree 30 having 115 terms in its standard support. Notice that the top two degrees cancel during the expansion of p_1. Similarly, if we choose

$h_{21} = x_1^8 x_2^5 x_3^2 \partial_1 \partial_2^2 \partial_3^4 + x_1^8 x_2^5 x_3^2 \partial_1 \partial_2^2 \partial_3^4 + x_1^8 x_2^2 x_3^{10} + x_1^2 x_2^4 x_3^{14}$
$\qquad + x_1 \partial_2 + \partial_2 \partial_3 + \partial_1^2 + x_3 + \partial_1 + 1$
$h_{22} = x_1^9 x_2^2 \partial_1^2 \partial_2^4 \partial_3^3 + x_1^3 x_2^6 x_3^2 \partial_1^2 \partial_2^4 \partial_3^3 + x_1^3 x_2^4 x_3^2 \partial_1^2 \partial_2^4 \partial_3^3 + x_1^2 x_2^4 x_3^2 \partial_1^6 \partial_2^4$
$\qquad + x_1 x_2^4 x_3^2 \partial_1^2 \partial_2^4 \partial_3^3 + \partial_2^3 + \partial_1 \partial_3 + x_1 \partial_1 + x_2^2 + \partial_1 + \partial_3 + 1$
$h_{23} = x_2^2 x_3^{10} \partial_1^6 \partial_2^4 + x_1 x_2^4 x_3^{10} \partial_1^2 \partial_2^4 \partial_3^3 + x_1 x_2 x_3 + \partial_1 \partial_2^2 + \partial_3^2 + x_1 + \partial_1 + \partial_2$

we get a polynomial p_2 of degree 30 having 146 terms in its standard support. Again the two top degrees cancel during the expansion of p_2. Finally, we choose

$h_{31} = x_1^4 x_2^2 x_3^{14} \partial_1^2 \partial_2^2 + x_1^4 x_3^{10} \partial_1^6 \partial_2^4 + x_1^4 x_2^2 x_3^{14} \partial_1^2 \partial_2^2 + x_1^6 x_3^{10} \partial_1^2 \partial_2^4$
$\qquad + x_1 \partial_3^3 + \partial_1 \partial_2 + \partial_2 \partial_3 + x_3 + \partial_1 + \partial_3$
$h_{32} = x_1^4 x_3^{10} \partial_3^8 + x_2^2 x_3^{14} \partial_1^6 + x_1^2 x_2^2 x_3^{14} \partial_3^4 + x_1^2 x_2^6 x_3^8 \partial_3^4 + x_1^2 x_2^2 x_3^{12} \partial_3^4 + x_1^2 x_2^2 x_3^{10} \partial_1^6$
$\qquad + x_3^2 \partial_1 + \partial_2^3 + \partial_2^2 \partial_3 + x_2 \partial_2 + x_3 \partial_1 + x_1 + \partial_3 + 1$
$h_{33} = x_1^4 x_2^2 \partial_1^6 \partial_2^4 \partial_3^4 + x_1^6 \partial_1^2 \partial_2^4 \partial_3^8 + x_2^2 \partial_1^{12} \partial_2^4 + x_1^4 \partial_1^2 \partial_2^4 \partial_3^8 + x_1^2 \partial_1^2 \partial_2^4 \partial_3^8 + x_2^2 \partial_1^6 \partial_2^4 \partial_3^4$
$\qquad + \partial_1^2 + \partial_1 \partial_3 + \partial_3^2 + x_1 + \partial_1 + \partial_2$
$h_{34} = x_1^2 x_3^4 \partial_1^2 \partial_2^4 \partial_3^8 + x_2^2 x_3^4 \partial_1^6 \partial_2^4 \partial_3^4 + x_1^2 \partial_1^6 \partial_2^4 \partial_3^2 + x_3^2 \partial_1^2 \partial_2^4 \partial_3^8$
$\qquad + x_1 \partial_1 + \partial_2 \partial_2 + \partial_1^2 + \partial_1 + \partial_2 + 1$

and we get a polynomial p_3 of degree 30 with 348 terms in its standard support. Here even the top four degrees cancel during the conversion of p_3 into standard form.

These public Weyl polynomials generate a two-sided ideal $J = \langle p_1, p_2, p_3 \rangle_\tau$ whose partial Gröbner bases cannot be computed to very high degrees. On our computing machine, we reach the end of its capabilities around degree 45. So, if we use ciphertext polynomials of degree 60 or higher, we are safe against partial Gröbner basis attacks.

Notice that the construction of p_1, p_2, p_3 does not yet satisfy all requirements of Remark 3.2. In a number of intermediate degrees they have no or very few standard terms. To create an actual working public key, we should add terms of intermediate degrees to the polynomials h_{ij}. This will enlarge the support of the p_i even more and points to the main disadvantage of two-sided WGBC: their message expansion appears to be worse than that of left WGBC.

Naturally, the size of the message expansion also depends on the size of the message space. In the case at hand, the Hilbert function of $\bar{A} = A_n / \operatorname{LT}_\sigma(G)$ grows until it stabilizes at $\operatorname{HF}_{\bar{A}}(i) = 28800$ for $i \geq 33$. Therefore we can safely use a set M consisting of 28000 elements. Then each ciphertext unit encrypts about 28 kbit, and even a substantial increase in the size of the support of the p_i is acceptable.

Further suggestions for secure instances of left and two-sided WGBC can be found in [2].

Acknowledgements. The authors thank Viktor Levandovskyy for valuable discussions and suggestions. This paper is based in part on the first author's dissertation [2]. The first author is grateful to HEC and DAAD for financial support. The second author wishes to thank Università di Genova, and in particular Lorenzo Robbiano, for the hospitality he enjoyed during the preparation of this paper.

References

[1] P. Ackermann and M. Kreuzer, *Gröbner basis cryptosystems*, Applicable Alg. in Eng., Commun. and Comput. **17** (2006), 173–194. MR2233780 (2007e:94048)

[2] R. Ali, *Weyl Gröbner basis cryptosystems*, dissertation, Universität Passau 2011.

[3] ApCoCoA team, *ApCoCoA: Applied Computations in Commutative Algebra*, available at http://www.apcocoa.org.

[4] A. Belov-Kanel and M. Kontsevich, *Automorphisms of the Weyl algebra*, Lett. Math. Phys. **74** (2005), 181–199. MR2191954 (2006k:16068)

[5] Boo Barkee et al., *Why you cannot even hope to use Gröbner bases in public key cryptology*, J. Symb. Comput. **18** (1994), 497–501. MR1334658

[6] S. Coutinho, *A Primer of Algebraic D-Modules*, Cambridge University Press, 1995. MR1356713 (96j:32011)

[7] M. Fellows and N. Koblitz, *Combinatorial cryptosystems galore!*, Contemp. Math. **168** (1994), 51–61. MR1291417 (95e:94028)

[8] D.R. Grayson and M.E. Stillman, *Macaulay 2, a software system for research in algebraic geometry*, available at http://www.math.uiuc.edu/Macaulay2.

[9] G.-M. Greuel, G. Pfister and H. Schönemann, *Singular 3.0. A computer algebra system for polynomial computations*, available at http://www.singular.uni-kl.de.

[10] D. Hofheinz and R. Steinwandt, *A "differential" attack on Polly Cracker*, Int. J. Inf. Secur. **1** (2002), 143–148.

[11] A. Kandri-Rody and V. Weispfenning, *Non-commutative Gröbner bases in algebras of solvable type*, J. Symb. Comput. **9** (1990), 1–26. MR1044911 (91e:13025)

[12] N. Koblitz, *Algebraic Aspects of Cryptography*, Alg. and Comput. in Math. **3**, Springer Verlag, 1998. MR1610535 (2000a:94012)

[13] M. Kreuzer, *Algebraic attacks galore!*, Groups - Complexity - Cryptology **1** (2009), 231–259. MR2598992 (2011c:94050)

[14] M. Kreuzer and L. Robbiano, *Computational Commutative Algebra 1*, Springer Verlag, 2000. MR1790326 (2001j:13027)

[15] M. Kreuzer and L. Robbiano, *Computational Commutative Algebra 2*, Springer Verlag, 2005. MR2159476 (2006h:13036)

[16] V. Levandovskyy, *Non-commutative computer algebra for polynomial algebras: Gröbner bases, applications and implementation*, dissertation, Universität Kaiserslautern 2005.

[17] V. Levandovskyy, *Plural, a noncommutative extension of Singular: past, present and future*, Reports on Computer Algebra **36**, Universität Kaiserslautern 2006.

[18] A. Leykin, *D-modules for Macaulay 2*, in: A.M. Cohen, X.-S. Gao and N. Takayama (eds.), *Mathematical Software (ICMS Beijing 2002)*, World Sci. Publishing 2002, pp. 169–179. MR1932609

[19] L. Ly, *Polly two – a new algebraic polynomial-based public-key scheme*, Applicable Alg. in Eng., Commun. and Comput. **17** (2006), 267–283. MR2233786 (2007c:94149)

[20] J.C. McConnell and J.C. Robson, *Noncommutative Noetherian Rings*, Graduate Studies in Math. **30**, Amer. Math. Soc., 1987. MR934572 (89j:16023)

[21] M. Noro, T. Shimoyama and T. Takeshima, *Risa/Asir, a computer algebra system*, available at ftp://archives.cs.ehime-u.ac.jp/pub/asir2000.

[22] T. Rai, *Infinite Gröbner bases and noncommutative Polly Cracker cryptosystems*, dissertation, Virginia Polytechnic Institute 2004.

[23] T. Rai and S. Bulygin, *Noncommutative Polly Cracker-type cryptosystems and chosen-ciphertext security*, Cryptology ePrint Archive, Report 2008/504, available at http://eprint.iacr.org.

[24] P. Revoy, *Algèbres de Weyl en charactéristique p*, Compt. Rend. Acad. Sci. Paris, Ser. A **276** (1973), 225–228. MR0335564 (49:345)

[25] N. Taslaman, *Private key extension of Polly Cracker cryptosystems*, Comp. Sci. J. of Moldova **16** (2008), 117–132. MR2418007 (2009f:94061)

[26] Y. Tsuchimoto, *Preliminaries on Dixmier conjecture*, Mem. Fac. Sci. Kochi Univ. (Math.) **24** (2003), 43–59. MR1967342 (2004k:16070)

FACULTY OF ENGINEERING, MOHAMMAD ALI JINNAH UNIVERSITY, ISLAMABAD, PAKISTAN
E-mail address: rashid.ali@jinnah.edu.pk

FAKULTÄT FÜR INFORMATIK UND MATHEMATIK, UNIVERSITÄT PASSAU, D-94030 PASSAU, GERMANY
E-mail address: Martin.Kreuzer@uni-passau.de

A New Look at Finitely Generated Metabelian Groups

Gilbert Baumslag, Roman Mikhailov, and Kent E. Orr

ABSTRACT. A group is metabelian if its commutator subgroup is abelian. For finitely generated metabelian groups, classical commutative algebra, algebraic geometry and geometric group theory, especially the latter two subjects, can be brought to bear on their study. The object of this paper is to describe some of the new ideas and open problems that arise.

1. Introductory remarks

There was a conference at the City College of New York on March 17 and 18 of 2011, funded by the National Science Foundation, entitled *Finitely Presented Solvable Groups*. The first named author of this paper discussed how geometric group theory and algebraic geometry might play a role in better understanding finitely generated metabelian groups, in particular the isomorphism problem. In the past 18 months the present authors have introduced some new ideas which more closely tie algebraic geometry to the isomorphism problem, in particular, and to the study of these finitely generated metabelian groups [**BMOa, BMOb**]. A key idea is that of *para-equivalence of finitely generated metabelian groups*, in which localization of modules plays a fundamental and simplifying role, as it does in algebraic geometry. The object of the present note is to provide a quick overview of the general subject and to explain how these ideas come into play. Subsequent papers will describe this in more detail.

Geometric group theory in the form of hyperbolic groups was introduced by Gromov in 1983 [**Gro84**]. Gromov's idea was to view infinite groups as geometric objects, and with his extraordinary power and insight he revolutionized part of combinatorial group theory. This founded the subject of *geometric group theory.* In fact, many of the ideas go back to Dehn in 1910, 1911, Tartakovski [**Tar49b**], [**Tar49c**], [**Tar49a**], Lyndon, Weinbaum, Greendlinger, Lipschutz, Schupp (see the book [**LS77**] for these and a host of references), and Rips [**Rip82**]. These hyperbolic groups are

---

2010 *Mathematics Subject Classification.* Primary 20F16, 20D10, 13B30, 13B35.

The research of the first author is supported by Grant CNS 111765.

The research of the second author is supported by the National Science Foundation under agreement No. DMS-0635607. Any opinions, findings and conclusions or recommendations expressed in this material are those of the authors and do not necessarily reflect the views of the National Science Foundation.

The third author thanks the National Science Foundation, Grant 707078, and the Simons Foundation, Grant 209082, for their support.

a far cry from solvable groups, and more closely resemble free groups. In fact solvable groups have been largely excluded from geometric group theory. Recently, some attention has been paid to metabelian groups - witness the work of Farb and Mosher [**FM00**], Eskin and Fisher and Whyte [**EFW07**], [**EFW**].

The hyperbolic groups of Gromov are finitely presented, unlike many finitely generated metabelian groups. However, as we will point out in due course, when viewed as objects in the category of metabelian groups, they are finitely presented and geometric ideas might, if properly developed, play a part in their study. Geometric ideas of a rather different kind do arise naturally in the case of metabelian groups through commutative algebra. This suggest that major progress may arise through ideas borrowed from algebraic geometry.

2. Philip Hall's approach

In 1954, Philip Hall began a systematic study of finitely generated metabelian groups [**Hal54**]. We recall his approach here. To this end, we shall use the following notation. Given a group G, we denote the conjugate $g^{-1}ag$ of an element $a \in G$ by the element g by a^g, the commutator $g^{-1}a^{-1}ga$ of g and a by $[g,a]$ and the subgroup of a group G generated by the commutators $[x,y] = x^{-1}y^{-1}xy$, where x and y range over the elements of G, by $[G,G]$. Then G is termed metabelian if $[G,G]$ is abelian. Thus a metabelian group G is an extension of an abelian normal subgroup A by an abelian group $G/A = Q$.

Hall first observed that one can view this subgroup A as a $\mathbb{Z}[Q]$-module. Precisely, Q acts on A on the right by conjugation:

$$a(gA) = g^{-1}ag \ (a \in A, g \in G).$$

So the integral group ring $R = \mathbb{Z}[Q]$ of Q acts on A which, writing it additively, then becomes a right R-module:

$$a(\Sigma_{i=1}^n c_i g_i Q) = \Sigma_{i=1}^n c_i(g_i^{-1} a g_i) \ (g_i \in G).$$

If G is finitely generated, then Q is a finitely generated abelian group. It then turns out that A is a finitely generated R-module and so one can avail oneself of the structure theory of classical commutative algebra. In particular, A is Noetherian.

3. A few properties of finitely generated metabelian groups

Philip Hall's approach to the study of finitely generated metabelian groups described above in §2 gave rise to an entire body of results. We describe some of these below. Items 1, 2, 3 and 6 are due to Hall. A general reference to much of what is described in this section can be found in the book [**LR04**].

(1) Finitely generated metabelian groups satisfy the maximal condition for normal subgroups, that is, any properly ascending chain of normal subgroups is finite. This follows from the fact, adopting the notation above, that A is Noetherian.

(2) Every finitely generated metabelian group G is finitely presentable in the category of metabelian groups, i.e., G can be defined in terms of finitely many generators and finitely many relations together with all relations of the form $[[x,y],[z,w]] = 1$. We write a group presentation in the category of metabelian groups using double angle brackets, as follows:

$$G = \langle\langle x_1, \ldots, x_m \mid r_1 = 1, \ldots, r_n = 1 \rangle\rangle.$$

(3) It follows from (2) that there are only countably many isomorphism classes of finitely generated metabelian groups.
(4) Algorithmically, finitely generated metabelian groups are well behaved. So for example, the word and conjugacy problem are solvable.
(5) There is an algorithm to decide if a finitely generated metabelian group is residually nilpotent, i.e., if the intersection of the terms of the lower central series is trivial.
(6) Finitely generated metabelian groups are residually finite, i.e., the normal subgroups of finite index have trivial intersection.
(7) Although finitely generated metabelian groups are not necessarily finitely presented, every finitely presented metabelian group can be embedded in a finitely presented metabelian group.
(8) Bieri-Strebel have introduced a geometric invariant which distinguishes the finitely presented metabelian groups from the others [**BS81**]. It is commonly believed that the computation of this invariant is algorithmic.
(9) Groves-Miller [**GM86**] have solved the isomorphism problem for free metabelian groups, i.e., for metabelian groups which can be presented in the form
$$G = \langle\langle x_1, \ldots, x_m \rangle\rangle.$$
(10) As of now the isomorphism problem for finitely generated metabelian groups remains, perhaps, the most intriguing open problem about these groups. At first sight it seems to be connected to Hilbert's Tenth problem which suggests that it is algorithmically undecidable. We shall not touch on this aspect of the isomorphism problem but go in a different direction. We will say more about the isomorphism problem in the course of further discussion.

4. The Bieri-Strebel invariant

Let Q be a finitely generated abelian group. As previously noted, for any extension G of a (not necessarily finitely generated) abelian group A by Q, the conjugation action of G on Q makes A into a $\mathbb{Z}[Q]$–module. It is easy to see that G is finitely generated if and only if A is finitely generated as a $\mathbb{Z}[Q]$–module. A homomorphism of Q to the additive group $\mathbb{R}$ of real numbers is called a *valuation* of Q. Associated to each such valuation v one has the submonoid of Q
$$Q_v = \{q \in Q \mid v(q) \geq 0\}.$$
For valuations v, v' we write $v \sim v'$ if and only if there exists $\lambda > 0$ such that $v(q) = \lambda v'(q)$ for all $q \in Q$. Let n be the torsion-free rank of Q. Then $Hom(Q, \mathbb{R}) \cong \mathbb{R}^n$, and there is an obvious identification between the set of equivalence classes of nontrivial valuations of Q and the $(n-1)$-sphere S^{n-1}.

Let A be a finitely generated $\mathbb{Z}[Q]$–module. We can view A as a module over the commutative ring $\mathbb{Z}[Q_v] \leq \mathbb{Z}[Q]$. Define Σ_A to be the set of $\sim$ classes of valuations v on Q such that A is finitely generated as a $\mathbb{Z}[Q_v]$–module.

The module A is said to be *tame* if $\Sigma_A \cup -\Sigma_A = S^{n-1}$, in other words, for every valuation v of Q, either A is finitely generated as a $\mathbb{Z}[Q_v]$–module, or else it is finitely generated as a $\mathbb{Z}[Q_{-v}]$–module.

Then Bieri and Strebel have proved the following theorems [**BS81**]:

(1) Submodules of tame modules are tame.

(2) If the finitely generated group G is an abelian extension of the abelian group Q, then G is finitely presented if and only if A is tame as a $\mathbb{Z}[Q]$-module.

5. The geometry of the Cayley graph and metabelian presentations

Gromov, in his paper in his paper [**Gro84**], introduced into group theory a view of infinite groups as geometric objects. What follows here is a very short account of Gromov's basic idea and possible applications to the study of finitely generated metabelian groups.

Suppose that G is a group equipped with a finite set X of generators. We turn G into a metric space $\Gamma = \Gamma(G, X)$ by defining the distance $d(g, h)$ between g and h to be the length of the shortest X-word equal to gh^{-1}. It turns out that Γ is an invariant of G, for if one changes the finite generating sets involved, the metric spaces turn out to be quasi-isometric (see [**BH99**] for more details and further explanation.) In the event that Γ has the geometry of a two-dimensional hyperbolic space, then the group G is said to be hyperbolic. These hyperbolic groups can be described in terms of the complexity of their word problems, i.e., in terms of what are called isoperimetric functions. Roughly speaking, given a finite presentation of a group, the isoperimetric function counts the number of times one uses defining relations to prove that a given relator of the group is the identity. This function is often referred to as the Dehn function. Hyperbolic groups turn out to be those finitely presented groups with linear Dehn functions. Little is known about isoperimetric functions of finitely presented solvable groups. Recently Kassabov and Riley [**KR**] showed that the example described in §11, below, has an exponential Dehn function, but when the extra relation $b^n = 1$ is added, the Dehn function becomes quadratic. This fascinating area needs further investigation. It would be interesting to explore whether any of Gromov's ideas can be adapted to the study of finitely generated metabelian groups.

Because finitely generated metabelian groups are finitely presentable in the variety of metabelian groups one can associate with them a corresponding relative isoperimetric function. The first named author computed some of these isoperimetric functions of metabelian groups in Rio many years ago. Cheng-Fen Fuh in her thesis at the Graduate Center of Cuny in 2000, computed the relative isoperimetric functions of a number of additional metabelian groups. In particular she proved that the group $\langle a, t \mid a^t = a^2 \rangle$ has a linear relative isoperimetric function. In the category of all groups, this group has an exponential isoperimetric function. It is not clear how much of Gromov's program can meaningfully be carried out for metabelian groups, but it is conceivable that some of these ideas can be used. This is a topic that has yet to be explored.

6. Hilbert functions

Let G be a finitely generated metabelian group, $\gamma_n(G)$ the n-th term of the lower central series of G. The Lazard Lie ring $L(G)$ of G is defined to be the additive abelian group

$$L(G) = \bigoplus_{n=1}^{\infty} \gamma_n(G)/\gamma_{n+1}(G)$$

with binary operation

$$[a\gamma_{m+1}(G), b\gamma_{n+1}(G)] = [a,b]\gamma_{m+n+1}(G).$$

$L(G)$ can be thought of as a linearization of G. Then it turns out that L(G) is metabelian and satisfies the maximal condition on ideals. Several authors have proven, following Hilbert and Serre, that $h(G) = \Sigma_{n=1}^{\infty} r_n t^n$, where r_n is the torsion-free rank of $\gamma_n(G)/\gamma_{n+1}(G)$, is a rational function, termed the Hilbert function. One can define other growth functions for these metabelian groups following some of the ideas that arise in algebraic geometry.

7. Where to go from here

As already noted, the major open problem about finitely generated metabelian groups is the isomorphism problem. The structure of finitely generated metabelian groups depends on the ideal theory of finitely generated commutative rings, and the corresponding submodule theory of the modules over these rings. Adjunction of inverses to rings and modules, that is, localization, results in simpler rings and modules. Localization allows one to focus attention on rings with exactly one maximal ideal, the so-called local rings. These rings need not be finitely generated, but they remain Noetherian, as do the modules attached to them. This powerful tool, localization, has been much used in algebraic geometry.

Localization, and the related tool of completion, provide the means for distinguishing one algebraic variety from another. They also provide an important way, not an easy one, to study finitely generated metabelian groups and in particular a possible way to attack the isomorphism problem. The major object of this paper is to sketch some of the ideas that will be used in our ongoing work to investigate this problem. We focus on the flexibility provided by localization to help deepen our understanding of finitely generated metabelian groups. Note, for instance, that the first named author used localization in the proof that finitely generated metabelian groups can be embedded in finitely presented metabelian groups. This will be discussed further in the sequel. However in order to better understand the point of view that we have used in investigating finitely generated metabelian groups, we first recall some notions from algebraic geometry.

8. Affine algebraic sets, localization and completions

The model used in our study of finitely generated metabelian groups is an affine algebraic set. We remind the reader that an affine algebraic set is a set X of points in k^n, where k is a field, consisting of the zeroes of a set of polynomials in $k[x_1,\ldots,x_n]$. These algebraic sets define a topology on k^n, in which they are the closed sets. This topology is known as the Zariski topology. An irreducible algebraic set, X, is called a variety, and an algebraic set X is a variety if the set of polynomials that vanish on X is a prime ideal. The set $A = A(X)$ of polynomial functions on X can be identified with $k[x_1,\ldots,x_n]/I(X)$, where $I(X)$ is the set of polynomials vanishing on X. A is called the coordinate ring of X.

Among the points P on X are the so-called non-singular ones. These are the points at which at least one of the partial derivatives does not vanish at P. The set of functions in A which vanish at such a non-singular point P is an ideal, $\mathfrak{m}$, of A and $A/\mathfrak{m} \cong k$. It follows that $\mathfrak{m}$ is a maximal ideal of A and so prime. $S = A - \mathfrak{m}$ is multiplicatively closed. Adjoin to A inverses of the elements of S - we will describe this construction in more detail in §10. The result is a ring denoted A_S containing A in which the elements of S are now invertible. Adjoining the inverses of the elements of S to $\mathfrak{m}$, denoted $\mathfrak{m}_S$, yields a maximal ideal of A_S, the unique maximal ideal of A_S. So A_S is a local ring and is called the local ring of the point P. A_S is called the field of fractions of this local ring. In a local ring A with maximal ideal $\mathfrak{m}$ the powers of $\mathfrak{m}$ have intersection $\{0\}$. It follows that A embeds in its completion using the powers of $\mathfrak{m}$. Two such local rings are isomorphic if and only if the varieties on which the given non-singular points lie are birationally equivalent. The latter holds only if the two fields of fractions are isomorphic. These ideas can be adapted to study finitely generated metabelian groups.

9. A new approach to the isomorphism problem for finitely generated metabelian groups

It seems that some of the ideas described above for studying the birational equivalence of affine algebraic sets can be carried over to studying finitely generated metabelian groups. To do so, we modify the techniques described above, namely localization and completion. In order to do so, we observe that a finitely generated metabelian group Γ contains a normal subgroup G of finite index which is residually nilpotent, see [**LR04**], p.73. So one can reduce the isomorphism problem for finitely generated metabelian groups to a problem involving finite extensions of finitely generated residually nilpotent groups and the isomorphism problem for finitely generated residually nilpotent metabelian groups. In the event that a finitely generated metabelian group G is residually nilpotent, both techniques, localization and completion, become available. In this section we will describe how this comes about.

To this end, we need to introduce some additional notation. Let G be a group. Then we define $\gamma_1(G) = G$ and inductively for $n > 1$, $\gamma_n(G) = [G, \gamma_{n-1}(G)]$. Then the series

$$G = \gamma_1(G) \geq \gamma_2(G) \geq \cdots \geq \gamma_n(G) \geq \ldots$$

is termed the lower central series of G. G is termed residually nilpotent if the intersection of the terms of the lower central series is trivial. Under these circumstances, we can avail ourselves of a completion, by analogy with the one we described in the discussion of algebraic geometry, the pro-nilpotent completion, the inverse limit, $\hat{G}$, of the quotients $G/\gamma_n(G)$. Some of the properties of G carry over to $\hat{G}$. In particular, if G is polycyclic, then we will prove that the finitely generated subgroups of $\hat{G}$ are also polycyclic. We will be more interested in our attempts to better understand the isomorphism problem, to investigate the implications when two groups G and H have the same lower central quotients, i.e. if

$$G/\gamma_n(G) \cong H/\gamma_n(H), \text{ for every } n \geq 1.$$

If G and H are residually nilpotent groups then we term H para-G if there exists a homomorphism ϕ from G to H which induces isomorphisms from $G/\gamma_n(G)$ to $H/\gamma_n(H)$ for $n > 0$. This notion will play an important role in our work. One of the objectives of our work is to prove that there are surprisingly close connections between H and G given the right hypothesis, as we will show in Theorem 10.2, our Telescope theorem, and Theorems 12.1, 12.2 and 12.3. As a simple sample of such a connection we note first the easily proved

THEOREM 9.1. *Suppose that G and H are residually nilpotent metabelian groups. If H is para-G and if H is finitely generated, then G is also finitely generated.*

It seems likely that given two residually nilpotent metabelian groups, one can decide algorithmically whether they have the same lower central quotients. Of course, if H is para-G, then G and H have the same lower central quotients. If G is free in some variety and G and H have the same lower central quotients, then H is para-G. In particular, if a group H has the same lower central quotients as the free group G, then H is para-G. These para-G groups are called parafree. There exists finitely generated parafree groups which are not free. In fact, it appears that for most groups G, there are para-G groups H which are not isomorphic to G [**Bau67, Bau69**]. In order to obtain some deeper connections between a group H which is para-G, and G, it turns out that localization can play a key role. Indeed, the use of localization in the study of finitely generated residually nilpotent metabelian groups parallels that of localization that arises in algebraic geometry detailed above. The basic idea here goes back to Levine although the work of Baumslag and Stammbach seem related [**Lev89b, Lev89a, BS77**].

We note in passing that a related construction to localization, which like localization is a smaller version of completion, arose from work in low dimensional topology. Jerome Levine defined what he called the *algebraic closure of a group* to study concordance of knots [**Lev89a, Lev89b**]. Let $f : G \to H$ be a group homomorphism which induces an isomorphisms of lower central quotients. Then f induces an isomorphism of the pro-nilpotent completions. These huge groups, often uncountable, can be unwieldy. Levine observed that a smaller, countable subgroup of the pro-nilpotent completion effectively replaces completion when studying finitely generated groups. Homomorphisms that induce isomorphisms on lower central series quotients become isomorphisms after taking closure, and the closure can be described as a direct limit through para-equivalencies of groups. The group closure functor $G \mapsto \overline{G}$ that appeared in Levine's work has played a significant role in low dimensional topology.

For metabelian groups, the localization tools which have their origins in commutative algebra, and are suggested by algebraic geometry, agree with Levine's topologically inspired group closure.

10. Localization of finitely generated, residually nilpotent, metabelian groups and the Telescope Theorem

The objective of this section is to describe, in part, the role localization plays in our efforts. Statement (3) below is our *Telescope Theorem*, which shows that in a sense, our group localization of a group G has the property of being *locally G*. That is, every finitely generated subgroup is contained in an isomorphic copy of the given group G. This has substantial consequences, and motivates our forthcoming definition of para-equivalence of groups.

We will need what can be viewed as an instantiation of the change of coefficients involving the second cohomology of a group, which we term here *an extension pushout*. In order to define such an extension pushout, suppose that H is a group and B is an abelian normal subgroup of H with quotient P. So H is the middle of a short exact sequence

$$0 \longrightarrow B \xrightarrow{\beta} H \longrightarrow P \longrightarrow 1.$$

Here β is the inclusion of B in H. H acts on B as a group of automorphisms by conjugation, as does P, and so we can form the semi-direct product $H \ltimes B$. We will denote the elements of $H \ltimes B$ simply as hb ($b \in B, h \in H$).

Now suppose that $S = 1 + \ker\{\mathbb{Z}[P] \to \mathbb{Z}\}$, and that $s \in S$. Consider a $\mathbb{Z}[P]$-module homomorphism $\gamma \colon B \to B$, given by $b \mapsto b \cdot s$. This allows us to form a new semi-direct product $H \ltimes B$. Then it follows that the subgroup K of $H \ltimes B$ generated by the elements $b^{-1}\gamma(b)$ ($b \in B$) is normal in $H \ltimes B$. We call $(H \rtimes B)/K$ the *extension pushout of B into B through H via γ*. Then, given the conditions above, the following lemma holds.

LEMMA 10.1. (1) *The extension pushout E of B into B through H via γ is isomorphic to H.*

(2) *The canonical homomorphsm ϕ of H into E mapping $h \in H$ to the coset hK in E is a monomorphism if H is residually nilpotent, and to a proper subgroup if γ is not onto.*

We will provide details of the proof of Lemma 10.1 in [**BMOa**].

We come now to the formulation of our Telescope Theorem, Theorem 10.2. To this end, let G be a finitely generated, residually nilpotent, metabelian group, and let A be the derived group of G, written additively as usual, and let $Q = G/A$. Let I be the augmentation ideal of the group ring R of Q, i.e., the elements of R with coefficient sum zero. A can be viewed as an R-module as already noted. It then follows from the residual nilpotence of G, that

$$\bigcap_{n=1}^{\infty} AI^n = 0.$$

Now set $S = 1 + I$. Then S is multiplicatively closed and contains the identity 1. Observe that if $a \in A, a \neq 0$ and $s \in S$, then $as \neq 0$. Suppose the contrary. Now $s = 1 - \alpha$, $\alpha \in I$, and $0 = as = a(1 - \alpha)$ implies $a = a\alpha$ and therefore $a = a\alpha^n$ for every n. Hence $a = 0$. We now form what is termed A *localized at S*, which we denote here by A_S, and which consists of the equivalence classes of $A \times S$ with respect to the equivalence relation $\sim$ defined as follows:

$$(a, s) \sim (b, t) \text{ if } (at - bs)u = 0 \text{ for some } u \in S.$$

We denote the equivalence class of (a, s) by a/s and turn A_S into an R-module in the obvious way.

Since S is countable, we can enumerate the elements of S:

$$S = \{s_0 = 1, s_1, s_2, \ldots, s_n, \ldots\}.$$

Now for each $i = 0, 1, \ldots$, define $A_i = A$, and define the $\mathbb{Z}[Q]$-module homomorphism $\gamma_n \colon A_n \to A_{n+1}$ given by $a \mapsto a \cdot s_n$. The homomorphism

$$\psi \colon A_0 \to A_S \text{ defined by } a \mapsto a/1$$

extends to a homomorphism $A_1 \to A_S$, and this inductively defines injective homomorphisms $A_n \to A_S$. Then A_S becomes a properly ascending union of its submodules A_n.

THEOREM 10.2. *(**The Telescope Theorem**)* Let G be an extension of A by Q. Then G is the middle term in a short exact sequence of groups

$$A \xhookrightarrow{\phi} G \twoheadrightarrow Q$$

where ϕ is the inclusion of $A = A_0$ in G. Now put $G_0 = G$. Notice that A_i is a proper submodule of A_{i+1}. Now let G_1 be the extension pushout of A_0 through $A_1 \rtimes G_0$ via γ_0, G_2 the extension pushout of A_1 through $A_2 \rtimes G_1$ via γ_1, ..., G_n the extension pushout of A_{n-1} through $A_n \rtimes G_{n-1}$ and so on. Then by Lemma 10.1 each of the G_n is isomorphic to G and the image of G_{n-1} under the canonical homomorphism of G_{n-1} into G_n is a proper subgroup of G_n. If we abuse the naming of the various groups involved in the sequence of groups above we obtain a properly ascending sequence of groups

$$G = G_0 < G_1 < \cdots < G_n < \cdots < G_\infty = \bigcup_{n \geq 0} G_n.$$

If we now define G_S to be the extension pushout of $A = A_0$ through $A_S \rtimes G$ via the map ψ defined above. Then it turns out that G_S is isomorphic to G_∞. We refer to G_S as G localized at S, or as the telescope of G.

We will give a complete proof of the Telescope Theorem in [**BMOa**]. As previously stated, this construction coincides with Levine's group closure for metabelian groups. The construction of G_S and A_S from G and A is functorial.

11. Embedding groups and localization

It turns out that localization not only provides a new tool for studying finitely generated metabelian groups, as we show in the sections that follow, but also allows for new embedding theorems which we hope will provide insights into this amazingly complex family of groups. Here is an illustration of one use of localization which sheds light on how one can embed a finitely generated metabelian group in a finitely presented metabelian group. We use a different multiplicative set than used in most of this paper, and to avoid possible confusion, we denote this new multiplicative set by K, and preserve the notation S in this paper for the multiplicative set $1 + \ker\{\mathbb{Z}[Q] \to Z\}$ used in our group telescope.

To this end, consider the group

$$G = \langle b, s, t \mid [s,t] = 1, b^s = bb^t, [b, b^t] = 1 \rangle.$$

It turns out that G is actually the result of localization. To see how this comes about, let W be the wreath product of one infinite cyclic group by another,

$$W = \langle b \rangle \wr \langle t \rangle.$$

It is worth explaining what this means. Here $W = AQ$, where $Q = \langle t \rangle$ and A is the free abelian group, freely generated by the conjugates of b by the powers of t. W is not a finitely presented metabelian group. Indeed it can be presented in the form

$$W = \langle b, t \mid [b, b^{t^i}] = 1, \ i = 1, 2, \dots \rangle$$

but it cannot be finitely presented. The problem here is that infinitely many of these commuting relations are needed in order to present W. Now let R be the integral group ring of T and let K be the set of powers of $1 + t$. Then

$$R_K = \{r/s \mid r \in R, s \in K\}$$

and

$$A_K = \{a/s \mid a \in A, s \in K\}.$$

A_K is then an R_K-module and $U = gp(t, s = 1+t)$ is a free abelian group on s and $1 + t$ and it acts on A_K by right multiplication. So we can form the semi-direct product $A_K \rtimes U$ of A_K and U. This then is our group G.

Note how these finitely many relations suffice to make all of the conjugates of b by the powers of t commute. The following simple observations show how this comes about. First we have $[b, b^t] = 1$. So

$$1 = [b, b^t] = [b, b^t]^s = [b^s, b^{ts}] = [b^s, b^{st}] = [bb^t, (b^s)^t] = [bb^t, b^t b^{t^2}]$$

and from this it follows that $[b, b^{t^2}] = 1$, and similarly, that all of the conjugates of b by the powers of t commute. It is this kind of trick that enables one to embed finitely generated metabelian groups in finitely presented metabelian groups [**Bau73**], [**Rem73**].

If we had chosen K to be the set of non-zero divisors of R, and U to be the multiplicative subgroup of R_K generated by K, then the semi-direct product $U \ltimes B_K$ is a metabelian group with an extremely interesting array of subgroups. We will use variants of this kind of group in constructing completions of finitely generated, residually nilpotent, metabelian groups.

A careful examination of the group G_S reveals surprising aspects which resemble the nature of the group G discussed above. In fact it allows us to construct a slightly different way of embedding a finitely generated metabelian group into a finitely presented metabelian group. Although this does not differ substantially from the first named author's original proof, it does lead to a slightly different view of that theorem. Indeed using the ideas involved in the discussion of theTelescope Theorem, Theorem 10.1, enables us to prove the following new theorem which is not easily proved along the lines of the earlier proof of the general embedding theorem.

Every finitely generated residually nilpotent metabelian group can be embedded in a finitely presented residually nilpotent metabelian group.

So the use of localization might well lead to a deeper understanding of finitely generated metabelian groups and their close connections with finitely presented metabelian groups.

12. Some of the implications of the use of localization

Now suppose that G and H are finitely generated residually nilpotent metabelian groups and that H is para-G. Then H comes equipped with a homomorphism $\phi: G \to H$ which induces an isomorphism on the corresponding quotients of their lower central series. The homomorphism ϕ is a monomorphism and in particular induces an isomorphism from $G_{ab} = G/[G, G]$ to $H_{ab} = H/[H, H]$. So we can identify H_{ab} with G_{ab} which we denote simply by Q. Let R be the integral group ring of Q, I the augmentation ideal of R and let $S = 1 + I$, as before. Then we have this crucial results linking the telescopes G_S and H_S and their submodules and some consequences of them as detailed below.

(1) The monomorphism ϕ induces an isomorphism

$$\phi_S : G_S \longrightarrow H_S.$$

(2) If we now put $A = [G, G]$ and $B = [H, H]$ then ϕ_S induces an isomorphism between the R-modules A_S and B_S and so ϕ_S^{-1} maps B_S to A_S. In particular ϕ_S^{-1} maps B into A_S.

(3) Now B is a finitely generated R-module and A_S is an ascending union of its R-submodules A_i (see §10), so there exists an integer ℓ such that ϕ_S^{-1} maps B into A_ℓ. As already noted, the R-submodules A_i are isomorphic to the R-module A. If now G is finitely presented, then A and also the A_i are tame R-modules and so too are their submodules (see §4). So B is a tame R-module. Consequently H is also finitely presented. Thus we have proved the rather surprising theorem:

THEOREM 12.1. *Suppose that G and H are finitely generated residually nilpotent metabelian groups and that H is para-G. Then H is finitely presented if and only if G is finitely presented.*

(4) Recall once again that the Telescope Theorem states that a countable group G_S is a union of an increasing, countable sequence of copies of groups isomorphic to G. If a finitely generated group H is para-G, then for any generator of H, there is some group G_n in this filtration of G_S that contains that generator of H. Thus we have unexpected theorem:

THEOREM 12.2. *Suppose G is a countable, residually nilpotent, metabelian group. If a finitely generated group H is para-G, then there are inclusions $G \leq H$ and $H \leq G$, both inducing isomorphisms on lower central series quotients.*

We call this *para-equivalence of groups*. That is, G and H are para-equivalent if H is para-G and G is para-H. Restated, the above theorem states that if H is finitely generated and para-G, then G and H are para-equivalent.

(5) Observe, as well, that for similar reasons, the Telescope Theorem implies another theorem:

THEOREM 12.3. *Suppose that H is a finitely generated, residually nilpotent, metabelian group such that H is para-G. If G is polycyclic, then so too is H. That is, a group para-equivalent to a polycyclic group is also polycyclic.*

(6) Finally we note that we can prove a variation of Theorem 12.3 which takes a somewhat different form.

THEOREM 12.4. *If G is a finitely generated subgroup of the pro-nilpotent completion of a residually nilpotent metabelian polycyclic group, then G is polycyclic.*

13. Classifying para-equivalent metabelian groups

he following theorem is mostly a summary of some of the results from the prior two sections, accumulating these results for the classification theorem and examples that follow. However, statement i) below holds some new information.

THEOREM 13.1. *Let G and H be residually nilpotent, metabelian groups with H finitely generated. Then*

i) *H is para-equivalent to G if and only if there is a homomorphism $G \to H$ inducing an isomorphism $G_S \to H_S$.*

ii) *The function $G \mapsto G_S$ is functorial. In particular, G and H are isomorphic if and only if there is a diagram as follows, where the vertical homomorphisms are inclusions:*

$$\begin{array}{ccc} G & \xrightarrow{\cong} & H \\ \downarrow & & \downarrow \\ G_S & \xrightarrow{\cong} & H_S \end{array}$$

In particular, the isomorphism class of the $\mathbb{Z}[G_{ab}]$-module $[G,G]_S$ is an invariant of the para-equivalence class of G.

The difference between statements $i)$ and $ii)$ create the foundation for our forthcoming Classification Theorem 13.5 for isomorphism classes of para-equivalent groups.

If $Ann([G,G])$ is the annihilator of $[G,G]$, viewed as a module over the quotient ring $\mathbb{Z}[G_{ab}]$, then it becomes a faithful module over the ring $R = \mathbb{Z}[G_{ab}]/Ann([G,G])$, i.e., each non-zero element of R acts non-trivially on $[G,G]$. Following our analogy with algebraic geometry, we call R the *coordinate ring of G*. The coordinate ring, too, is an invariant of the para-equivalence class of G.

THEOREM 13.2. *Let G be a finitely generated, residually nilpotent group, and suppose H is a para-G group. Then the coordinate ring of G is isomorphic to the coordinate ring of H.*

We classify isomorphism classes of groups para-equivalent to G using properties $i)$ and $ii)$ of Theorem 13.1, and some basic observations about submodules of $[G,G]_S$. Toward that goal, we first consider submodules $C \leq [G,G]_S$. Following the theory of ideal classes in Dedekind domains, we call such a sub-module an *S-fractional submodule* if it is finitely generated and the inclusion of C in $[G,G]_S$ induces an isomorphism between C_S and $[G,G]_S$. We denote the set of S-fractional submodules by $\mathcal{F}([G,G]_S)$.

LEMMA 13.3. *Each S-fractional submodule $C \subset [G,G]_S$ determines a para-G group.* We'll call this the para-G group determined by C.

In order to see how to prove Lemma 13.3, we need to use our Telescope Theorem. Since $[G,G]$ is a finitely generated module, there is some element $s \in S$ such that $s \cdot [G,G] \subset C$. The resulting homomorphism $[G,G] \to C$ induces a homomorphism of second cohomology groups $H^2(G_{ab};[G,G]) \to H^2(G_{ab};C)$ and thereby gives rise to an extension of C by G_{ab}. The resultant extension is then what we have termed the para-G group determined by C. Note then an automorphism of G_S determines an automorphism of $[G,G]_S$, and therefore an action of $Aut(G_S)$ on $\mathcal{F}([G,G]_S)$. We term two fractional S-modules equivalent if such an induced automorphism of $[G,G]_S$ maps one onto the other.

DEFINITION 13.4. *We define the* ideal class monoid of G, $\mathcal{Cl}(G)$ *to be the set of equivalence classes*

$$\mathcal{Cl}(G) = \frac{\mathcal{F}([G,G]_S)}{Aut(G_S)}.$$

That is, $\mathcal{Cl}(G)$ is the set of S-fractional modules where two are equivalent if an induced automorphism of $[G,G]_S$ maps one onto the other.

Some examples in the remaining sections, and especially §14, will help to justify using the term *ideal class monoid*. Then we have the following important theorem.

THEOREM 13.5. *Isomorphism classes of groups para-equivalent to G lie in one-to-one correspondence to elements of $\mathcal{C}\ell(G)$.*

14. Examples

We give a number applications of Theorem 13.5, with minimal explanation. Details will appear in forthcoming papers.

For the remainder of this paper we always assume that G is a finitely generated, residually nilpotent, metabelian group. *These hypotheses will not be repeated.*

The coordinate ring of G plays a central role in calculating the ideal class monoid $\mathcal{C}\ell(G)$. In this section and the next, we consider the simplest cases, where the coordinate ring is a principal ideal domain or a Dedekind domain. The module theory of Dedekind domains is highly structured, and should admit a deeper analysis than given here.

Most of our examples in this section lie in a rich, but on the face of it, a relatively simple class of groups. Throughout the following discussion, we denote the infinite cyclic group with generator t by T and, as usual, use multiplicative notation for T. So the integral group ring $\mathbb{Z}[T]$ of T is simply the ring of polynomials in t and t^{-1}, i.e., consists of finite Laurent polynomials in t. We assume that the groups G that we consider here take the form $T \ltimes A$, where A is an abelian group, and that G_{ab} modulo the torsion subgroup of G_{ab} is infinite cyclic. We denote the coordinate ring of G by R and by $Aut_R(A_S)$ the group of R-homomorphisms of A_S. Now $Aut_R(A_S)$ acts on $\mathcal{F}(A_S)$ which again gives rise to an equivalence relation on $\mathcal{F}(A_S)$. We denote the set of equivalence classes under this equivalence relation by $\mathcal{C}\ell_S(A)$. It then turns out, for groups of the form $T \ltimes A$ as above, that

LEMMA 14.1.
$$\mathcal{C}\ell(G) = \mathcal{C}\ell_S(A).$$

If A is the additive group of the coordinate ring R of G then the group U of units in the ring R acts on A by right multiplication and we can form the semi-direct product $U \ltimes A$. Many of our examples will either take this form or variations of it.

Now Lemma 14.1 states that for groups which split over an infinite cyclic group, the action of $Aut(G_S)$ on the set of S-fractional modules is equivalent to the action of the group of module automorphisms of A_S. The simplest case is that in which the coordinate ring R is a principal ideal domain. We have the following theorem:

THEOREM 14.2. *Let A be the additive group of the ring $R = \mathbb{Z}[T]/J$, where J is an ideal of $\mathbb{Z}[T]$ and let $G = T \ltimes A$. If R is a principal ideal domain, then any finitely generated para-G group is isomorphic to G.*

For instance, consider the groups $G = T \ltimes \mathbb{Z}[1/n]$, $n \neq 2$, where we view $\mathbb{Z}[1/n]$ is the additive subgroup of the subring of the rational numbers with denominators a power of n. (The example $n = 2$ is not residually nilpotent.) Here t^n acts on $\mathbb{Z}[1/n]$ by multiplication by n. Then the coordinate ring of G is $\mathbb{Z}[1/n]$, a principal ideal domain. Hence any para-G is isomorphic to G by Theorem 14.2.

Another interesting example is the Lamplighter group, the wreath product of the group $C_2 = \langle a \mid a^2 = 1\rangle$ of order 2 and the infinite cyclic group $T = \langle t \rangle$:
$$L = C_2 \wr T = \langle a, t \mid a^2, [a, t^{-k}at^k] \text{ for all } k \in \mathbb{Z}\rangle.$$
This coordinate ring of L is $\mathbb{Z}[T]$, a Laurent polynomial ring over a field, and thus a principal ideal domain. Again Theorem 14.2 applies.

One can construct numerous other interesting examples. Here we give an example of a group G such that any group para-equivalent to G is isomorphic to G but for which the coordinate ring for G is not a principal ideal domain.

EXAMPLE 14.3. *Consider the group* $G = C_2 \ltimes A$, *where now A is the additive group of the integral group ring of C_2 and C_2 acts on A by right multiplication. Then the coordinate ring of G is $\mathbb{Z}[C_2]$. One can show that an ideal $J \leq \mathbb{Z}[C_2]$ is S-fractional if and only if $J \cap S \neq \emptyset$. But in the ring $\mathbb{Z}[C_2]$, every S-fractional ideal is a principal ideal, and hence, every group para-equivalent to G is isomorphic to G.*

We conclude with an example where $\mathcal{C}\ell(G)$ is infinite. We give only one example. To this end, consider the wreath product:
$$W = \langle b\rangle \wr \langle t\rangle = \langle t, b \mid [b, t^{-k}bt^k] \text{ for all } k \in \mathbb{Z}\rangle.$$
Then, for example, if J is the ideal of $\mathbb{Z}[T]$ generated by $2t-1$ and $2-t$. The groups $T \ltimes J$ and W are para-equivalent. However J is not a principal ideal and therefore the two groups are not isomorphic. Contrast this with the Lamplighter example, a quotient group of the above example, and one for which para-equivalence implies isomorphism.

15. Further examples and Dedekind domains

We consider the next simplest example, where the coordinate ring is a Dedekind domain.

Recall that a Dedekind domain is an integral domain such that non-zero proper ideals factor uniquely into products of prime ideals. Such rings are integrally closed, i.e., they contain the roots of monic polynomials over the integers. The ring of algebraic numbers in a number field is a Dedekind domain, as well as the coordinate ring of a nonsingular, geometrically integral, affine algebraic curve over a field $\Bbbk$.

We will consider special Dedekind domains, that is, Dedekind domains which are quotient rings of a Laurent polynomial ring on a single variable. We call such a Dedekind domain a *Laurent domain*.

We first consider examples of the form $G = T \ltimes D$ where D is a ring of algebraic integers over a real quadratic extension of the rationals. More precisely we consider fields of the form $\mathbb{Q}(\sqrt{d})$ where d is a square free positive integer. In any such number field, the ring of algebraic integers is the field of the roots of monic polynomials with coefficients in $\mathbb{Z}$. This is always a Dedekind domain, and for quadratic extensions has the form

$$D = \mathbb{Z}[\sqrt{d}] \text{ for } d \equiv_4 2, 3 \quad \text{and} \quad D = \mathbb{Z}\left[\frac{1+\sqrt{d}}{2}\right] \text{ for } d \equiv_4 1.$$

Such rings have a norm which can be used to compute units in the ring. Given a unit in the ring D, this defines a homomorphism $T \to D$ by sending t to that unit. D is Laurent if and only if ι is onto for some choice of unit. We made this computation for all $d < 100$, and the ring of algebraic integers in $\mathbb{Q}(\sqrt{d})$ is a Laurent domain for the following values of d:

$$d = 2, 3, 10, 13, 15, 23, 26, 29, 35, 53, 77, 82, 85.$$

The S-ideal class monoid $\mathcal{C}\ell_S(D)$ forms an abelian group when D is a Dedekind domain, and elements in this group lie in one-to-one correspondence to isomorphism classes of groups para-equivalent to $T \ltimes D$, by Lemma 14.1. In these cases, the coordinate ring of $T \ltimes D$ is precisely D.

The ring D is a principal ideal domain for $d = 2, 3, 13, 23, 29, 53$, and 77.

For the remaining cases above, one easily computes these abelian groups using elementary techniques from number theory, and there are exactly two isomorphism classes of groups in each para-equivalence class for each remaining value of d above, with the exception of $d = 82$. In this last case there are 4 such groups, and the S-ideal class group is cyclic of order four. For every Laurent domain we have computed, the homomorphism $\mathcal{C}\ell_S(D) \to \mathcal{C}\ell(D)$ is an isomorphism, but we see little reason to believe this holds in general.

To illustrate these results concretely, we give two isomorphism classes of groups which are para-equivalent.

If R is a ring, then we denote the ideal of R generated by the elements $a, b, \ldots$ by $(a, b, \ldots)$. Consider the ideal J of $A = \mathbb{Z}[T]/(t^2 - 6t - 1)$ generated by the image of $(3, t - 2)$. Standard arguments show that this is a non-principal ideal. (It's square is principal.) The ring homomorphism $\mathbb{Z}[T] \to \mathbb{Z}[\sqrt{10}]$ defined by $t \mapsto 3 + \sqrt{10}$ induces an isomorphism $A \cong \mathbb{Z}[\sqrt{10}]$, and sends the ideal J to $(3, 1 + \sqrt{10})$. One easily shows that both J and A are free abelian groups of rank two. We can write these as such, and the inclusion $J \subset A$ is the homomorphism

$$J = \mathbb{Z}^2 \xrightarrow{\begin{bmatrix} 3 & -2 \\ 0 & 1 \end{bmatrix}} \mathbb{Z}^2 = A.$$

Here, t acts on the domain and range, respectively, via the matrices:

$$\begin{bmatrix} 2 & 3 \\ 3 & 4 \end{bmatrix} \text{ and } \begin{bmatrix} 0 & 1 \\ 1 & 6 \end{bmatrix}.$$

This determines a para-equivalence of non-isomorphic groups

$$T \ltimes J \to T \ltimes A.$$

We have the following theorem:

THEOREM 15.1. *For D a Laurent domain, the S-ideal class group $\mathcal{C}\ell_S(D)$ is finite, and a subgroup of the usual ideal class group of D.*

Another class of interesting Laurent domains arise from the ring of cyclotomic integers, $\mathbb{Z}[\zeta_n] \cong \mathbb{Z}[t, t^{-1}]/(\phi_n(t))$, where $\phi_n(t)$ is the n-th cyclotomic polynomial. These rings are also Dedekind domains, in fact, the ring of algebraic integers in $\mathbb{Q}(\zeta_n)$, and hence, Laurent domains.

THEOREM 15.2. *Let $G = T \ltimes \mathbb{Z}[\zeta_n]$, where the action of a generator t of T on $\mathbb{Z}[\zeta_n]$ is multiplication by ζ_n.*

   i) *G is residually nilpotent if and only if $n = p^k$ for some prime p and positive integer k.*

ii) *In each of these cases, isomorphism classes of groups para-equivalent to G lie in one-to-one correspondence with*

$$\mathcal{C}\ell_S(D) \cong \mathcal{C}\ell(D).$$

Recall that the first is the group of S-fractional ideals of D under the operation of multiplication of ideals, and modulo principal ideals. The latter group, $\mathcal{C}\ell(D)$, is the classical ideal class group of D, that is, one considers all ideals in D.

$D = \mathbb{Z}[\zeta_n]$ is a principal ideal domain for $n < 23$, and thus any group para-equivalent to $G = T \ltimes \mathbb{Z}[\zeta_{p^k}]$ is isomorphic to G for prime powers $p^k < 23$. Exactly three remaining groups, $D \rtimes T$, are determined by their lower central series quotients. These are $p^k = 25, 27$, and 32.

The first interesting case occurs for $n = 23$. In this case the following is a para-equivalence of non-isomorphic groups.

$$T \ltimes \left(2, \frac{1+\sqrt{-23}}{2}\right) \subset T \ltimes \mathbb{Z}[\zeta_{23}].$$

References

[Bau67] Gilbert Baumslag, *Groups with the same lower central sequence as a relatively free group. I. The groups*, Trans. Amer. Math. Soc. **129** (1967), 308–321. MR0217157 (36:248)

[Bau69] _____, *Groups with the same lower central sequence as a relatively free group. II. Properties*, Trans. Amer. Math. Soc. **142** (1969), 507–538. MR0245653 (39:6959)

[Bau73] _____, *Subgroups of finitely presented metabelian groups*, J. Austral. Math. Soc. **16** (1973), 98–110, Collection of articles dedicated to the memory of Hanna Neumann, I. MR0332999 (48:11324)

[BH99] Martin R. Bridson and André Haefliger, *Metric spaces of non-positive curvature*, Grundlehren der Mathematischen Wissenschaften [Fundamental Principles of Mathematical Sciences], vol. 319, Springer-Verlag, Berlin, 1999. MR1744486 (2000k:53038)

[BMOa] G. Baumslag, R. Mikhailov, and K. E. Orr, *Localization, completions and metabelian groups*.

[BMOb] G. Baumslag, R. Mikhailov, and K.E. Orr, *Ideal class theory and metabelian groups*.

[BS77] G. Baumslag and U. Stammbach, *On the inverse limit of free nilpotent groups*, Comment. Math. Helv. **52** (1977), no. 2, 219–233. MR0463304 (57:3257)

[BS81] Robert Bieri and Ralph Strebel, *A geometric invariant for modules over an abelian group*, J. Reine Angew. Math. **322** (1981), 170–189. MR603031 (82f:20017)

[EFW] Alex Eskin, David Fisher, and Kevin Whyte, *Coarse differentiation of quasi-isometries II: Rigidity for Sol and Lamplighter groups*, no. arXiv:0706.0940.

[EFW07] _____, *Quasi-isometries and rigidity of solvable groups*, Pure Appl. Math. Q. **3** (2007), no. 4, part 1, 927–947. MR2402598 (2009b:20074)

[FM00] Benson Farb and Lee Mosher, *Problems on the geometry of finitely generated solvable groups*, Crystallographic groups and their generalizations (Kortrijk, 1999), Contemp. Math., vol. 262, Amer. Math. Soc., Providence, RI, 2000, pp. 121–134. MR1796128 (2001j:20064)

[GM86] J. R. J. Groves and Charles F. Miller, III, *Recognizing free metabelian groups*, Illinois J. Math. **30** (1986), no. 2, 246–254. MR840123 (87i:20057)

[Gro84] Mikhael Gromov, *Infinite groups as geometric objects*, Proceedings of the International Congress of Mathematicians, Vol. 1, 2 (Warsaw, 1983) (Warsaw), PWN, 1984, pp. 385–392. MR804694 (87c:57033)

[Hal54] P. Hall, *Finiteness conditions for soluble groups*, Proc. London Math. Soc. (3) **4** (1954), 419–436. MR0072873 (17:344c)

[KR] Martin Kassabov and Tim Riley, *The Dehn function of Baumslag's metabelian group*, no. arXiv:1008.1966.

[Lev89a] Jerome P. Levine, *Link concordance and algebraic closure. II*, Invent. Math. **96** (1989), no. 3, 571–592. MR91g:57007

[Lev89b] _____, *Link concordance and algebraic closure of groups*, Comment. Math. Helv. **64** (1989), no. 2, 236–255. MR91a:57016

[LR04] John C. Lennox and Derek J. S. Robinson, *The theory of infinite soluble groups*, Oxford Mathematical Monographs, The Clarendon Press Oxford University Press, Oxford, 2004. MR2093872 (2006b:20047)

[LS77] Roger C. Lyndon and Paul E. Schupp, *Combinatorial group theory*, Springer-Verlag, Berlin, 1977, Ergebnisse der Mathematik und ihrer Grenzgebiete, Band 89. MR0577064 (58:28182)

[Rem73] V.R. Remeslennikov, *On finitely presented groups*, Proc. Fourth All-Union Symposium on the Theory of Groups, Novosibirsk (1973), 164–169.

[Rip82] E. Rips, *Generalized small cancellation theory and applications. I. The word problem*, Israel J. Math. **41** (1982), no. 1-2, 1–146. MR657850 (83m:20047)

[Tar49a] V. A. Tartakovskiĭ, *Application of the sieve method to the solution of the word problem for certain types of groups*, Mat. Sbornik N.S. **25(67)** (1949), 251–274. MR0033815 (11:493b)

[Tar49b] _____, *The sieve method in group theory*, Mat. Sbornik N.S. **25(67)** (1949), 3–50. MR0033814 (11:493a)

[Tar49c] _____, *Solution of the word problem for groups with a k-reduced basis for $k > 6$*, Izvestiya Akad. Nauk SSSR. Ser. Mat. **13** (1949), 483–494. MR0033816 (11:493c)

Dept. of Computer Science, City College of New York, Convent Avenue and 138th Street, New York, New York 10031
E-mail address: gilbert.baumslag@gmail.com

Steklov Mathematical Institute, Gubkina 8, 119991 Moscow, Russia and Institute for Advanced Study, Princeton, New Jersey
E-mail address: romanvm@mi.ras.ru
URL: http://www.mi.ras.ru/~romanvm/pub.html

Dept of Mathematics, Indiana University, Bloomington Indiana 47405
E-mail address: korr@indiana.edu

IA-Automorphisms of Groups with Almost Constant Upper Central Series

Marianna Bonanome, Margaret H. Dean, and Marcos Zyman

ABSTRACT. Let G be any group for which there is a least j such that $Z_j = Z_{j+1}$ in the upper central series. Define the group of j-central automorphisms as the kernel of the natural homomorphism from $Aut(G)$ to $Aut(G/Z_j)$. We offer sufficient conditions for $IA(G)$ to have a useful direct product structure, and apply our results to certain finitely generated center-by-metabelian groups.

1. Introduction

The IA-group of a group G, denoted by $IA(G)$, consists of those automorphisms that induce the identity on G/G', where G' is the commutator subgroup of G. The investigation of the IA-group has been of interest in different contexts. P. Hall, for example, has shown that the IA-group of a nilpotent group of class c is nilpotent of class $c-1$ (see [5]); and M. Zyman has remarked that if G is finitely generated nilpotent, so too is $IA(G)$ (see [8]).

A result of particular interest here is due to S. Bachmuth (see [1]), who has shown that the IA-group of a free metabelian group of rank two is equal to its inner automorphism group. He has also shown that this is not the case when the rank is larger than two. In this paper we study the IA-group of a group G for which the upper central series stalls at some point. This means that there exists a least positive integer j for which $Z_j = Z_{j+1}$. We refer to these groups as $\mathcal{H}_j$-groups, whose examination was inspired by an example of P. Hall in [6]. We discuss this example in §4.

We define the group of *j-central automorphisms* of G, denoted by $Aut_{c_j}(G)$, as the kernel of the natural homomorphism from $Aut(G)$ to $Aut(G/Z_j)$, where Z_j is a term of the upper central series. In Lemma 3.1 we prove that for G an $\mathcal{H}_j$-group, the subgroup of $Aut(G)$ generated by the inner automorphisms together with the j-central automorphisms is a direct product of these two groups, modulo $Aut_{c_{j-1}}(G)$. Let $Aut_{z_j}(G) = IA(G) \cap Aut_{c_j}(G)$. In Theorem 3.2, we give sufficient conditions

---

2010 *Mathematics Subject Classification.* Primary 20F14, 20F16, 20F22, 20F28.

Key words and phrases. Automorphism groups, center-by-metabelian groups, central automorphisms, $\mathcal{H}_j$-groups, IA-automorphisms, inner automorphisms, j-central automorphisms, solvable groups, upper central series.

The third-named author was supported by The City University of New York PSC-CUNY research award program (grant # 60093-38 39).

The second and third-named authors were supported by a BMCC faculty development grant.

©2012 American Mathematical Society

for the group of IA-automorphisms of an $\mathcal{H}_j$-group to equal the direct product of the inner automorphisms with $Aut_{z_j}(G)$, modulo $Aut_{z_{j-1}}(G)$. The main result is:

Theorem 3.3. *Let G be a group for which there is a least j such that $Z_j = Z_{j+1}$ in the upper central series. If $IA\,(G/Z_j) = Inn\,(G/Z_j)$, then*
$$IA(G)/Aut_{z_{j-1}}(G) = \left(Inn(G)Aut_{z_{j-1}}(G)/Aut_{z_{j-1}}(G)\right) \times \left(Aut_{z_j}(G)/Aut_{z_{j-1}}(G)\right).$$

When $Z_1 = Z_2$, we get the very pleasing result:

Corollary 3.4 *Let G be a group such that $Z_1 = Z_2$ in the upper central series. If $IA(G/\zeta(G)) = Inn(G/\zeta(G))$, then*
$$IA(G) = Inn(G) \times Aut_{z_1}(G).$$

The paper is organized as follows: In §2 we provide definitions, notation, and a preliminary lemma; while §3 is devoted to the proof of our theorem. In §4 we consider some examples, motivated by the existence of a family of finitely generated metabelian groups whose IA-group is not finitely generated [2]. As further evidence of the complexity of solvable groups, we end §4 with the construction of continuously many finitely generated center-by-metabelian groups whose group of IA-automorphisms is not finitely generated.

2. Preliminary discussion

We write $G = gp(A)$ to express the fact that A is a generating set for G. We write a typical commutator as $[g, h] = g^{-1}h^{-1}gh$.

For any group G define the group of *IA-automorphisms* (or *IA-group*) to be the kernel of the natural homomorphism from $Aut(G)$ to $Aut\,(G/G')$. Thus,
$$IA(G) = \left\{\alpha \in Aut(G) : g^{-1}(g\alpha) \in G'\ (g \in G)\right\}.$$

For any group G, $Inn(G) \leq IA(G)$ and $G/\zeta(G) \cong Inn(G)$. Throughout this paper, the inner automorphism induced by $x \in G$ will be denoted as α_x.

For any positive integer j, we define the group of *j-central automorphisms* of G, denoted by $Aut_{c_j}(G)$, to be the kernel of the natural homomorphism from $Aut(G)$ to $Aut\,(G/Z_j)$. A j-central automorphism acts as the identity on G modulo Z_j. Thus:
$$Aut_{c_j}(G) = \left\{\alpha \in Aut(G) : g^{-1}(g\alpha) \in Z_j\ (g \in G)\right\}.$$

We use the notation $Aut_c(G) = Aut_{c_1}(G)$ and we refer to $Aut_c(G)$ as the group of *central automorphisms* of G. An alternate definition of $Aut_{c_j}(G)$ is given by the following lemma:

LEMMA 2.1. *For any group G,*
$$Aut_{c_j}(G) = \left\{\varphi \in Aut(G) : [\varphi, \alpha_x] \in Aut_{c_{j-1}}(G)\ \text{for all}\ \alpha_x \in Inn(G)\right\}.$$

PROOF.
$$\begin{aligned}
\varphi \in Aut_{c_j}(G) &\Leftrightarrow g^{-1}(g\varphi) \in Z_j\ \text{for all}\ g \in G \\
&\Leftrightarrow x^{-1}\left(x\alpha_{g^{-1}(g\varphi)}\right) = x^{-1}\left(x\alpha_g^{-1}\alpha_g^{\varphi}\right) \in Z_{j-1}\ \text{for all}\ x, g \in G \\
&\Leftrightarrow \alpha_g^{-1}\alpha_g^{\varphi} = [\alpha_g, \varphi] \in Aut_{c_{j-1}}(G)\ \text{for all}\ g \in G.\ \square
\end{aligned}$$

As stated in §1, a group G is an $\mathcal{H}_j$-group if there exists a least positive integer j for which $Z_j = Z_{j+1}$.

Since there is a natural correspondence between $Aut_{z_j}(G) = Aut_{c_j}(G) \cap IA(G)$ and $G' \cap Z_j$, we can also characterize $Aut_{z_j}(G)$ as

$$Aut_{z_j}(G) = \{\alpha \in AutG : g^{-1}(g\alpha) \in Z_j \cap G' \ (g \in G)\}.$$

We remark that every nilpotent group of class c is trivially an $\mathcal{H}_c$-group, since $G = Z_c = Z_{c+1} = \cdots$.

A group G is termed *center-by-metabelian* if $G/\zeta(G)$ is metabelian. G is center-by-metabelian if and only if $G'' \leq \zeta(G)$, where $G'' = [G', G']$. The variety of center-by-metabelian groups contains interesting examples of $\mathcal{H}_1$-groups, some of which will be discussed in §4.

3. Proof of the main result

LEMMA 3.1. *Let G be an $\mathcal{H}_j$-group. Then, modulo $Aut_{c_{j-1}}(G)$,*

$$gp\left(Inn(G), Aut_{c_j}(G)\right) \cong \left(Inn(G) Aut_{c_{j-1}}(G)\right) \times Aut_{c_j}(G).$$

PROOF. By Lemma 2.1,

$$[Inn(G), Aut_{c_j}(G)] \leq Aut_{c_{j-1}}(G)$$

for any group G.

To prove that $Inn(G) \cap Aut_{c_j}(G) \leq Aut_{c_{j-1}}(G)$, suppose that $\alpha_x \in Inn(G) \cap Aut_{c_j}(G)$, where $g\alpha_x = g[g,x]$. Then for all $g \in G$, $[g,x] \in Z_j$. This implies $x \in Z_{j+1}$. Since $Z_j = Z_{j+1}$, we have $x \in Z_j$, and thus $[g,x] \in Z_{j-1}$ for all $g \in G$.

Hence, $Inn(G) Aut_{c_{j-1}}(G) \cap Aut_{c_j}(G) = 1$ modulo $Aut_{c_{j-1}}(G)$, and our result follows. □

Note that if G is not an $\mathcal{H}_j$-group, then it is not necessarily the case that $Inn(G) \cap Aut_{c_j}(G) \leq Aut_{c_{j-1}}(G)$. For example, let

$$G = \langle a, t_0, t_1, \ldots; t_0 = 1, [a, t_i] = t_{i-1} \ (i = 1, 2, \ldots), [t_i, t_j] = 1\rangle.$$

Observe that $Z_j = gp(t_1, \ldots, t_j)$ and G has a strictly ascending upper central series. In particular G is not an $\mathcal{H}_j$-group for any j. Let $T = gp(t_1, t_2, \ldots)$. Any element of G can be written uniquely as $g = a^m t$, for some $m \in \mathbb{Z}$ and some $t \in T$. Then, $g\alpha_{t_{j+1}} = g[g, t_{j+1}] = g[a^m, t_{j+1}]$. Since $[a, t_{j+1}] = t_j \in Z_j$, $[a^m, t_{j+1}] \in Z_j$. Hence $\alpha_{t_{j+1}} \in Aut_{c_j}(G)$. Moreover, since $a\alpha_{t_{j+1}} = at_j$, then $\alpha_{t_{j+1}} \notin Aut_{c_{j-1}}(G)$.

Our main result is:

THEOREM 3.2. *Let G be a group for which there is a least j such that $Z_j = Z_{j+1}$ in the upper central series. If $IA(G/Z_j) = Inn(G/Z_j)$, then, modulo $Aut_{z_{j-1}}(G)$,*

$$IA(G) = (Inn(G) Aut_{z_{j-1}}(G)) \times Aut_{z_j}(G).$$

PROOF. It suffices to show that $IA(G) = Inn(G) Aut_{z_j}(G)$. Then, by Lemma 2.1 and the proof of Lemma 3.1, $IA(G) = Inn(G) \times Aut_{z_j}(G)$, modulo $Aut_{z_{j-1}}(G)$.

Let $\varphi \in IA(G)$. Then $x^{-1}(x\varphi) \in G' \leq G'Z_j$ for all $x \in G$. Thus,

$$x^{-1}(x\varphi) G' Z_j = G' Z_j \text{ for all } x \in G.$$

By hypothesis $IA(G/Z_j) = Inn(G/Z_j)$, so there exists $g \in G$ such that for all $x \in G$,

$$(x\varphi) Z_j = x^g Z_j.$$

Hence, $x^{-g}(x\varphi) = \left(x^{-1}(x\varphi)^{g^{-1}}\right)^g \in Z_j$, and consequently, $x^{-1}\left(x\varphi\alpha_g^{-1}\right) \in Z_j$. I.e., $\varphi\alpha_g^{-1} \in Aut_{c_j}(G) \cap IA(G) = Aut_{z_j}(G)$. Then $\varphi = \left(\varphi\alpha_g^{-1}\right)\alpha_g \in Aut_{z_j}(G)Inn(G)$. □

COROLLARY 3.3. *Let G be a group such that $Z_1 = Z_2$ in the upper central series. If $IA(G/\zeta(G)) = Inn(G/\zeta(G))$, then $IA(G) = Inn(G) \times Aut_{z_1}(G)$.*

4. Examples

For the remainder of the paper, we will be considering $\mathcal{H}_1$-groups; for simplicity, we will refer to them as $\mathcal{H}$-groups, and we will denote $Aut_{c_j}(G)$ and $Aut_{z_j}(G)$ as $Aut_c(G)$ and $Aut_z(G)$, respectively. Accordingly, The 1-central automorphisms will be simply termed *central automorphisms*.

Although the group of IA-automorphisms of a two-generator metabelian group is metabelian [4], it need not be finitely generated. There is a family of examples of finitely generated metabelian groups whose group of IA-automorphisms is not finitely generated [2].

As further evidence of the complexity of solvable groups, we provide continuously many finitely generated center-by-metabelian groups whose group of IA-automorphisms is not finitely generated.

The following results play a central role in the discussion of our examples.

LEMMA 4.1. *Let $G = gp(x_1, \ldots, x_n)$ be a finitely generated $\mathcal{H}$-group such that $\zeta(G) \leq G'$. Suppose that there exists a presentation for G in which every relator is a product of commutators. Assume also that $\zeta(G) = gp(z_1, z_2, \ldots)$. Then each map of the form*

$$\varphi_{ij} : x_i \mapsto x_i z_j$$
$$x_k \mapsto x_k \ (k \neq i)$$

lifts to a central automorphism of G, and the set of all φ_{ij} generates $Aut_c(G)$. Furthermore, if $\zeta(G)$ is freely generated by $\{z_1, z_2, \ldots\}$, then $Aut_c(G)$ is freely generated by the φ_{ij}.

PROOF. Let $G = gp(x_1, \ldots, x_n)$ be a finitely generated $\mathcal{H}$-group, and let $\{z_1, z_2, \ldots\}$ be a generating set for $\zeta(G)$. Since G comes with a presentation in which every relator is a product of commutators, each φ_{ij} extends to a map on G that sends defining relators to 1. Hence, each φ_{ij} lifts to a homomorphism. Since $G' \geq \zeta(G)$, the inverse for φ_{ij} can be constructed; therefore, φ_{ij} is a central automorphism.

Next we show that $\{\varphi_{ij}\}$ generates $Aut_c(G)$. Let $\varphi \in Aut_c(G)$ and write

$$x_i\varphi = x_i \prod_{j=1}^{m} z_j^{\alpha_{ij}}.$$

Since central automorphisms act trivially on G' and $G' \geq \zeta(G)$,

$$x_i \prod_{k,j=1}^{n,m} \varphi_{kj}^{\alpha_{kj}} = x_i \prod_{j=1}^{m} \varphi_{ij}^{\alpha_{ij}} = x_i \prod_{j=1}^{m} z_j^{\alpha_{ij}} = x_i\varphi$$

for each x_i. It is easy to show that if the φ_{ij} do not freely generate $Aut_c(G)$, then the z_i do not freely generate $\zeta(G)$. This completes the proof. □

LEMMA 4.2. *Let $G = gp(x_1, \ldots, x_n)$ be a finitely generated $\mathcal{H}$-group such that $\zeta(G) \leq G'$. Suppose that there exists a presentation for G in which every relator is a product of commutators. If $Aut_c(G)$ is finitely generated, then $\zeta(G)$ is finitely generated.*

PROOF. Suppose $\{\alpha_1, \ldots, \alpha_m\}$ generates $Aut_c(G)$. Then
$$x_i \alpha_j = x_i z_{ij}$$
with $i = 1, \ldots, n$; $j = 1, \ldots, m$; and $z_{ij} \in \zeta(G)$.

Suppose now that there exists a $z \in \zeta(G)$ such that $z \notin gp(z_{11}, \ldots, z_{nm})$. Then we can define a central automorphism
$$\begin{aligned} \alpha_z : x_1 &\mapsto x_1 z \\ x_i &\mapsto x_i \ (i \neq 1) \end{aligned}$$
which is not generated by $\{\alpha_1, \ldots, \alpha_m\}$, a contradiction. □

Example 1. Let
$$G = F/[F'', F]$$
be the free center-by-metabelian group of rank 2, where $F = gp(x, y)$ is the free group on two generators. Then $\zeta(G) = G''$ and $G/\zeta(G)$ is free metabelian of rank 2. It is well known that a free metabelian group has trivial center. Hence, G is an $\mathcal{H}$-group, with $Aut_c(G) = Aut_z(G)$.

By a theorem of Bachmuth (see [1]), $IA(G/\zeta(G)) = Inn(G/\zeta(G))$, so we may apply Theorem 3.2 to conclude that $IA(G) = Inn(G) \times Aut_c(G)$. Ridley has shown in [7] that $\zeta(G)$ is free abelian of countably infinite rank. Thus, it follows from Lemmas 4.1 and 4.2 that $Aut_c(G)$ is itself free abelian of infinite rank. In particular, we have exhibited a finitely generated center-by-metabelian group whose group of IA-automorphisms is not finitely generated.

Before proceeding, we state an important lemma. We are unaware of an existing proof in the literature.

LEMMA 4.3. *If $W = \langle a \rangle \wr \langle t \rangle$ is the wreath product of an infinite cyclic group by another, then $IA(W) = Inn(W)$.*

PROOF. There is a standard expanded presentation for W that is convenient for us to use:
$$W = \langle a, t, a_i \ (i \in \mathbb{Z}); a_0 = a, a_i^t = a_{i+1}, [a_i, a_j] = 1 \ (i, j \in \mathbb{Z}) \rangle.$$
By examining this presentation we deduce that W' is free abelian, freely generated by
$$\left\{ [t, a]^{t^i} \ (i \in \mathbb{Z}) \right\}.$$
It follows that W' can also be regarded as a free module over $\mathbb{Z}[t, t^{-1}]$, generated by the single element $\mu = [t, a]$. The module structure allows us to show that every IA-automorphism is an inner automorphism.

Let $\varphi \in IA(G)$. Then
$$\begin{aligned} \varphi : a &\mapsto a\mu^{r_a} \\ t &\mapsto t\mu^{r_t}, \end{aligned}$$
where $r_a, r_t \in \mathbb{Z}[t, t^{-1}]$.

Applying the commutator calculus and using the module structure of W' we have:

(i) $[t,\mu] = [t,[t,a]] = [t,a^{-t}a] = [t,a][t,a^{-t}] = \mu\mu^{-t} = \mu^{1-t}$, and hence:
(ii) $\mu\varphi = [t,a]\varphi = [t\mu^{r_t}, a\mu^{r_a}] = [t,a][t,\mu^{r_a}] = [t,\mu]^{r_a}[t,a] = \mu^{(1-t)r_a+1}$.

Similarly, if
$$\varphi^{-1} : a \mapsto a\mu^{s_a}$$
$$t \mapsto t\mu^{s_t},$$
then $\mu\varphi^{-1} = \mu^{(1-t)s_a+1}$. Thus,
$$a(\varphi^{-1}\varphi) = (a\mu^{s_a})\varphi = a\mu^{r_a+[(1-t)r_a+1]s_a} = a.$$

Consequently,
$$r_a + [(1-t)r_a + 1]s_a = 0.$$

Therefore r_a and s_a are mutual divisors, so they are associates in $\mathbb{Z}[t, t^{-1}]$. As a result, $(1-t)r_a + 1$ is a unit in $\mathbb{Z}[t, t^{-1}]$. The units of $\mathbb{Z}[t, t^{-1}]$ are of the form $\pm t^\gamma$.

If $1 + r_a - r_a t = -t^\gamma$, then $r_a(t-1) = t^\gamma + 1$. This implies that 1 is a root of $t^\gamma + 1$, a contradiction.

Hence $1 + r_a - r_a t = t^\gamma$, and $r_a(1-t) = t^\gamma - 1$. Thus,
$$r_a = \begin{cases} -(1 + t + \cdots + t^{\gamma-1}) & \text{if } \gamma > 0, \\ t^{-1} + \cdots + t^{\gamma+1} + t^\gamma & \text{if } \gamma < 0. \end{cases}$$

Finally, suppose $r_t = n_1 t^{i_1} + n_2 t^{i_2} + \cdots + n_k t^{i_k}$, where $n_j, i_j \in \mathbb{Z}$. A straightforward computation shows that φ is the inner automorphism induced by $w = t^\gamma a_{i_1}^{n_1} \cdots a_{i_k}^{n_k} \in W$. □

Example 2. The following example is given by P. Hall in [6]. Let B be the free nilpotent group of class 2, freely generated by $\{b_i : i \in \mathbb{Z}\}$. Thus
$$B = \langle \ldots, b_{-1}, b_0, b_1, \ldots ; [b_i, b_j, b_k] \ (i,j,k \in \mathbb{Z}) \rangle.$$

Consider the quotient group
$$N = \langle B ; [b_i, b_j] = [b_{i+1}, b_{j+1}] \ (i,j \in \mathbb{Z}) \rangle.$$

For each positive integer r, set $d_r = [b_0, b_r]$. In N, $d_r = [b_i, b_{i+r}]$ for all $i \in \mathbb{Z}$. In fact, the effect of the relations $[b_i, b_j] = [b_{i+1}, b_{j+1}]$ is that N' is free abelian on $\{d_1, d_2, \ldots\}$.

Let
$$a : b_i \mapsto b_{i+1}$$
be the translation map on the generators of N. By the nature of the defining relations of N, a induces an automorphism of N. Set $b = b_0$ and form the semidirect product
$$G = N \rtimes \langle a \rangle = gp(b, a).$$

It is straightforward to verify that $\zeta(G) = N'$. In particular, $\zeta(G)$ is a free abelian group of countably infinite rank, contained in G'. As a consequence of the defining relations for G, it is the case that
$$G/\zeta(G) \cong \mathbb{Z} \wr \mathbb{Z}.$$

Since $\mathbb{Z} \wr \mathbb{Z}$ has trivial center, G is a two-generator group satisfying that
$$\zeta(G/\zeta(G)) = 1.$$

Thus, G is a two-generator $\mathcal{H}$-group with $Aut_c(G) = Aut_z(G)$, which is center-by-metabelian. G can also be viewed as a matrix group (see [3], page 45, for details). Since $\zeta(G)$ is not finitely generated, it follows from Lemma 4.2 that $Aut_c(G)$ is not finitely generated. By Lemma 4.3 and Theorem 3.2 we conclude that the IA-group of G is not finitely generated.

Now, both Examples 1 and 2 lead to a collection of continuously many finitely generated center-by-metabelian groups whose IA-group is not finitely generated. Suppose G is either of the groups discussed in the above examples. Then $\zeta(G)$ is free abelian of countably infinite rank. Let X be a free generating set for $\zeta(G)$. It is known that there are continuously many subgroups G_σ of $\zeta(G)$, each generated by a non-empty subset A_σ of X, and that the quotient groups G/G_σ fall into continuously many isomorphism classes (see [3], page 3). For each G_σ,

$$\zeta(G/G_\sigma) = \{\bar{x} \in G/G_\sigma : [x,g] \in G_\sigma \ (g \in G)\}.$$

Since G is an $\mathcal{H}$-group and $G_\sigma \leq \zeta(G)$, we conclude that

(4.1) $$\zeta(G/G_\sigma) = \zeta(G)/G_\sigma.$$

By (4.1), $\zeta(G/G_\sigma)$ is finitely generated only if the complement of A_σ is finite. Discard the subsets of X whose complement is finite. A continuous family of subsets $A_\sigma \subset X$ still remains, the corresponding quotient groups G/G_σ fall into continuously many isomorphism classes, and each $\zeta(G/G_\sigma)$ is free abelian of countably infinite rank.

It follows again from (4.1) that $\zeta(G/G_\sigma) \leq (G/G_\sigma)'$, and

$$G/G_\sigma / \zeta(G/G_\sigma) = G/G_\sigma / \zeta(G)/G_\sigma \cong G/\zeta(G).$$

By Theorem 3.2 each G/G_σ is a 2-generator center-by-metabelian group whose IA-group is not finitely generated.

5. Acknowledgements

We would like to express our gratitude to Professor Peter Neumann for his helpful discussions. This work was inspired by a suggestion of his. We would also like to extend our thanks to Professor Gilbert Baumslag for his guidance and helpful comments; and to Professor Katalin Bencsáth for her generous encouragement. Finally, we thank the referee for offering a shorter version of the proof of Theorem 3.2, and for several other insightful comments.

References

1. S. Bachmuth, Automorphisms of free metabelian groups, Trans. Amer. Math. Soc. 118 (1965), 93-104. MR0180597 (31:4831)
2. S. Bachmuth, G. Baumslag, J. Dyer, and H. Y. Mochizuki, Automorphism groups of two generator metabelian groups, J. London Math. Soc. (2) 36 (1987) no. 3, 393-406. MR918632 (89a:20040)
3. K. Bencsáth, M. C. Bonanome, M. H. Dean, and M. Zyman, Lectures on Finitely Generated Solvable Groups, soon to appear in the SpringerBriefs Series.
4. C. K. Gupta, IA-automorphisms of two generator metabelian groups, Arch. Math. 37 (1981), 106-112. MR640795 (83b:20042)
5. P. Hall, The Edmonton notes on nilpotent groups, Queen Mary College Mathematics Notes, 1969. MR0283083 (44:316)
6. P. Hall, Finiteness conditions for soluble groups, Proc. London Math. Soc. 4 (1954), 419-436. MR0072873 (17:344c)
7. J. N. Ridley, The free centre-by-metabelian group of rank two, Proc. London Math. Soc. (3) 20 (1970), 321-347. MR0255650 (41:310)
8. M. Zyman, IA-automorphisms and localization of nilpotent groups, Ph.D. Thesis-City University of New York, 2007. MR2711065

Department of Mathematics, The City University of New York-CityTech, 300 Jay Street, Brooklyn, New York 11201
E-mail address: mbonanome@citytech.cuny.edu

Department of Mathematics, The City University of New York-BMCC, 199 Chambers Street, New York, New York 10007
E-mail address: mdean@bmcc.cuny.edu

Department of Mathematics, The City University of New York-BMCC, 199 Chambers Street, New York, New York 10007
E-mail address: mzyman@bmcc.cuny.edu

A Proposed Alternative to the Shamir Secret Sharing Scheme

Chi Sing Chum, Benjamin Fine, Gerhard Rosenberger, and Xiaowen Zhang

ABSTRACT. We propose an alternative to the Shamir (t,n) secret sharing scheme based on the closest vector theorem. We also give a brief introduction to the Shamir method.

1. Introduction

A **secret sharing protocol** is a method to distribute a secret among a group of participants by giving a share of the secret to each. The secret can be recovered only if a sufficient number of participants combine their pieces.

Formally we have the following. We have a secret $\mathcal{K}$ and a group of n participants. This group is called the **access control group**. A **dealer** allocates shares to each participant under given conditions. If a suffucent number of participants combine their shares then the secret can be recovered. If $t \leq n$ then an (t,n)-**threshold scheme** is one with n total participants and in which any t participants can combine their shares and recover the secret but not fewer than t. The number t is called the **threshold**. It is a **secure secret sharing scheme** if given less than the threshold there is no chance to recover the secret. If a measure is placed on the set of secrets, and on the set of shares, security can be made precise by saying that given less than the threshold all secrets are equally likely but given the threshold there is a unique secret. We refer to [**Sh**] and [**Bl**] for more on the security. Secret sharing is an old idea but was formalized mathematically in independent papers in 1979 by Adi Shamir [**Sh**] and George Blakley [**Bl**].

In the 1979 paper Shamir [**Sh**] proposed a beautiful (t,n) threshold scheme, based on polynomial interpolation, that has many desirable properties. We describe this in the next section. It is now the standard method for solving the (t,n) secret sharing problem although there are modifications for different situations (see [**CFZ**]). Blakely in his original paper proposed a geometric solution based on hyperplanes that is less space efficient for computer storage than Shamir's. In Blakely's scheme the distributed shares are larger than the secret, whereas in Shamir's scheme they are the same size. In this short note we propose an alternative to the Shamir scheme that has the same desirable properties but is somewhat simpler to adjust. It is geometric in nature and based on the closest vector theorem so it has some of the geometric content of the Blakely scheme, but with the shares and secret the

---

2010 *Mathematics Subject Classification.* Primary 94A60.

©2012 American Mathematical Society

same size. In format it follows the Shamir scheme. We also consider this note as a nice introduction, in general, to the secret sharing problem.

There are many different motivations for the secret sharing problem. One of the most important is the the problem of maintaining sensitive information. There are two cruicial issues here: availability and secrecy. If only one person keeps the entire secret, then there is a risk that the person might lose the secret or the person might not be available when the secret is needed. Hence it is often useful to utilize several people in order to access a secret. On the other hand, the more people who can access the secret, the higher the chance the secret will be leaked. By sharing a secret in a threshold scheme the availability and reliability issues can be addressed. The paper by Chum,Fine and Zhang [**CFZ**] contains a wealth of information on secret sharing in general and managing an access control group.

2. The Shamir Secret Sharing Scheme

We suppose that we want to develop an (t,n) secret sharing scheme. The beautiful general idea in the Shamir Scheme is the following. Let F be any field and $(x_0, y_0), ..., (x_n, y_n)$ be $n+1$ points in F^2 with distinct x_i. We say that a polynomial $P(x)$ over F **interpolates** these points if $P(x_i) = y_i$ for $i = 0, ..., n$. The relevant theoretical result that we need is the following (we can see [**A**] as a reference and for a proof):

THEOREM 2.1. *Let F be any field and $x_0, x_1, ..., x_n$ be $n+1$ distinct elements of F and $y_0, y_1, ..., y_n$ any elements of F. Then there exists a **unique** polynomial of degree $\leq n$ that interpolates the $n+1$ points $(x_i, y_i), i = 0, 1, ..., n$.*

Using this theorem the Shamir (t,n) scheme is roughly this. We choose a field F. The secret is $K \in F$ and we choose a polynomial $P(x)$ of degree $t-1$ with K as its constant term. We randomly choose distinct $x_1, ..., x_n$ with no $x_i = 0$ and distribute to each of the n participants a point $(x_i, P(x_i)), i = 1, ..., n$. By the theorem above any t people can determine the interpolating polynomial $P(x)$ and hence recover the secret K. Given less that t people then any $K \in F$ can be the constant term of the polynomial. Therefore every secret, given less shares than the threshold, is equally likely. Hence a member of the access control group has no more information than an outsider until the threshold is reached.

It is suggested that the secret be frequently changed or altered and hence the shares must be redistributed (see [**CFZ**]).

For an explicit version of the Shamir scheme, Shamir suggests using a finite field $\mathbb{Z}_q$ where q is a large prime. By working with a finite field he places a finite distribution on the set of secrets and shows that this scheme has perfect secrecy. See the original paper [**Sh**] or [**CFZ**] for a more complete discussion of this.

3. An Alternative Based on the Closest Vector Theorem

In this section we present an (t,n) secret sharing scheme that could provide an alternative to the Shamir Scheme. The scheme we present is also perfect in the sense that any t people out of the size n access control group can recover the secret but given less than t each possible secret is equally likely.

As described above the Shamir scheme depends on the uniqueness of interpolating polynomials. The present scheme depends on the **closest vector theorem** (see [**A**]).

THEOREM 3.1. **Closest Vector Theorem.** *Let W be a real inner product space and V a subspace of finite dimension t. Suppose that $w \in W$, with w not in V, and $e_1, e_2, ..., e_t$ is an orthonormal basis of V. Then the unique vector $w* \in V$ closest to w is given by*

$$w* = <w, e_1> e_1 + <w, e_2> e_2 + ... + <w, e_t> e_t$$

where $<\ ,\ >$ is the inner product on W.

Notice that given any basis for the subspace V the Gram-Schmidt orthonormalization procedure (see [**A**]) can be used to find an orthonormal basis for V. Hence given $w \in W$ we can algorithmically always find $w*$, the unique vector in V closest to w. If a basis for V is not known and we have only have knowledge or information on proper subspace spans in V of dimension less than t we cannot do this procedure and any point in W can be the secret. That is if we do not have complete knowledge of a basis for V we cannot apply the closest vector theorem. Further since given a subspace of dimension less than t there are infinitely many subspaces of dimension t properly containing it, there is a probability of zero of obtaining the subspace V with only partial knowledge.

THE SECRET SHARING SCHEME

We are given an inner product space W of dimension greater than n and an access control group of size n. We assume that the dimesnion of W is much greater than n. There is a hidden subspace V of dimension t. The secret is $v \in V$ a vector in V.

The dealer gives each of the n members of the access control group, $i = 1, ..., n$, two vectors v_i, w where $v_i \in V$, and w is a vector in the big space W with the property that $w \notin V$ and v is the vector in V closest to w. The set $\{v_1, v_2, ..., v_n\}$ has the property that any subset of size t is independent. Hence any subset of size t determines a basis for V.

Suppose t valid users get together. They can determine a basis for V and hence using the Gram-Schmidt procedure (see [**A**]) determine an orthonormal basis. Since w is given they can determine v by the closest vector theorem and recover the secret.

Given a subset of size less than t they generate a subspace of V of dimension less than t and hence in W there are infinitely many extensions to subspaces of dimension t so determining V has probability zero.

As suggested by Shamir the secret should be altered periodically. Here very easily we can change the secret v and just send each user a new w.

References

[A]   K. Atkinson "An Introduction to Numerical Analysis" Wiley, 2nd edition (1989).
[Bl]   G. Blakley "Safeguarding cryptographic keys" **Proceedings of the National Computer Conference**, 48, 313-317
[CFZ]   C.Chum,B.Fine, X.Zhang "Shamir's Threshold Scheme and Its Enhancements" to appear
[Sh]   A. Shamir. How to share a secret. Communications of the ACM, 22(11):612–613, 1979. MR549252 (80g:94070)

[St]   D. Stinson. An explication of secret sharing schemes. Design, Codes and Cryptology, 2:357–390, 1992. MR1194776 (93m:94021)

Computer Science Department, Graduate Center CUNY New York, New York

Department of Mathematics, Fairfield University, Fairfield, Connecticut 06430

Fachebreich Mathematik, University of Hamburg, Bundesstrasasse 55, 20146 Hamburg, Germany

Computer Science Department, College of Staten Island, Staten Island, New York

Improving Latin Square Based Secret Sharing Schemes

Chi Sing Chum and Xiaowen Zhang

ABSTRACT. The Latin square is a good candidate in a secret sharing scheme to represent a secret, because of the huge number of the Latin squares for a reasonably large order. This makes it difficult for outsiders to reveal the secret due to the tremendous number of possibilities. We can improve the efficiency by distributing the shares of the critical set, instead of the full Latin square, to the participants. However, finding a critical set of a large order Latin square is very difficult. This makes the implementation of Latin square based secret sharing scheme hard. We explore these limitations, then we propose to apply herding hash technique to overcome them. We also show how to make the implementation practical.

1. Introduction

How to set up an effective procedure to keep a secret is important. However, how to represent the secret is equally important. If we can reveal the secret by an exhaustive search, then we can bypass the secret sharing scheme, no matter how good it is. Also, it would be efficient to keep the secret short, and difficult to be revealed at the same time. Latin square is a good candidate in a secret sharing scheme. We can use a Latin square to represent the secret, because of the huge number of different Latin squares for a reasonably large order. For example, there are about 10^{37} different Latin squares of order 10. This makes it difficult for outsiders to reveal the secret without any knowledge due to the tremendous number of possibilities. We can even improve the efficiency by distributing the shares of the critical set, instead of the full Latin square, to the participants. Whenever any subset of the participants joins together to form any critical set, the original Latin square and hence the secret can be recovered.

There are Latin square based secret sharing schemes in the literature [2, 3, 8]. However, there are practical limitations to implement such secret sharing schemes due to the limited knowledge about Latin squares and their critical sets.

In order to conquer the aforementioned limitations, we propose to apply cryptographic hash functions, herding attack idea to Latin square based secret sharing schemes. This applies to any general access structure. We further show how to set up a verifiable or proactive or multi-secrets secret sharing scheme. The flexibility and security of our proposed schemes are dramatically improved.

2010 *Mathematics Subject Classification.* Primary 94A60.

Key words and phrases. Secret sharing scheme, Latin square, critical set, hash function, herding and Nostradamus attack.

©2012 American Mathematical Society

The rest of the paper is organized as follows. In Sections 2 and 3 we review cryptographic hash functions and secret sharing schemes. From Section 4 to Section 6, we talk about Latin squares, critical sets and their applications to secret sharing, and limitations, respectively. In Section 7 we discuss how to apply hash function to Latin square based schemes for improvements. We conclude the paper with future research in Section 8.

2. Cryptographic hash functions

2.1. Iterative hash functions. A *cryptographic hash function* H takes an input message M of arbitrary length and outputs a fixed-length string h. The output h is called the hash or message digest of the message M. It should be fast, preimage, second preimage and collision resistant. Please refer to the textbooks, such as [18, 19], for the details.

An iterative hash function H is basically built from iterations of a compression function C using the *Merkle-Damgård construction* [9, 14]. Briefly, the construction repeatedly applies the compression function as follows. (a) Pad the arbitrary length message M into multiple v-bits blocks: $m_1, m_2, \ldots, m_b$. (b) Iterate the compression function $h_i = C(h_{i-1}, m_i)$, where h_i and h_{i-1} are intermediate hashes of u-bits strings, h_0 is the initial value (or initial vector) IV, and i ($1 \leq i \leq b$) is an integer index. (c) Output h_b as the hash of the message M, i.e., $H(M) = h_b = C(h_{b-1}, m_b)$.

2.2. Herding and Nostradamus attack. Kelsey and Kohno [12] have a detailed analysis of this attack. Stevens, Lenstra and Weger [17] applied the technique to predict the winner of the 2008 US Presidential Elections using a Sony PlayStation 3 in November 2007. We first build a large set of intermediate hashes at the first level: $h_{11}, h_{12}, \ldots, h_{1w}$. Then message blocks are generated, so that they are linked and each intermediate hash at level 1 can reach the final hash, say h. This is called the *diamond structure* (see Fig. 1). We claim we can predict that something will happen in the future by announcing the final hash to the public. When the result is available, we construct a message as follows:

(2.1) $$M = \text{Prefix} \| M^* \| \text{Suffix},$$

where "Prefix" contains the results that we claimed we knew before it happens. M^* is a message block which links the "Prefix" to one of the intermediate hashes at level 1. "Suffix" is the rest of message blocks which linked M^* to the final hash. In the example of Fig. 1, $M = \text{Prefix} \| M^* \| \text{Suffix}$, Suffix $= m_{15} \| m_{23} \| m_{32}$, and $H(M) = h_{41} = h$.

3. Secret sharing schemes

When we examine the problem of maintaining sensitive information, we will consider two issues: *availability and secrecy*. If only one person keeps the entire secret, then there is a risk that the person might lose it or the person might not be available when it is needed. We can solve the availability and reliability issues by letting more than one person keep the same secret. But the more people who can access the secret, the higher the chance the secret will be leaked. By splitting and distributing a secret among a set of participants, each of whom receives a share of the secret, a secret sharing scheme [18, 19] is designed to address these issues. The secret can only be recovered when the participants join together to combine their shares.

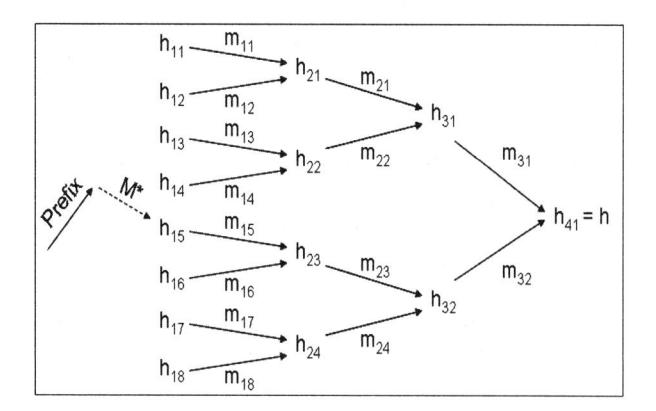

FIGURE 1. A simplified diamond structure to illustrate Nostradamus attack.

3.1. A $(t+1, n)$ threshold scheme. In 1979 Shamir [16] proposed a $(t+1, n)$ *threshold scheme*, in which each of the n participants $P_1, P_2, \ldots, P_n$ receives a share of the secret and any subset of $t + 1$ or more participants ($t \leq n - 1$) can recover the secret. Any subset of fewer than $t + 1$ participants cannot recover the secret. The concept used by Shamir is based on Lagrange polynomial interpolation. The dealer generates a polynomial of degree at most t over $\mathbb{Z}_q$, where q is a large prime number ($q > n \geq t + 1$). The coefficients, $a_t, \ldots, a_1 \in \mathbb{Z}_q$, are generated randomly and $a_0 \in \mathbb{Z}_q$ is the secret.

$$(3.1) \qquad P(x) = a_t x^t + a_{t-1} x^{t-1} + \ldots + a_1 x^1 + a_0 \pmod{q}.$$

The dealer arbitrarily chooses different $x_i \in \mathbb{Z}_q - \{0\}$, $i = 1, 2, \ldots, n$, and stores them in a public area. The corresponding shares $P(x_i) \pmod{q}$, $i = 1, 2, \ldots, n$ are then calculated and distributed to the participants privately, so that each participant gets a share of the secret. By the polynomial interpolation given any $t + 1$ points the polynomial coefficients can be recovered, hence the constant term a_0 which is the secret. Note that we want the n points to be all different to each other and the coefficients must be from the field $\mathbb{Z}_q$ to make sure we can recover the original polynomial. Here, we do not want to give out the point $P(0)$, because $P(0)$ is the secret itself.

We will use the $(t+1, n)$ threshold scheme to illustrate the following concepts.

3.2. Access structure. It is reasonable to assume that any subset of more than $t + 1$ participants can always recover the secret. We call this property *monotone*. A subset of participants, which can recover the secret when they join together, is called an *authorized subset*. On the other hand, any subset of participants that cannot recover the secret is called an *unauthorized subset*. An *access structure* Γ is the set of all authorized subsets.

Given an access structure Γ, $A \in \Gamma$ is called a *minimal authorized subset* if $B \subset A$ then $B \notin \Gamma$.

We use Γ_0 to denote the set of the *minimal authorized subsets* of Γ. In a $(t+1, n)$ threshold scheme, let P be the set of the participants:

$$(3.2) \qquad \Gamma = \{A | A \subseteq P \ \& \ |A| \geq t+1\}, \ \Gamma_0 = \{A | A \subseteq P \ \& \ |A| = t+1\}$$

In secret sharing, we first define the access structure. Then we realize it by a secret sharing scheme. Ito, Saito, and Nishizeki [11] proved that any general access structure can be realized by a secret sharing scheme.

3.3. Perfect and ideal scheme. Shamir's scheme allows no partial information to be given out even up to t participants joined together [18]. In other words, any subset of up to t participants cannot get more information about the secret than any outsider. A secret sharing scheme with this property is called a *perfect scheme*.

Based on the information theory, the length of any share must be at least as long as the secret itself in order to have perfect secrecy. The argument is that up to t participants have zero information about the secret under a perfect sharing scheme, but when one extra participant joins the subset, the secret can be recovered. This means that any participant has his share at least as long as the secret. If the shares and the secret come from the same domain, we call it an *ideal scheme*. In this case, the shares and the secret have the same size.

4. Latin square

A *Latin square* of order n is a 2-dimensional array that consists of n rows and n columns such that for any row and any column only one out of the n given symbols is filled in exactly once. For simplicity, we usually use $0, \ldots, n-1$ to represent the symbols so that each entry in a Latin square can be represented as a triple (i, j, k), where $0 \leq i, j, k \leq n-1$, and i, j, k are the row, the column and the symbol, respectively. For any order n, there exists a Latin square of this order. The addition table of the additive group $\mathbb{Z}/n\mathbb{Z}$ of integers mod n is an example [15].

4.1. Use a Latin square as a secret. Suppose we use a Latin square to represent a secret and its order, n, is made public.

If the order n is increased by 1, the number of Latin squares will grow exponentially [13, 15]. For a reasonably large n, say $n \geq 10$, there are sufficient Latin squares of that order which makes it very difficult for an outsider to figure out the secret itself without having any prior knowledge.

4.2. Partial Latin square and extension of a partial Latin square. A *partial Latin square* of order n is a 2-dimensional array that consists of n rows and n columns such that for any row and any column no symbol occurs more than once and one or more cell(s) can be empty. That is, there exists one or more pairs (i, j) such that there is no symbol in the cell of row i and column j.

Some partial Latin squares can be extended to Latin squares of the same order, while others cannot be extended. In the following example (see Tab. 1), the partial Latin square on the left can be extended into a Latin square in the middle, whence the partial Latin square on the right cannot be extended to a Latin square.

4.3. Critical set and strong critical set. A *critical set* of a Latin square is a partial Latin square that can be extended to a full Latin square uniquely. Also, after deletion of any entry of a critical set, the unique completion property does not hold any more. For a given Latin square, there may exist critical sets of different sizes.

TABLE 1. Partial Latin square extendibility.

0		3	
	2		
		1	
			3

0	1	3	2
3	2	0	1
2	3	1	0
1	0	2	3

0		3	1
	2		

By definition, we know we can recover the original Latin square from one of its critical sets and the completion is unique. However, whether we can complete a Latin square from an arbitrary partial Latin square is an NP-complete problem [6]. That means that the recovery of the Latin square from one of its critical sets may be time-consuming. We really need some criteria to speed up the process.

Donovan, Cooper, Nott and Seberry [10] suggested a notion of a *strong critical set*. Let L be a Latin square of order n and C one of its critical sets. Let $|C|$ be the size of C, i.e., the number of nonempty cells in C. If there is a sequence of partial Latin squares $\{P_0, P_1, \ldots, P_m\}$ such that

1) $C = P_0 \subset P_1 \subset \ldots \subset P_m = L$, where $m = n^2 - |C|$;
2) for any $i, 0 \leq i \leq m-1, P_i \cup \{(r_i, c_i, k_i)\} = P_{i+1}$ and $P_i \cup \{(r_i, c_i, k)\}$ is not a partial Latin square if $k \neq k_i$.

It means that we start from the critical set C and enter an entry at a time to an empty cell until we finish the extension to a full Latin square L. Each time when we get a new partial Latin square P_{i+1} $(0 \leq i \leq m-1)$, there always exists a cell (r_i, c_i) that can be filled with only one symbol k_i. We call such critical set a strong critical set if it has the above properties. In other words, the "forcing out" process makes a strong critical set to be extended to a full Latin square easily.

5. Application of critical set in secret sharing

Cooper, Donovan and Seberry [8] proposed to form a collection of critical sets of a Latin square, say S. Elements of S are distributed to participants. Any subset of participants is an authorized subset if their shares form one of the critical sets in S.

1) For example: A $(2,3)$ threshold scheme is shown in Tab. 2.

TABLE 2. A $(2,3)$ threshold secret sharing scheme.

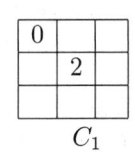

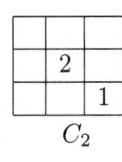

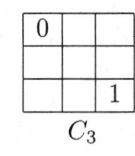

0		
	2	

C_1

	2	
		1

C_2

0		
		1

C_3

0	1	2
1	2	0
2	0	1

L

We can easily verify that all the partial Latin squares C_1, C_2, C_3 are critical sets. They can be extended uniquely to the full Latin square L. This unique completion property does not hold any more if any entry of C_1, C_2, C_3 is deleted.

Let S be the union of the three critical sets C_1, C_2, C_3. Then

(5.1) $$S = \{(0,0,0), (1,1,2), (2,2,1)\}.$$

We distribute a triple to a participant as a share. Any two participants can recover the full Latin square. So we have a (2, 3) threshold scheme.

2) The above simple example can be extended to the following general case. Let $C_1, C_2, C_3, \ldots, C_n$ be the critical sets of a given Latin square of size $s_1, s_2, \ldots, s_n$. Each C_i consists of a set of triples as follows:

(5.2) $$C_1 = \{(x_{11}, y_{11}, k_{11}), \ldots, (x_{1s_1}, y_{1s_1}, k_{1s_1})\}$$

(5.3) $$C_2 = \{(x_{21}, y_{21}, k_{21}), \ldots, (x_{2s_2}, y_{2s_2}, k_{2s_2})\}$$

$$\ldots \quad \ldots \quad \ldots$$

(5.4) $$C_n = \{(x_{n1}, y_{n1}, k_{n1}), \ldots, (x_{ns_n}, y_{ns_n}, k_{ns_n})\}$$

A triple (x_{ij}, y_{ij}, k_{ij}) is interpreted as follow: x_{ij} is the row index of the jth element in C_i, y_{ij} is the column index of the jth element in C_i, and k_{ij} is the symbol of the jth element in C_i.

In general, we make S as a union of some critical sets of a given Latin square L which represents a secret. Then, the dealer distributes a share in S, in this case a triple of the Latin square, to each participant. Whenever a subset of participants joins together to form a critical set, the original Latin square, and hence the secret, can be recovered.

6. Limitations of Latin square based schemes

Additional research has been done since the original secret sharing ideas of Shamir [16] and Blakley [1] in 1979. Latin square was suggested as a good candidate for secret sharing schemes. However, there are certain limitations as follows.

1) By just distributing shares of a critical set to participants, partial information will be available to any unauthorized subset. So, the schemes proposed by Cooper, Donovan and Seberry [8] and Chaudhry and Seberry [3] are not perfect.

2) The scheme proposed by Chaudhry, Ghodosi and Seberry [2] is not flexible if there is only one authorized subset. In this case it is just a secret splitting scheme. If more than one authorized subset exists, the secret sharing scheme is not ideal. Each participant needs to have different shares for different authorized subsets he/she belongs to.

3) As we know, distributing shares of a critical set instead of a Latin square is definitely more efficient. However, there are two issues need to be considered:

a) Even getting all the shares about a critical set, it may not be easy to get back the original Latin square, i.e., the shared secret. In order to speed up the recovery process, we should use a strong critical set.

b) However, if the participants of an authorized subset join together, it will be much easier for them to figure out the shared secret if the chosen critical set is a strong one.

4) Given a Latin square of large order (say ≥ 10), there are many critical sets of different sizes. It is very difficult to verify or find such critical sets [7].

a) Control: Let S be a collection of critical sets C_1, C_2, C_3 of a Latin square L. We would like to design a secret sharing scheme such that any authorized subset of participants can recover C_1, C_2, C_3. But there is a possibility that S contains another critical set C_4. If individuals of any unauthorized subset (in the sense that they cannot recover C_1, C_2 or C_3) can pool their shares to form C_4, then they can recover L. Hence some careful controls are needed especially when dealing with critical sets of a large order Latin square. Unfortunately for a partial Latin square of a given Latin square of large order, there is no efficient method to determine/verify whether it is a critical set. This is the major limitation of the scheme.

Example: As shown in the example in Section 5 (see Tab. 2), C_1, C_2, C_3 are critical sets of L. Suppose the dealer does not notice that C_3 is a critical set. He assigns $(0, 0, 0)$ to A, $(1, 1, 2)$ to B, and $(2, 2, 1)$ to C. So A and B can come up with C_1 to recover L; B and C can come up with C_2 to recover L. So two participants need to recover the secret and B is more important in the sense he or she must be present. However, A and C (an unauthorized subset in the dealer's mind) can come up with C_3 to recover the secret. □

b) Implementation: The knowledge about the critical sets of Latin squares of a large order is very limited. These hinder the implementation of various schemes based on critical sets.

7. Apply hash function to Latin square based schemes

In [4], we discuss how to apply the cryptographic hash functions to improve the Latin square based secret sharing schemes. The idea is to build a diamond structure based on Herding and Nostradamus attack [12]. Every authorized subset can then be herded into a final hash, which is the secret. We also discuss how to store a Latin square (secret) with a minimal number of bits. [5] is a detailed version of [4] with some improvements, for example, how to store the secret more effectively.

In [4, 5], we started with a new idea to apply the cryptographic hash functions to Latin square based secret sharing schemes, but there are limitations. Based on the birthday attack, it needs $2^{n/2}$ (n is the number of bits in the output of the hash function) to find a collision. So the complexity of building a diamond structure is too expensive to implement. In this paper, while we still follow the herding idea, we will show how to avoid such costly procedure and make the scheme efficient and practical. Since the proposed scheme is efficient, we can further implement other secret sharing schemes such as proactive, multi-secrets practically. We also make the improvement to further cut down the number of bits needed to store the secret.

7.1. Store Latin square in a hash. If we want to use the hash to store a Latin square of order 10, we need to store 81 numbers (since the last row and last column are not necessary). We proceed in the following way to minimize the number of bits needed (see Fig. 2 and Tab. 3). This shows an improvement over [4].

All the A's, B's, C's, and D in any row or column are different because of Latin property. A's, B's, C's, and D have different encoding/decoding methods. There is a one-to-one correspondence between the remaining decimals and number of bits in each stage.

1) The 0th and 1st row (Fig. 2: 1-2):

TABLE 3. Store a Latin square of order 10 in fewer bits.

A	A	B	B	B	B	C	C	D	
A	A	B	B	B	B	C	C	D	
B	B	B	B	B	B	C	C	D	
B	B	B	B	B	B	C	C	D	
B	B	B	B	B	B	C	C	D	
B	B	B	B	B	B	C	C	D	
C	C	C	C	C	C	C	C	D	
C	C	C	C	C	C	C	C	D	
D	D	D	D	D	D	D	D	D	

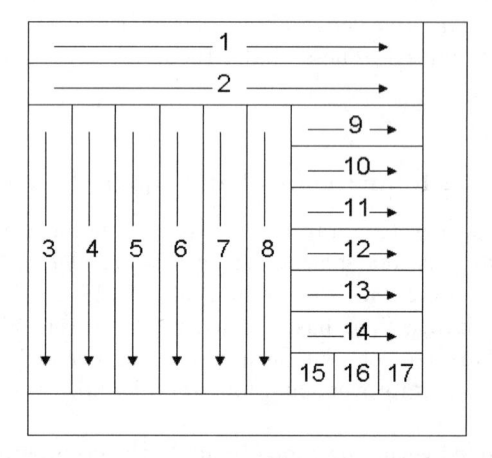

FIGURE 2. The order to encode/decode Latin square of order 10.

a) Use 7 bits to represent the first two A's. The maximal number for the first two A's is 98, and $2^7 = 128$. Therefore 7 bits are enough to represent the two A's.
b) For the next four B's, use 3 bits each.
   For example, if the first two digits (A's) are 7 and 3, then the remaining decimals will be decoded as:

(7.1)
$$9 \leftrightarrow 111;\ 8 \leftrightarrow 110;\ 6 \leftrightarrow 101;\ 5 \leftrightarrow 100;\ 4 \leftrightarrow 011;\ 2 \leftrightarrow 010;\ 1 \leftrightarrow 001;\ 0 \leftrightarrow 000.$$

c) For the next two C's, use 2 bits each.
   For example if four B's are 4, 6, 1, and 2, then the remaining numbers will be encoded as

(7.2)
$$9 \leftrightarrow 11;\ 8 \leftrightarrow 10;\ 5 \leftrightarrow 01;\ 0 \leftrightarrow 00.$$

d) For the next D, use 1 bit.
   For example, if two C's are 8 and 5, then the remaining numbers will be encoded as

(7.3)
$$9 \leftrightarrow 1;\ 0 \leftrightarrow 0.$$

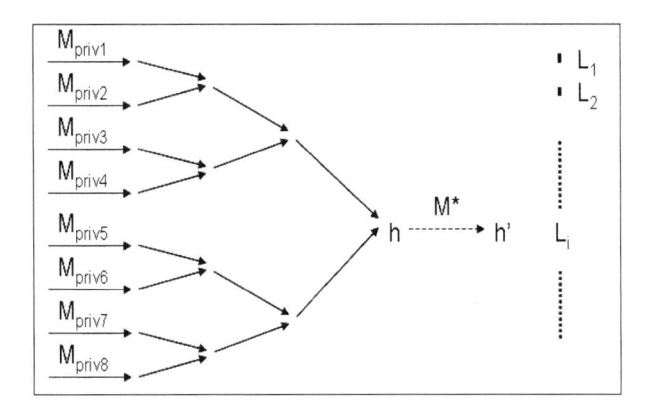

FIGURE 3. A modified diamond structure. $C(h, M^*) = h' =$ one of the Latin squares $L_1, L_2, \ldots$.

We need $7 + 4 \times 3 + 2 \times 2 + 1 = 24$ bits to encode one row. So we need $2 \times 24 = 48$ bits.

2) Remaining columns from 0th to 5th (Fig. 2: 3-8):
It takes $4 \times 3 + 2 \times 2 + 1 = 17$ bits for each column. So we need $6 \times 17 = 102$ bits.

3) Remaining rows from 2nd to 7th (Fig. 2: 9-14):
It takes $2 \times 2 + 1 = 5$ bits for each row. So we need $6 \times 5 = 30$ bits.

4) Remaining columns 6, 7, 8 (Fig. 2: 15-17): It needs 3 bits.

In total we need $48 + 102 + 30 + 3 = 183$ bits.

7.2. A modified diamond structure. We set up one message M_{priv} for one authorized subset. After building a diamond structure, all the M_{priv}'s will be herded to a final hash h. However, h may not be a Latin square. So we need to generate a long list of Latin squares: $L_1, L_2, \ldots$, and then find a linking block M^* to link h to one of these Latin squares h'. Building the list of $L_1, L_2, \ldots$ Latin squares can be done in parallel (see Fig. 3).

Based on birthday attack, we need $2^{n/2}$ (where n is the length of hash) steps just to find a pair of colliding messages. So it is infeasible to build such a diamond structure for our purpose. In the next subsection we will show how to avoid such complexity and make the scheme efficient and practical.

7.3. Newly proposed scheme. We just need to consider the size of the secret (Latin square) if it is less than that of the hash. On the other hand, if the secret is larger than the size of the hash, we will need to use multiple hashes to represent it. For simplicity, we assume here that the size of the hash is the same as that of the Latin square that represents the secret.

1) Setup

a) We randomly generate a share of the same size as that of the hash for each participant. Suppose there are n participants, then share s_i will be assigned to participant P_i, $i = 1, \ldots, n$.

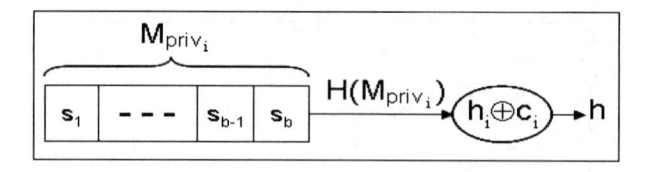

FIGURE 4. Secret recovery by combination of private and public information.

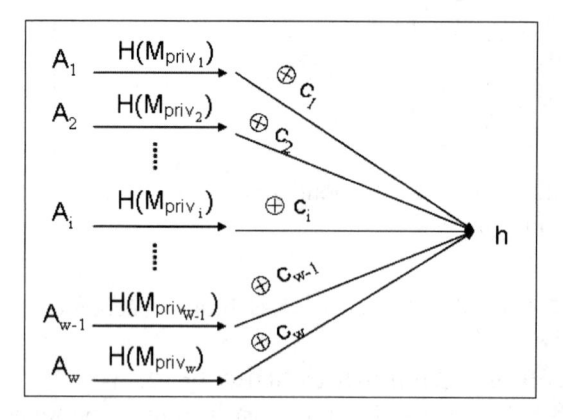

FIGURE 5. Secret recovery for any authorized subset.

b) We determine all the minimal authorized subsets. Suppose we have $A_1, \ldots, A_w$ minimal authorized subsets. Each participant holds a share and combination of the shares of any one of these w authorized subsets will form a private message M_{priv}. The combination will be the concatenation of the shares in participant sequence. For example, if an authorized subset consists of P_1, P_3 and P_4, then $M_{priv} = s_1 || s_3 || s_4$.

c) Calculate the hashes for the following

(7.4) $$H(M_{priv_i}) = h_i, \ i = 1, \ldots, w.$$

Let h be the secret and of the same size of h_i. We continue to generate a control c_i as follows (here $\oplus$ is bitwise exclusive OR):

(7.5) $$c_i = h_i \oplus h, \ i = 1, \ldots, w.$$

To summarize, after the setup process each participant P_i gets a random share $s_i, i = 1, \ldots, n$. Public information c_i, where $i = 1, \ldots, w$, is generated. Control area c_i's help to herd all the intermediate hashes h_i's to the final hash h. This eliminates the complexity of building a diamond structure.

2) Secret recovering

Suppose authorized subset A_i consists of participants $P_1, \ldots, P_b$. Joining together they can recover the secret as follows, see Fig. 4.

a) Get the public information c_i.
b) $H(s_1 || s_2 || \ldots || s_b) = h_i$, and $h_i \oplus c_i = h$.

This applies to any authorized subset, see Fig. 5.

3) Performance

In the setup step the operations involved are a generation of random shares $s_1, \ldots, s_n$, a calculation of hashes $h_i = H(M_{priv_i})$, and a generation of control area $c_i, i = 1, 2, \ldots, w$. In the secret recovering step, assuming participants of authorized subset A_i join together, we just need to calculate the hash (i.e., secret) by $h = H(M_{priv_i}) \oplus c_i$. All the operations during the setup and secret recovering are efficient. This makes the proposed scheme practical.

4) Properties of the proposed scheme

a) Perfect: Based on randomness of a hash function, any participant cannot figure out any information about the hash from his/her share. Suppose a participant in a minimal authorized subset is missing, the randomness property makes it impossible to recover his/her share directly. Brute force is the only way to determine the share of the missing participant. The rest of the participants cannot rule out any possibility of the value of the share, as they cannot figure out which guessed value cannot come up to a valid hash (Latin square) until they tried. So, in the worst case, they need to try $2^{|s|}$ times. On the other hand, an outsider can eliminate those invalid hashes (non-Latin squares), which means that the number of trials is less than $2^{|h|}$. If we choose the size of the share s as same as that of the hash h, any authorized subset with just one participant missing does not have any additional information to help them do better than any outsider. Consider the following example. Suppose there are three minimal authorized subsets: $\{(P_1, P_2), (P_1, P_3), (P_1, P_4)\}$. It means that we need P_1 and another participant to recover the secret. Obviously P_1 is more important than anyone else for secret recovering because he/she must be present. However, P_1 by himself/herself cannot recover the secret and he/she does not have more information than any other participants, as we never set up a corresponding control area for him/her.

b) Ideal: Each participant holds one share which has the same size of the hash.

c) Fast setup and recovery of the secret: The calculation of hash function is fast. No complicated or intensive computation is needed, for example, polynomial evaluation and interpolation.

d) Application of minimal authorized subset: As we explained earlier, we can speed up the whole process by considering the minimal authorized subset only.

e) General access structure: We can setup any M_{priv} for any authorized subset and herd them together. This applies to any general access structure.

f) Flexible: A hash function can handle any message of arbitrary length so there is no limit to the number of participants. We can always change to a new and better hash function should it become available. For example, we use SHA-2 now; when SHA-3 is available we can switch to it.

g) Control: Control areas are set up to make sure that only authorized subsets can recover the secret. This is a great improvement compared to 4) a) in Section 6.

h) No special hardware or software is required: For example, no need to handle a large number or find a large prime, etc.

From the properties of the proposed scheme, we can see that the limitations discussed in Section 6 have been eliminated.

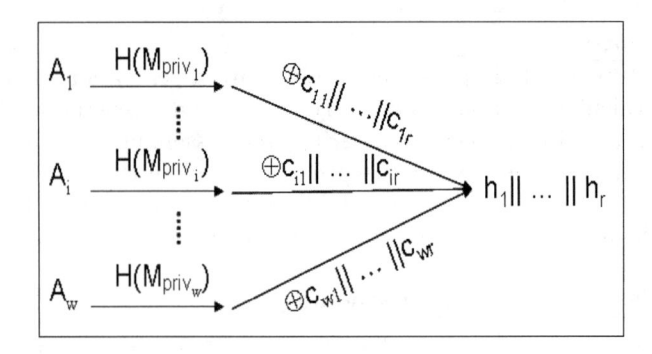

FIGURE 6. A multi-secrets scheme.

7.4. Set up a verifiable scheme for general access structure. Let f, g be cryptographic hash functions. The dealer generates shares $s_1, s_2, \ldots$, and distributes each share to each participant and then publishes the hashes (by hash function g) of each share as commitments: $g_1, g_2, \ldots$. Participant i verifies his or her share by checking if $g(s_i) = g_i$ holds. If all participants confirm that taking his or her share as input to the hash function g, he or she gets the hash value equal to one of the commitments published by the dealer, we conclude the dealer sends out consistent shares. Likewise, when the participants return their shares, the dealer can verify in the same way.

Hash function g is used to make the scheme *verifiable*. Hash function f is used as H in Subsection 7.3 for the scheme. Partial information was given out here, however, if g is preimage resistant, it would be infeasible to find the original share s_i from g_i. Participant i can fool the party if he or she can find s'_i such that $g(s_i) = g(s'_i) = g_i$. However, this is also extremely difficult to achieve if g is second preimage resistant.

7.5. Set up a proactive scheme. We pick up any authorized subset to recover the secret h, then repeat the process to generate and re-distribute new shares $s'_1, s'_2, \ldots$. Based on the secret h and the newly generated shares $s'_1, s'_2, \ldots$, we determine and update the new public control information $c'_1, c'_2, \ldots$. Finally we delete the secret h. So shares are refreshed and the secret remains unchanged.

7.6. Set up a multi-secrets scheme. We can extend the above to a *multi-secrets scheme*. Everything would be the same except for that we generate a set of control areas for each secret as follows, see Fig. 6.

In Fig. 7, we show how to control the accessibility of the secrets for different authorized subsets. Any subset of participants will be prohibited to recover a secret if their corresponding control area is not generated.

8. Conclusion and Future Research

In this paper, we use cryptographic hash functions to improve the security and performance of Latin square based secret sharing schemes.

We believe that it is an interesting area of research to extend the herding attack idea of cryptographic hash function to any secret sharing schemes and the effective implementation of such schemes.

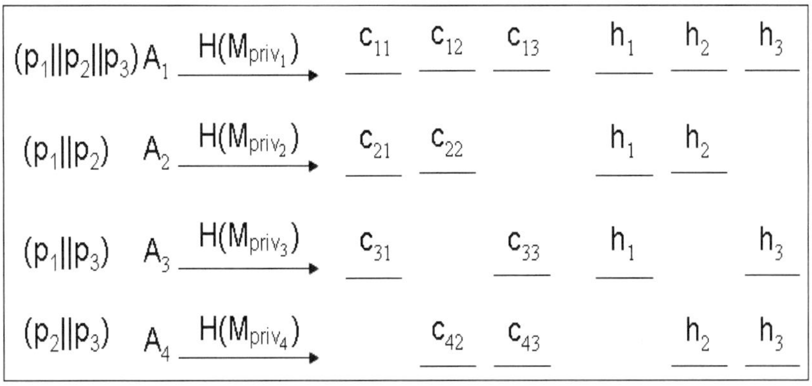

$$
\begin{array}{l}
(p_1\|p_2\|p_3)A_1 \xrightarrow{H(M_{priv_1})} \quad c_{11} \quad c_{12} \quad c_{13} \quad h_1 \quad h_2 \quad h_3 \\[4pt]
(p_1\|p_2)\ A_2 \xrightarrow{H(M_{priv_2})} \quad c_{21} \quad c_{22} \quad\quad\ \ h_1 \quad h_2 \\[4pt]
(p_1\|p_3)\ A_3 \xrightarrow{H(M_{priv_3})} \quad c_{31} \quad\quad\ c_{33} \quad h_1 \quad\quad\ h_3 \\[4pt]
(p_2\|p_3)\ A_4 \xrightarrow{H(M_{priv_4})} \quad\quad\ \ c_{42} \quad c_{43} \quad\quad\quad h_2 \quad h_3
\end{array}
$$

FIGURE 7. Control of accessibility of authorized subsets in a multi-secrets scheme.

Acknowledgments. Authors would like thank Joseph Vaisman for getting us many useful references, and Vincent Falco for proofreading the draft. We greatly appreciate the referees for their valuable suggestions and comments.

References

1. G.R. Blakley, *Safeguarding cryptographic keys*. Proc. of the National Computer Conference, American Federation of Information Processing Societies Proceedings, **48** (1979), 313–317.
2. G. Chaudhry and H. Ghodosi and J. Seberry, *Perfect Secret Sharing Schemes from Room Squares*. Journal of Combinatorial Mathematics and Combinatorial Computing, **28** (1998), 55–61. MR1668498 (99j:94040)
3. G. Chaudhry and J. Seberry, *Secret sharing schemes based on room squares*. Proc. of DMTCS'96 - Combinatorics, Complexity and Logic, (1996), 158–167.
4. C. Chum and X. Zhang, *Applying hash functions in the Latin square based secret sharing schemes*. Proc. of The 2010 International Conference on Security and Management (SAM'10), (2010), 197–203.
5. C. Chum and X. Zhang, *The Latin squares and the secret sharing schemes*. J. of Groups - Complexity - Cryptology, **2** (2010), 175–202. MR2747148 (2012b:05052)
6. C.J. Colbourn, *The Complexity of Completing Partial Latin Squares*. Discrete Applied Mathematics, **8** (1984), 25–30. MR739595 (85d:05055)
7. C.J. Colbourn and M.J. Colbourn and D.R. Stinson, *The computational complexity of recognizing critical sets*. Proc. of Graph Theory Singapore 1983, Lecture Notes in Mathematics, **1073** (1983), 248–253. MR761025 (85k:68035)
8. J.A. Cooper and D. Donovan and J. Seberry, *Secret sharing schemes arising from Latin squares*. Bulletin of the ICA, **12** (1994), 33–43. MR1301402 (95j:05043)
9. I. Damgård, *A design principle for hash functions*. Proc. of CRYPTO 1989, LNCS, **435** (1989), 416–427. MR1062248
10. D. Donovan and J.A. Cooper and D.J. Nott and J. Seberry, *Latin squares: Critical sets and their lower bounds*. Ars Combinatoria, **39** (1995), 33–48. MR1328482 (95k:05030)
11. M. Ito and A. Saito and T. Nishizeki, *Secret sharing scheme realizing general access structure*. Proc. of IEEE GLOBECOM'87, (1995), 99–102.
12. J. Kelsey and T. Kohno, *Herding hash functions and the Nostradamus attack*. IACR Cryptology ePrint Archive, Report 2005/281, (2005), 1–18.
13. B. D. McKay and I. M. Wanless, *On the number of Latin squares*. Ann. Combin., **9** (2005), 335–344. MR2176596 (2006f:05027)
14. R.C. Merkle, *One way hash function and DES*. Proc. of CRYPTO'89, LNCS, **435** (1989), 428–446. MR1062249 (91c:94023)
15. G. Mullen and C. Mummert, *Finite Fields and Applications (Student Mathematical Library)*. American Mathematical Society, (2007). MR2435301 (2009h:11004)

16. A. Shamir, *How to share a secret.* Communications of the ACM, **22(11)** (1979), 612–613. MR549252 (80g:94070)
17. M. Stevens and A.K. Lenstra and B. Weger, *Predicting the winner of the 2008 US presidential elections using a Sony PlayStation 3.* Available online http://www.win.tue.nl/hashclash/Nostradamus, (2007).
18. D. Stinson, *Cryptography, Theory and Practice.* 3rd Ed., Chapman and Hall/CRC, (2006). MR2182472 (2007f:94060)
19. W. Trappe and L. Washington, *Introduction to Cryptography with Coding Theory.* 2nd Ed., Prentice Hall, (2006). MR2372272 (2008k:94055)

DEPARTMENT OF COMPUTER SCIENCE, GRADUATE CENTER, CITY UNIVERSITY OF NEW YORK, 365 FIFTH AVENUE, NEW YORK, NEW YORK 10016
E-mail address: `CChum@gc.cuny.edu`

DEPARTMENT OF COMPUTER SCIENCE, COLLEGE OF STATEN ISLAND, CITY UNIVERSITY OF NEW YORK, 2800 VICTORY BOULEVARD, STATEN ISLAND, NEW YORK 10314
E-mail address: `Xiaowen.Zhang@csi.cuny.edu`

A Hand-Computation Involving Surface Groups, the Reidemeister-Schreier Rewriting Process and Kurosh Subgroup Theorem

Anthony E. Clement

ABSTRACT. The object of this paper is to give a hand-computational proof of Kurosh Subgroup Theorem, using as tools, surface groups and the Reidemeister-Schreier Rewriting Process.

1. Introduction

An interesting question that arises about the nature of any group is what are the structures of all of its subgroups. Given a presentation of a group G, the Reidemeister-Schreier Rewriting Process shows *how* to compute a presentation for a given subgroup H defined by generators and relations in terms of certain knowledge of its right (or left) cosets; whereas given $G = \prod * A_i (i \in I)$ a free product of groups A_i, the Kurosh Subgroup Theorem enables one to determine the *structure* of a given subgroup H of G. There are several different proofs of Kurosh Subgroup Theorem that have been published. The original [5] is combinatorial in its nature but utilizes a complicated double transfinite induction. A topological proof was given by R. Bear and F. Levi [1]. M. Takahasi [9] and M. Hall [3] gave simplified proofs but still make use of cancelation arguments. H. Kuhn [4] and A. Weir [10] introduced methods which gave the subgroups of a free product in terms of generators and relations.

The purpose of this paper is to give a step-by-step hand-computational proof that if $G = \prod * A_i (i \in I)$ a free product of surface groups A_i, each of which is presented in terms of generators and relations, then a presentation for a subgroup H of G reflects the nature of H as a free product of certain subgroups of G - in particular, a free product of a free group F and the intersection of H with certain conjugates of G. This result coincides exactly with Kurosh Subgroup Theorem which states that every subgroup H of a free product of groups $G = \prod * A_i (i \in I)$ is itself a free product of a free group F and the intersection of H with certain conjugates of $G = \prod * A_i (i \in I)$.

2. Surface Groups

A surface group is a special kind of one-relator group. One-relator groups have always played an important role in combinational group theory. Motivation

---

2010 *Mathematics Subject Classification.* Primary 20F05; Secondary 20E06, 20E07.

©2012 American Mathematical Society

for studying surface groups come from topology. The fundamental group of an orientable surface of genus g, has presentation $G_g = \langle a_1, b_1, ..., a_g, b_g; [a_1, b_1]...[a_g, b_g] = 1 \rangle$. If $g = 2$, $G_2 = \langle a_1, b_1, a_2, b_2; [a_1, b_1][a_2, b_2] = 1 \rangle$. In this paper we will present G_2 with genus 2 as $G_2 = \langle a, b, c, d; [a, b][c, d] = 1 \rangle$ which is a free product with amalgamation $F_{(u) \underset{=(v)}{*}} \widehat{F}$, where F is the free group with basis $\{a, b\}$, and $\widehat{F}$ is the free group on $\{c, d\}$, isomorphic to F via the isomorphism which sends a to d and b to c, with $u = [a, b]$ is not a proper power in F, and v is the image of u under this isomorphism. The group $G = \langle a, b, c, d, A, B, C, D; [a, b][c, d] = [A, B][C, D] = 1, \rangle$ is a free product of two surface groups $G_2 = \langle a, b, c, d; [a, b][c, d] = 1 \rangle$ and $G_2 = \langle A, B, C, D; [A, B][C, D] = 1 \rangle$.

3. The Reidemeister-Schreier Rewriting Process

Recall that a presentation $G = \langle X; R \rangle$
(i) comes equipped with a presentation map $\phi : X \longrightarrow G$;
(ii) if F denotes the free group on X, then ϕ extends to a unique homomorphism $\phi_* : F \twoheadrightarrow G$;
(iii) if K stands for the kernel of ϕ_*, then $K = gp_F(R) = gp(frf^{-1} \mid f \in F, r \in R)$ and by the first isomorphism theorem $\phi_* : F/K \xrightarrow{\sim} G$.

Now suppose that H is a subgroup of G, then H can be presented by generators and defining relations as follows: The pre-image of H in F/K is isomorphic to E/K, for some subgroup E in F. We seek generators for the free subgroup E. To this end, let T be a right Schreier transversal of E in F. Then E is free on Y, where $Y = \{tx(\overline{tx})^{-1} \neq 1 \mid t \in T, x \in X\}$. Notice that due to the isomorphism between E/K and H, $\phi|_Y$ is a presentation map of H, so $H = \langle Y; W \rangle$, with W still to be determined (in terms of Y-words). We know that $K = gp(frf^{-1} \mid f \in F, r \in R) = gp_E(\{trt^{-1} \mid t \in T, r \in R\})$. Since T is a right Schreier transversal of E in F, the cosets of E partition F, so $F = \bigcup\{et \mid e \in E, t \in T\}$. Every element of F can be written in the form $f = et$, $e \in E$, $t \in T$. Thus $K = gp(frf^{-1} \mid f \in F, r \in R)$
$= gp(\{(et)r(et)^{-1} \mid e \in E, t \in T, r \in R\})$
$= gp(\{e(trt^{-1})e^{-1} \mid e \in E, t \in T, r \in R\})$
$= gp_E(\{trt^{-1} \mid t \in T, r \in R\})$.
We still face the problem of re-expressing the relators in W, a subset of F, in terms of the generators of E above. The rewriting process provides for methodical replacement of the X-words in relation specifying elements of H by Y-words. Let $\varrho(trt^{-1})$ be the rewrite of trt^{-1} as a reduced Y-word. So W above can be written as $W = \{\varrho(trt^{-1}) \mid t \in T, r \in R\}$. Now H can be presented: $H = \langle \{tx(\overline{tx})^{-1} \neq 1 \mid t \in T, x \in X\}; \varrho(trt^{-1}) \mid t \in T, r \in R\} \rangle$.
Now we can describe the rewriting process in general using symbol by symbol replacement.

If we were to rewrite the word $z_1 z_2 z_3 \cdots z_n$, we proceed as follows:
z_1 is replaced by, $\quad \overline{1} z_1 \cdot (\overline{1 z_1})^{-1}$
z_2 is replaced by, $\quad \overline{z_1} z_2 \cdot (\overline{z_1 z_2})^{-1}$
z_3 is replaced by, $\quad \overline{z_1 z_2} z_3 \cdot (\overline{z_1 z_2 z_3})^{-1}$
$$\vdots$$
z_n is replaced by, $\quad \overline{z_1 z_2 z_3 \cdots z_{n-1}} z_n \cdot (\overline{z_1 z_2 z_3 \cdots z_{n-1} z_n})^{-1}$.
Observe that in general, $\overline{\overline{z_{n-1}} z_n} = \overline{z_{n-1} z_n}$.
Notice that if we were to replace x in $z_1 z_2 \cdots z_{i-1} x$, $z_i \in X \cup X^{-1}$, it

becomes $\overline{z_1 z_2 \cdots z_{i-1}x} \cdot (\overline{z_1 z_2 \cdots z_{i-1}x})^{-1}$. If x has exponent -1, then the rewrite
of $z_1 z_2 \cdots z_{i-1}x^{-1}$ will be $\left(\overline{z_1 z_2 \cdots z_{i-1}x^{-1}x} \cdot (\overline{z_1 z_2 \cdots z_{i-1}x^{-1}x})^{-1} \right)^{-1}$
$= \left(\overline{z_1 z_2 \cdots z_{i-1}x^{-1}}x \cdot (\overline{z_1 z_2 \cdots z_{i-1}})^{-1} \right)^{-1}$
$= \overline{z_1 z_2 \cdots z_{i-1}} \cdot x^{-1} (\overline{z_1 z_2 \cdots z_{i-1}x^{-1}})^{-1}$.

The proof of the following theorem gives a step-by-step hand-computational procedure and an appreciation of the usefulness of the Reidemeister-Schreier Rewriting Process in revealing the nature of Kurosh Subgroup Theorem.

THEOREM 1. *The group*
$G = \langle a,b,c,d,A,B,C,D; [a,b][c,d] = [A,B][C,D] = 1 \rangle$ *which is a free product of surface groups* $G = \langle a,b,c,d; [a,b][c,d] = 1 \rangle$ *and* $G = \langle A,B,C,D; [A,B][C,D] = 1 \rangle$ *has subgroup* $H = \langle b_0, c_i, d_i, A_i, B_i, C_i, D_i ; [A_i, B_i][C_i, D_i] = 1, \forall i \in \mathbb{Z} \rangle$ *where H is the free product of a free group* $\langle b_0, c_i, d_i \rangle$ *(of infinite rank) and free abelian groups* $\langle A_i, B_i; [A_i, B_i] = 1 \rangle$ *and* $\langle C_i, D_i; [D_i, C_i] = 1 \rangle$ *respectively.*

PROOF. $G = \langle a,b,c,d,A,B,C,D; [a,b][c,d] = [A,B][C,D] = 1 \rangle$ and in this presentation let $X = \{a,b,c,d,A,B,C,D\}$.

Suppose $U = \langle u \rangle$ and $\gamma: X \twoheadrightarrow \langle u \rangle$ be a map given by $a \to u$, and $b,c,d,A,B,C,D \to 1$, then γ can be extended to an epimorphism from G into U by Von Dyck's theorem. Notice that under γ, $[a,b][c,d] \to [u,1][1,1] \to 1$, $[A,B][C,D] \to [1,1][1,1] \to 1$, i.e., relators goes to relators. So $\gamma(w(a,b,c,d,A,B,C,D)) \to w(u,1,1,1,1,1,1,1)$.

Suppose $H = gp_G(b,c,d,A,B,C,D)$. Then $G/H = \langle a \rangle = gp(a^i H \mid i \in \mathbb{Z})$. $G/ker(\gamma) \cong im(\gamma) = \langle u \rangle \cong \langle a \rangle$. Hence $Ker(\psi) \cong H = gp_G(b,c,d,A,B,C,D)$.

We can find a right Schreier transversal for H in G, say, $S = \{a^i \mid i, \in \mathbb{Z}\}$. Using the Reidemeister-Schreier Rewriting Process mentioned above, the generators for H are as follows:
$\{sx \cdot (\overline{sx})^{-1} \neq 1 | s \in S, x \in X\}$.
$a^i b \cdot (\overline{a^i b})^{-1} = a^i b \cdot (\overline{a^i})^{-1} = a^i b a^{-i} = b_i$,
$a^i c \cdot (\overline{a^i c})^{-1} = a^i c \cdot (\overline{a^i})^{-1} = a^i c a^{-i} = c_i$,
$a^i d \cdot (\overline{a^i d})^{-1} = a^i d \cdot (\overline{a^i})^{-1} = a^i d a^{-i} = d_i$,
$a^i A \cdot (\overline{a^i A})^{-1} = a^i A \cdot (\overline{a^i})^{-1} = a^i A a^{-i} = A_i$,
$a^i B \cdot (\overline{a^i B})^{-1} = a^i B \cdot (\overline{a^i})^{-1} = a^i B a^{-i} = B_i$,
$a^i C \cdot (\overline{a^i C})^{-1} = a^i C \cdot (\overline{a^i})^{-1} = a^i C a^{-i} = C_i$,
$a^i D \cdot (\overline{a^i D})^{-1} = a^i D \cdot (\overline{a^i})^{-1} = a^i D a^{-i} = D_i$.
So the generations for H are,
$\{b_i, \quad c_i, \quad d_i, \quad A_i, \quad B_i, \quad C_i, \quad D_i\}$.
The relators for H are as follows:
$\{\varrho(srs^{-1}) \mid s \in S, r \in R\}$, where ϱ stands for the rewrite.
$\varrho([a,b][c,d]) = a^i [a,b][c,d] a^{-i} = a^i a^{-1} b^{-1} a b c^{-1} d^{-1} c d a^{-i}$
$= \underbrace{a^{i-1} b^{-1} a^{-(i-1)}}_{} a^{i-1} a b c^{-1} d^{-1} c d a^{-i}$
$= b_{i-1}^{-1} \underbrace{a^i b a^{-i}}_{} a^i c^{-1} d^{-1} c d a^{-i}$
$= b_{i-1}^{-1} b_i \underbrace{a^i c^{-1} a^{-i}}_{} a^i d^{-1} c d a^{-i}$
$= b_{i-1}^{-1} b_i c_i^{-1} \underbrace{a^i d^{-1} a^{-i}}_{} a^i c d a^{-i}$

$$= b_{i-1}^{-1}b_i c_i^{-1}d_i^{-1}\underbrace{a^i c a^{-i}}\underbrace{a^i d a^{-i}}$$
$$= b_{i-1}^{-1}b_i c_i^{-1}d_i^{-1}c_i d_i$$
$$= b_{i-1}^{-1}b_i[c_i, d_i].$$

And $\varrho([A,B][C,D]) = a^i[A,B][C,D]a^{-i} = a^i A^{-1}B^{-1}ABC^{-1}D^{-1}CDa^{-i}$,

$$= \underbrace{a^i A^{-1}a^{-i}}a^i B^{-1}ABC^{-1}D^{-1}CDa^{-i}$$
$$= A_i^{-1}\underbrace{a^i B^{-1}a^{-i}}a^i ABC^{-1}D^{-1}CDa^{-i}$$
$$= A_i^{-1}B_i^{-1}\underbrace{a^i A a^{-i}}a^i BC^{-1}D^{-1}CDa^{-i}$$
$$= A_i^{-1}B_i^{-1}A_i \underbrace{a^i B a^{-i}}a^i C^{-1}D^{-1}CDa^{-i}$$
$$= A_i^{-1}B_i^{-1}A_i B_i \underbrace{a^i C^{-1}a^{-i}}a^i D^{-1}CDa^{-i}$$
$$= [A_i, B_i]C_i^{-1}\underbrace{a^i D^{-1}a^{-i}}a^i CDa^{-i}$$
$$= [A_i, B_i]C_i^{-1}D_i^{-1}\underbrace{a^i Ca^{-i}}\underbrace{a^i Da^{-i}}$$
$$= [A_i, B_i]C_i^{-1}D_i^{-1}C_i D_i$$
$$= [A_i, B_i][C_i, D_i].$$

Thus, H can be presented as follows:
$H = \langle b_i, c_i, d_i, A_i, B_i, C_i, D_i \,;\, b_{i-1}^{-1}b_i[c_i, d_i] = [A_i, B_i][C_i, D_i] = 1, \forall i \in \mathbb{Z}\rangle. \,(*)$

Consider the relation $b_{i-1}^{-1}b_i[c_i, d_i] = 1$ or equivalently $b_{i-1} = b_i[c_i, d_i]$ for $i \leq 0$,

$b_{-1} = b_0[c_0, d_0]$ for $i = 0$,
$b_{-2} = b_{-1}[c_{-1}, d_{-1}]$ for $i = -1$,
$\quad = b_0[c_0, d_0][c_{-1}, d_{-1}]$
$\quad\quad \vdots$

$b_{-i} = b_0[c_0, d_0][c_{-1}, d_{-1}]\cdots[c_{-i+1}, d_{-i+1}]$.
Since $b_{i-1} = b_i[c_i, d_i]$, then equivalently $b_i = b_{i-1}[d_i, c_i]$.

Consider $b_i = b_{i-1}[d_i, c_i]$ for $i \geq 1$,

$b_1 = b_0[d_1, c_1]$ for $i = 1$,
$b_2 = b_1[d_2, c_2]$ for $i = 2$,
$\quad = b_0[d_1, c_1][d_2, c_2]$
$\quad\quad \vdots$

$b_i = b_0[d_1, c_1][d_2, c_2]\cdots[d_i, c_i]$.

Now we make use of Tietze transformations $T1, T1', T2, T2'$.
By Tietze transformation $T2$ we can add the relations $b_{i-1} = b_i[c_i, d_i]$ for $i \leq 0$ and $b_i = b_{i-1}[d_i, c_i]$ for $i \geq 1$ to our relations the presentation $(*)$ above.
By $T2'$ we can delete $b_{i-1}^{-1}b_i[c_i, d_i] = 1, \forall i \in \mathbb{Z}$ in the presentation $(*)$.
By $T1'$ we can delete the b_i's in the generators, and at the same time delete the b_i's in the relations of $(*)$.
Thus in doing so, we see that all the b_i's in the relation is deleted and all but b_0 in the generators are deleted. Thus we now have a new presentation for H.
$H = \langle b_0, c_i, d_i, A_i, B_i, C_i, D_i \,;\, [A_i, B_i][C_i, D_i] = 1, \forall i \in \mathbb{Z}\rangle$

$$= \langle b_0, c_i, d_i \rangle * \langle A_i, B_i; , [A_i, B_i] = 1 \rangle * \langle C_i, D_i; [D_i, C_i] = 1 \rangle.$$

H is the free product of a free group $\langle b_0, c_i, d_i \rangle$ (of infinite rank) and free abelian groups $\langle A_i, B_i; [A_i, B_i] = 1 \rangle$ and $\langle C_i, D_i; [D_i, C_i] = 1 \rangle$ respectively; thus, providing a computational proof of Kurosh subgroup theorem for free products of surface groups. □

References

[1] R. Baer and F. Levi, *Freie Producte und ihre Untergruppen*, Compositio Math., 3 (1936), 391-398. MR1556952

[2] Gilbert Baumslag, *Topics in Combinatorial Group Theory*, Birkhäuser Verlag: Basel; Boston; Berlin, (1993). MR1243634 (94j:20034)

[3] M. Hall, *Subgroups of free products*, Pacific Journal of Math., 3 (1953), 115-120. MR0055346 (14:1060e)

[4] H. Kuhn, *Subgroup Theorems for Groups Presented by Generators and Relations*, Annals of Math., 56 (1952), 22-46. MR0049888 (14:241h)

[5] A. G. Kurosh, *Die Untergruppen der freien Produkte von beliebigen Gruppen.* Mathematische Annalen., vol. 109 (1934), 647-660. MR1512914

[6] A. G. Kurosh, *Lectures on general algebra*, Fizmatgiz, Moscow, 1962; English transl., Chelsea, New York, 1963. MR0141700 (25:5097)

[7] A. G. Kurosh, *The theory of groups*, Vol. 2, Chelsea, New York, 1960. MR0109842 (22:727)

[8] Wilhelm Magnus, Abraham Karrass, and Donald Solitar, *Combinatorial Group Theory: Presentations of Groups in Terms of Generators and Relations*, Interscience Publishers., New York/London/Sydney, (1966). MR0207802 (34:7617)

[9] M. Takahashi, *Bemerkungen tiber den Untergruppensatz in freien Produkten*, Proc. Japan Acad., 20 (1944), 589-594. MR0014090 (7:239e)

[10] A. J. Weir, *The Reidemeister-Schreier and Kurosh subgroup theorems*, Mathematika., 3 (1956), 47-55. MR0080662 (18:280b)

DEPARTMENT OF MATHEMATICS, BROOKLYN COLLEGE, THE CITY UNIVERSITY OF NEW YORK, 2900 BEDFORD AVENUE BROOKLYN, NEW YORK 11210

E-mail address: `aclement@brooklyn.cuny.edu`

Adjunction of Roots in Exponential A-Groups

Margaret H. Dean, Stephen Majewicz, and Marcos Zyman

ABSTRACT. The purpose of this note is to prove that if G is an exponential A-group with $g \in G$ and $\beta \in A$, then G can be partially A-embedded in a partial exponential A-group where g has a βth root. This is a variation of a theorem by B. H. Neumann.

1. Introduction

An *exponential A-group* is a group G, equipped with an action by a commutative ring A with unity, such that for all $g \in G$ and for all $\alpha \in A$, the element $g^\alpha \in G$ is uniquely defined and the following axioms hold:
(1) $g^1 = g$, $g^\alpha g^\beta = g^{\alpha+\beta}$, and $(g^\alpha)^\beta = g^{\alpha\beta}$ for all $g \in G$ and $\alpha, \beta \in A$;
(2) $(h^{-1}gh)^\alpha = h^{-1}g^\alpha h$ for all $g, h \in G$ and $\alpha \in A$;
(3) If g and h are commuting elements of G, then $(gh)^\alpha = g^\alpha h^\alpha$ for all $\alpha \in A$.

Examples of exponential A-groups include A-modules, R. Lyndon's free $\mathbb{Z}[x]$-group (see [2]), and P. Hall's nilpotent A-powered groups ([1], pages 27-28).

A group G is called a *partial A-group* if for some elements $g \in G$ and $\alpha \in A$, g^α is defined, and whenever it is defined it is uniquely defined; and each equality in the axioms of an exponential A-group holds whenever both sides of the axiom are defined.

In [6], B. H. Neumann considered the problem of adjoining roots to elements in groups. Using amalgamated free products, he showed that for any group G, an arbitrary element $g \in G$, and a positive integer m, G embeds in a group H where $h^m = g$ for some $h \in H$.

In this note, we carry out a similar construction in the context of exponential A-groups. The difficulty here is that the amalgamated free product of two exponential A-groups need not be an exponential A-group, not even a partial one (see [5]). However, the particular construction given in the proof of Theorem 3.2 admits the structure of a partial A-group, where A is assumed to be a PID.

2. Preliminaries

We begin by recalling some terminology and results that will be used throughout the paper.

2010 *Mathematics Subject Classification.* Primary 20E06, 20E99; Secondary 20F99.
Key words and phrases. Exponential A-group, partial A-group, adjunction of roots, free product with amalgamation.

The collection of exponential A-groups forms a category. Thus, notions such as A-subgroup, normal A-subgroup, and A-homomorphism are defined in the obvious way (see [5]). If H is an A-subgroup (a normal A-subgroup) of an exponential A-group G, then we write $H \leq_A G$ ($H \trianglelefteq_A G$).

DEFINITION 2.1. Let G be an exponential A-group, and let $S = \{s_1, \ldots, s_j\}$ be a subset of G. Then

$$gp_A(s_1, \ldots, s_j) = \bigcap_{S \subset H_i \leq_A G} \{H_i\}$$

is called the A-subgroup of G which is A-generated by S.

Given a subset $S = \{s_1, \ldots, s_j\} \subseteq G$, if $gp_A(s_1, \ldots, s_j) = G$, then G is termed *finitely A-generated*. In case $j = 1$, we say that G is a *cyclic A-group*.

DEFINITION 2.2. Let G be an exponential A-group. An element $g \in G$ ($g \neq 1$) is called a *torsion element* if $g^\alpha = 1$ for some non-zero $\alpha \in A$. The *annihilator* or *order ideal* of g is the ideal

$$Ann(g) = \{\alpha \in A \,|\, g^\alpha = 1\}.$$

If A is a PID, then $Ann(g) = \langle \mu \rangle$ for some $\mu \in A$. We define μ to be the *order* of g, and denote it as $|g| = \mu$. Note that any other generator of $Ann(g)$ is an associate of μ in A, and the relation

$$\mu_1 \sim \mu_2 \text{ if and only if } \mu_1 = a\mu_2 \text{ for some unit } a \in A$$

is an equivalence relation on A. Thus, for any torsion element $g \in G$, $|g|$ is unique up to associates in A. This definition of $|g|$ is a slight revision of the one given in [4]. By convention, if $A = \mathbb{Z}$, the choice of generator for the order ideal of g is always a positive integer.

We remark that if g has torsion in the usual sense as a group element, say $|g| = m$ ($m \in \mathbb{Z}^+$), then $m \in Ann(g)$ and m is the least positive integer in $Ann(g)$.

Next we mention some well-known notions and facts about free products with amalgamation (see [3] for details). Let $P = G *_H T$ be the free product of G and T with H amalgamated. A sequence $c_1, c_2, \ldots, c_n$ in the ambient free product $G * T$ is termed *reduced* if

(1) each c_i lies in either factor G or T,
(2) for each i, c_i and c_{i+1} lie in distinct factors,
(3) if $n > 1$, no c_i lies in H, and
(4) if $n = 1$, $c_1 \neq 1$.

Given a reduced sequence $c_1, \ldots, c_n$, the product $w = c_1 \cdots c_n$ is an element of P. We call n the *reduced length* of w and write $l(w) = n$. It is clear that every element of P is equal to a product of elements in a reduced sequence.

THEOREM 2.3. *Let $w = c_1 \cdots c_n$ be an element in $G * T$, where $c_1, \ldots, c_n$ is a reduced sequence. Then $w \neq 1$ in P.*

It follows that G and T are embedded in P by their canonical maps.

3. Results

It is mentioned in [5] that a free product with amalgamation of two exponential A-groups may not admit the structure of a partial A-group. Theorem 3.1 provides an instance when such a partial A-action is definable. This result plays a central role in the proof of Theorem 3.2.

THEOREM 3.1. *Let G and T be exponential A-groups with $g_0 \in G$ and $t_0 \in T$, and suppose $gp_A(g_0) \cong_A gp_A(t_0)$. Let $P = (G*T)/N$, where N is the normal closure of the set $\{g_0^\alpha t_0^{-\alpha} \mid \alpha \in A\}$ in $G*T$. Then P is a partial A-group and the canonical maps $G \to P$ and $T \to P$ are partial A-embeddings.*

PROOF. Consider the isomorphism from $gp_A(g_0)$ to $gp_A(t_0)$ defined by $g_0^\alpha \mapsto t_0^\alpha$ for every $\alpha \in A$. Observe that this isomorphism can be realized as an A-isomorphism via the mapping $g_0 \mapsto t_0$. If $P = (G*T)/N$ is defined as in the statement of the theorem, then P can be realized as a free product with amalgamation $G *_H T$, where

$$H \cong gp_A(g_0) \cong gp_A(t_0).$$

Let $\overline{G} = GN/N$ and $\overline{T} = TN/N$ be the embedded copies of G and T in P, respectively. We define a partial A-action on P by defining an A-action on both $\overline{G}$ and $\overline{T}$ (but on no elements of P that lie outside of $\overline{G}$ or $\overline{T}$) as follows:

If $\overline{g} \in \overline{G}$, $\overline{t} \in \overline{T}$, and $\alpha \in A$, then

$$\overline{g}^\alpha = \overline{g^\alpha} \quad \text{and} \quad \overline{t}^\alpha = \overline{t^\alpha}.$$

To see that this is a well-defined partial action of A on P, note first that $\overline{g_0}^\alpha \overline{t_0}^{-\alpha} = g_0^\alpha t_0^{-\alpha} N = N = \overline{1}$. Hence, $\overline{g_0}^\alpha = \overline{t_0}^\alpha$.

Now consider $\overline{G}$. Since G embeds in P, the action is well-defined on $\overline{G}$. Let $\overline{g}, \overline{g_1}, \overline{g_2} \in \overline{G}$ and $\alpha, \beta \in A$. The following facts are inherited from the assumption that G is an exponential A-group:

(i) $\overline{g}^1 = \overline{g}$, $\overline{g}^\alpha \overline{g}^\beta = \overline{g}^{\alpha+\beta}$, and $(\overline{g}^\alpha)^\beta = \overline{g}^{\alpha\beta}$;
(ii) if $\overline{g_1}$ and $\overline{g_2}$ commute, then $(\overline{g_1}\,\overline{g_2})^\alpha = \overline{g_1}^\alpha \overline{g_2}^\alpha$.

It suffices to check the conjugation axiom. Let $\overline{g}$ be a nontrivial element of $\overline{G}$. In order for this axiom to make sense we assume, in turn, that either $w^{-1}\overline{g}w \in \overline{G}$, or $w^{-1}\overline{g}w \in \overline{T}$ for some $w \in P$. Suppose first that $w^{-1}\overline{g}w = \overline{x} \in \overline{G}$. Then $w \in \overline{G}$. Otherwise, w can be written as the product of a reduced sequence of one of the following forms:

(i) $w = \overline{t_1}\, v$;
(ii) $w = \overline{g_1}\,\overline{t_1}\, v$;

where $\overline{t_1} \in \overline{T} \setminus H$, $\overline{g_1} \in \overline{G} \setminus H$, and v is either the identity or the product of a reduced sequence beginning with an element from $\overline{G} \setminus H$. If w begins with $\overline{t_1}$ as in the first case, it is easy to see that $w^{-1}\overline{g}w\overline{x}^{-1}$ is a reduced product equal to 1, contradicting Theorem 2.3. Suppose then that w is of form (ii). We get

$$w^{-1}\overline{g}w\overline{x}^{-1} = v^{-1}\overline{t_1}^{-1}\,\overline{g_1}^{-1}\,\overline{g}\,\overline{g_1}\,\overline{t_1}\,v\overline{x}^{-1} = 1.$$

The only way that this product could collapse is if $\overline{g_1}^{-1}\overline{g}\,\overline{g_1} = 1$, hence, $\overline{g} = 1$; a contradiction. If it does not collapse, we get yet another reduced product equal to 1, again contradicting Theorem 2.3. Thus, $w \in \overline{G}$.

Since $\overline{G}$ is an exponential A-group, we know that

(3.1) $$(w^{-1}\overline{g}w)^\alpha = w^{-1}\overline{g}^\alpha w.$$

Next assume that $w^{-1}\overline{g}w \in \overline{T} \setminus H$; that is, $w^{-1}\overline{g}w = \overline{t}$ for some $\overline{t} \in \overline{T} \setminus H$. Applying Theorem 2.3 again, we conclude in this case that both $\overline{g}$ and w must lie in $\overline{T}$. Otherwise, since w clearly cannot lie in $\overline{G}$, we can write w as the product of a reduced sequence either of the form $w = \overline{t_1}\,\overline{g_1}\,v$ or of the form $w = \overline{g_1}\,\overline{t_1}\,v$, where $\overline{t_1} \in \overline{T} \setminus H$, $\overline{g_1} \in \overline{G} \setminus H$, and v is either the identity or the product of a reduced sequence. It is easy to see that if w begins with $\overline{t_1}$ and $\overline{g}$ is nontrivial, then $w^{-1}\overline{g}w\overline{t}^{-1}$ is a reduced product equal to 1, contradicting Theorem 2.3. If $w = \overline{g_1}\,\overline{t_1}\,v$, then we can write

$$w^{-1}\overline{g}w\overline{t}^{-1} = v^{-1}\overline{t_1}^{-1}\overline{g_1}^{-1}\overline{g}\,\overline{g_1}\,\overline{t_1}\,v\overline{t}^{-1} = 1.$$

Although $\overline{t_1}\,v\,\overline{t}^{-1}$ could collapse at the end, $v^{-1}\overline{t_1}^{-1}$ at the beginning is not trivial. Again we get a contradiction to Theorem 2.3; and we conclude that $w \in \overline{T}$. It immediately follows that $\overline{g} \in \overline{T}$. Since $\overline{T}$ is an exponential A-group, (3.1) also holds.

Clearly, then, the embedding $G \hookrightarrow P$ is a partial A-embedding. By the same reasoning, the action of A on $\overline{T}$ is well-defined, satisfies the required axioms, and the embedding $T \hookrightarrow P$ is a partial A-embedding. □

THEOREM 3.2. *Let G be an exponential A-group, where A is a PID, and suppose $g \in G$ and $\beta \in A$ are arbitrary. Then G can be partially A-embedded in a partial A-group in which g has a βth root.*

PROOF. Let G, g, and β be as stated in the hypothesis. Define T to be a cyclic A-group A-generated by t, where $T \cap G = 1$ and

$$|t| = \begin{cases} \alpha\beta & \text{if } |g| = \alpha \neq \infty \\ \infty & \text{if } |g| = \infty. \end{cases}$$

Observe that $gp_A(g)$ is A-isomorphic to $gp_A(t^\beta)$. (Note that in particular, g and t^β have the same order, in the usual group sense.) If N denotes the normal closure of the set

$$\{g^\gamma t^{-\beta\gamma} \mid \gamma \in A\}$$

in $G * T$, then Theorem 3.1 shows that $P = (G * T)/N$ is a partial A-group in which both G and T may be partially A-embedded in such a way that t^β and g are identified, and therefore t is a βth root of g in P. □

References

[1] P. Hall, *Nilpotent Groups*, (Canad. Math. Congress, Summer Seminar, University of Alberta, 1957)

[2] R. C. Lyndon, *Groups with parametric exponents*, (Trans. of the Amer. Math. Soc., 1960), 96, 518 - 533 MR0151502 (27:1487)

[3] R. C. Lyndon and P. E. Schupp, *Combinatorial Group Theory*, (Springer-Verlag, Berlin-Heidelberg-New York, 1977) MR0577064 (58:28182)

[4] S. Majewicz and M. Zyman, *Power-commutative nilpotent R-powered groups*, (Groups-Complexity-Cryptology, 2009), Vol. 1, No. 2, 297 - 309 MR2598996 (2011e:20049)

[5] A. Myasnikov and V. Remeslennikov, *Groups with Exponents. I: Fundamentals of the theory and tensor completions*, (Siberian Math. Journal, 1994), 35 (5), 986 - 996 MR1308240 (95m:20047)

[6] B. H. Neumann, *Adjunction of elements to groups*, (J. Lond. Math. Soc., 1943), 18, 4 - 11 MR0008808 (5:58s)

Department of Mathematics, The City University of New York-BMCC, 199 Chambers Street, New York, New York 10007
E-mail address: `mdean@bmcc.cuny.edu`

Department of Mathematics, The City University of New York-KCC, Brooklyn, New York 11235
E-mail address: `smajewicz@kbcc.cuny.edu`

Department of Mathematics, The City University of New York-BMCC, 199 Chambers Street, New York, New York 10007
E-mail address: `mzyman@bmcc.cuny.edu`

Logspace Computations in Coxeter Groups and Graph Groups

Volker Diekert, Jonathan Kausch, and Markus Lohrey

ABSTRACT. Computing normal forms in groups (or monoids) is computationally harder than solving the word problem (equality testing), in general. However, normal form computation has a much wider range of applications. It is therefore interesting to investigate the complexity of computing normal forms for important classes of groups.

For Coxeter groups we show that the following algorithmic tasks can be solved by a deterministic Turing machine using logarithmic work space, only: 1. Compute the length of any geodesic normal form. 2. Compute the set of letters occurring in any geodesic normal form. 3. Compute the Parikh-image of any geodesic normal form in case that all defining relations have even length (i.e., in even Coxeter groups.) 4. For right-angled Coxeter groups we can actually compute the shortlex normal form in logspace.

Next, we apply the results to right-angled Artin groups. They are also known as free partially commutative groups or as graph groups. As a consequence of our result on right-angled Coxeter groups we show that shortlex normal forms in graph groups can be computed in logspace, too. Graph groups play an important rôle in group theory, and they have a close connection to concurrency theory. As an application of our results we show that the word problem for free partially commutative inverse monoids is in logspace. This result generalizes a result of Ondrusch and the third author on free inverse monoids. Concurrent systems which are deterministic and co-deterministic can be studied via inverse monoids.

1. Introduction

The study of group theoretical decision problems, like the word problem: "Is a given word equal to 1 in the group?", the conjugacy problem: "Are two given words conjugate in the group?", and the isomorphism problem: "Do two given group presentations yield isomorphic groups?", is a classical topic in combinatorial group theory with a long history dating back to the beginning of the 20th century (e.g. Tietze 1908, Dehn 1911). More details are in the survey [33].

With the emergence of computational complexity theory, the complexity of these decision problems in various classes of groups has developed into an active research area, where algebraic methods as well as computer science techniques complement one another in a fruitful way.

2010 *Mathematics Subject Classification.* Primary 20F55, 20F10.

©2012 American Mathematical Society

In this paper we are interested in group theoretical problems which can be solved efficiently in parallel. More precisely, we are interested in *deterministic logspace*, called simply *logspace* in the following. The class logspace is at a lower level in the NC-hierarchy[1] of parallel complexity classes:

$$\mathsf{NC}^1 \subseteq \mathsf{LOGSPACE} \subseteq \mathsf{NC}^2 \subseteq \mathsf{NC}^3 \subseteq \cdots \subseteq \mathsf{NC} = \bigcup_{i \geq 1} \mathsf{NC}^i \subseteq \mathsf{P}$$

For more details about circuit complexity we refer to [40]. It is a standard conjecture in complexity theory that NC and P are different, just like a standard conjecture says $\mathsf{P} \neq \mathsf{NP}$. But this is debatable, of course.

A fundamental result in this context was shown in [29, 38]: The word problem of finitely generated linear groups belongs to logspace. In [29], Lipton and Zalcstein proved this result for fields of characteristic 0. The case of a field of prime characteristic was considered in [38] by Simon. The class of groups with a word problem in logspace is further investigated in [41]. Another important result is Cai's NC^2 algorithm for the word problem of a hyperbolic group [6]. In [31] this result was improved to LOGCFL, which is the class of all languages that are logspace-reducible to a context-free language. LOGCFL is a subclass of NC^2 and hence in the intersection of the class of problems which can be decided in polynomial time and the class of problems which can be decided in space $\log^2(n)$. As a parallel complexity class LOGCFL coincides with the (uniform) class SAC^1.

Often, it is not enough to solve the word problem, but one has to compute a normal form for a given group element. Fix a finite generating set Γ (w.l.o.g. closed under inverses) for the group G. Then, a *geodesic* for $g \in G$ is a shortest word over Γ that represents g. By choosing the lexicographical smallest (w.r.t. a fixed ordering on Γ) word among all geodesics for g, one obtains the *shortlex normal form* of g. The problem of computing geodesics and various related problems were studied in [18, 19, 21, 27, 35, 37]. It turned out that there are groups with an easy word problem (in logspace), but where simple questions related to geodesics are computationally hard. For instance, every metabelian group embeds (effectively) into a direct product of linear groups; hence its word problem can be solved in logspace. On the other hand, it is shown in [18], that the question whether a given element x of the wreath product $\mathbb{Z}/2\mathbb{Z} \wr (\mathbb{Z} \times \mathbb{Z})$ (a metabelian group) has geodesic length at most n is NP-complete. A corresponding result was shown in [35] for the free metabelian group of rank 2. Clearly, these results show that in general one cannot compute shortlex normal forms in metabelian groups in polynomial time (unless $\mathsf{P} = \mathsf{NP}$). On the positive side, for *shortlex automatic groups* [22] (i.e., automatic groups, where the underlying regular set of representatives is the set of shortlex normal forms) shortlex normal forms can be computed in quadratic time. Examples of shortlex automatic groups are Coxeter groups, Artin groups of large type, and hyperbolic groups. So for all these classes, shortlex normal forms can be computed in quadratic time. In [35], it is also noted that geodesics in nilpotent groups (which are in general not automatic) can be computed in polynomial time.

In this paper, we deal with the problem of computing geodesics and shortlex normal forms in logspace. A function can be computed in logspace, if it can be

---

[1] The notation NC refers to Nicholas "Nick" Pippinger. Nick's class NC^i denotes the class of languages that can be accepted by (uniform) Boolean circuits of polynomial size and polylogarithmic depth $O(\log^i(n))$, where all gates have fan-in ≤ 2.

computed by a deterministic *logspace transducer*. The latter is a Turing machine with three tapes: (i) a read-only input tape, (ii) a read/write work tape of length $\mathcal{O}(\log n)$, and (iii) a write-only output tape. The output is written sequentially from left to right onto the output tape. Every logspace transducer can be transformed into an equivalent deterministic polynomial time machine. Still better, it can be simulated by a Boolean circuit of polynomial size and $\mathcal{O}(\log^2 n)$ depth. Although it is not completely obvious, the class of logspace computable functions is closed under composition, see e.g. the textbook [36].

Recently, the class of groups, where geodesics and shortlex normal forms can be computed in logspace, attracted attention, see [20], where it was noted among other results that shortlex normal forms in free groups can be computed in logspace. (Implicitly, this result was also shown in [32].) In this paper, we deal with the problem of computing shortlex normal forms for Coxeter groups. Coxeter groups are discrete reflection groups and play an important role in many parts of mathematics, see [3, 12]. Every Coxeter group is linear and therefore has a logspace word problem. Moreover, as mentioned above, Coxeter groups are shortlex automatic [5, 7]. Therefore shortlex normal forms can be computed in quadratic time. However, no efficient parallel algorithms are known so far. In particular, it is open whether shortlex normal forms in Coxeter groups can be computed in logspace. We do not solve this problem in this paper, but we are able to compute in logspace some important invariants of geodesics. More precisely, we are able to compute in logspace (i) the length of the shortlex normal form of a given element (Theorem 3.7) and (ii) the alphabet of symbols that appear in the shortlex normal form of a given element (Theorem 3.9). The proofs for these results combine non-trivial results for Coxeter groups with some advanced tools from computational algebra. More precisely, we use the following results:

- The Chinese remainder representation of a given number m (which is the tuple of remainders $m \bmod p_i$ for the first k primes $p_1, \ldots, p_k$, where $m < p_1 p_2 \cdots p_k$) can be transformed in logspace into the binary representation of m [8, 24]. This result is the key for proving that iterated multiplication and division can be computed in logspace.
- Arbitrary algebraic constants can be approximated in logspace up to polynomially many bits. This result was recently shown in [11, 26].

For the case of even Coxeter groups, i.e., Coxeter groups where all defining relations have even length, we can combine Theorem 3.7 and Theorem 3.9 in one more general result, saying that the Parikh-image of the shortlex normal form can be computed in logspace (Theorem 3.12). The Parikh-image of a word $w \in \Sigma^*$ is the image of w under the canonical homomorphism from Σ^* to $\mathbb{N}^{|\Sigma|}$.

As mentioned above, it remains open, whether shortlex normal forms in Coxeter groups can be computed in logspace. In the second part of this paper, we prove that for the important subclass of *right-angled Coxeter groups* shortlex normal forms can be computed in logspace (Theorem 5.4). A right-angled Coxeter group is defined by a finite undirected graph (Σ, I) by taking Σ as the set of group generators and adding the defining relations $a^2 = 1$ for all $a \in \Sigma$ and $ab = ba$ for all edges $(a, b) \in I$. We use techniques from the theory of Mazurkiewicz traces [13]. More precisely, we describe right-angled Coxeter groups by strongly confluent length-reducing trace rewriting systems. Moreover, using the geometric representation of right-angled Coxeter groups, we provide an elementary proof that the alphabet of symbols that appear in a geodesic for g can be computed in logspace from g (Corollary 5.3).[2] In contrast to general Coxeter groups, for right-angled Coxeter groups this alphabetic information suffices in order to compute shortlex normal forms in logspace.

Right-angled Coxeter groups are closely related to graph groups, which are also known as *free partially commutative groups* or as *right-angled Artin groups*. A graph group is defined by a finite undirected graph (Σ, I) by taking Σ as the set of group generators and adding the defining relations $ab = ba$ for all edges $(a, b) \in I$. Hence, a right-angled Coxeter group is obtained from a graph group by adding all relations $a^2 = 1$ for all generators a. Graph groups received in recent years a lot of attention in group theory because of their rich subgroup structure [2, 10, 23]. On the algorithmic side, (un)decidability results were obtained for many important group-theoretic decision problems in graph groups [9, 16]. There is a standard embedding of a graph group into a right-angled Coxeter group [25]. Hence, also graph groups are linear and have logspace word problems. Using the special properties of this embedding, we can show that also for graph groups, shortlex normal forms can be computed in logspace (Theorem 5.4). We remark that this is an optimal result in the sense that logspace is the smallest known complexity class for the word problem in free groups already. Clearly, computing shortlex normal forms is at least as difficult than solving the word problem.

Finally, we apply Theorem 5.4 to *free partially commutative inverse monoids*. These monoids arise naturally in the context of deterministic and co-deterministic concurrent systems. This includes many real systems, because they can be viewed as deterministic concurrent systems with *undo*-operations. In [15] it was shown that the word problem for a free partially commutative inverse monoid can be solved in time $\mathcal{O}(n \log(n))$. (Decidability of the word problem is due to Da Costa [39].) Using our logspace algorithm for computing shortlex normal forms in a graph group, we can show that the word problem for a free partially commutative inverse monoid can be solved in logspace (Theorem 6.1). Again, with state-of-the art techniques, this can be viewed as an optimal result. It also generalizes a corresponding result for free inverse monoids from [32]. Let us emphasize that in order to obtain Theorem 6.1 we have to be able to compute shortlex normal forms in graph groups in logspace; knowing only that the word problem is in logspace would not have been sufficient for our purposes.

Let us remark that for all our results it is crucial that the group (resp., the free partially commutative inverse monoids) is fixed and not part of the input. For instance, it is not clear whether for a given undirected graph (Σ, I) and a word w over $\Sigma \cup \Sigma^{-1}$ one can check in logspace whether $w = 1$ in the graph group defined by the graph (Σ, I).

The work on this paper started at the AMS Sectional Meeting, Las Vegas, May 2011, and was motivated by the lecture of Gretchen Ostheimer [20]. A preliminary version of our results appeared as a conference abstract at the Latin American Symposium on Theoretical Informatics (LATIN 2012), [14]. In contrast to the conference abstract this paper provides full proofs and it contains new material about even Coxeter groups and how to compute geodesic lengths in all Coxeter groups.

---

[2] In contrast, the proof of Theorem 3.9, which generalizes Corollary 5.3 to all Coxeter groups, is more difficult in the sense that it uses geometry and more facts from [3].

2. Notation

Throughout Σ (resp. Γ) denotes a finite *alphabet*. This is a finite set, sometimes equipped with a linear order. An element of Σ is called a *letter*. By Σ^* we denote the free monoid over Σ. For a word $w \in \Sigma^*$ we denote by $\alpha(w)$ the *alphabet of w*: it is the set of letters occurring in w. With $|w|$ we denote the length of w. The *empty word* has length 0; and it is denoted by 1 as other neutral elements in monoids or groups.

All groups and monoids M in this paper are assumed to be finitely generated; and they come with a monoid homomorphism $\pi : \Sigma^* \to M$. Frequently we assume that M comes with an involution[3] $x \mapsto \bar{x}$ on M, and then we require that $\pi(\Sigma) \cup \overline{\pi(\Sigma)}$ generates M as a monoid.

If the monoid M is a group G, then the involution is always given by taking inverses, thus $\bar{x} = x^{-1}$. Moreover, G becomes a factor group of the *free group* $F(\Sigma)$ thanks to $\pi : \Sigma^* \to G$.

Let $\overline{\Sigma}$ be a disjoint copy of Σ and $\Gamma = \Sigma \cup \overline{\Sigma}$. There is a unique extension of the natural mapping $\Sigma \to \overline{\Sigma} : a \mapsto \bar{a}$ such that Γ^* becomes a monoid with involution: We let $\bar{\bar{a}} = a$ and $\overline{a_1 \cdots a_n} = \overline{a_n} \cdots \overline{a_1}$. Hence, we can lift our homomorphism $\pi : \Sigma^* \to M$ to a surjective monoid homomorphism $\pi : \Gamma^* \to M$ which respects the involution, i.e., $\pi(\bar{x}) = x^{-1}$.

Given a surjective homomorphism $\pi : \Gamma^* \to M$ and a linear order on Γ we can define the geodesic length and the shortlex normal form for elements in M as follows. For $w \in M$, the *geodesic length* $\|w\|$ is the length of a shortest word in $\pi^{-1}(w)$. The *shortlex normal form* of w is the lexicographical first word in the finite set $\{u \in \pi^{-1}(w) \mid |u| = \|w\|\}$. By a *geodesic* we mean any word in the finite set $\{u \in \pi^{-1}(w) \mid |u| = \|w\|\}$.

3. Coxeter groups

A *Coxeter group* G is given by a generating set $\Sigma = \{a_1, \ldots, a_n\}$ of n generators and a symmetric $n \times n$ matrix $M = (m_{i,j})_{1 \leq i,j \leq n}$ over $\mathbb{N}$ such that $m_{i,j} = 1$ if and only if $i = j$. The defining relations are $(a_i a_j)^{m_{i,j}} = 1$ for $1 \leq i,j \leq n$. In particular, $a_i^2 = 1$ for $1 \leq i \leq n$. Traditionally, one writes the entry ∞ instead 0 in the Coxeter matrix M and then $m_{i,j}$ becomes the order of the element $a_i a_j$.

A Coxeter group is called *even*, if all $m_{i,j}$ are even numbers for $i \neq j$. It is called *right-angled*, if $m_{i,j} \in \{0, 1, 2\}$ for all i,j. The defining relations of a right-angled Coxeter group can be rewritten in the following form: $a_i^2 = 1$ for $1 \leq i \leq n$ and $a_i a_j = a_j a_i$ for $(i,j) \in I$ where I denotes a symmetric and irreflexive relation $I \subseteq \{1, \ldots, n\} \times \{1, \ldots, n\}$. Thus, one could say that a right-angled Coxeter group is a *free partially commutative* Coxeter group. Readers interested only in right-angled Coxeter groups or in the application to graph groups (i.e., right-angled Artin groups) may proceed directly to Section 4.

3.1. Computing the geodesic alphabet and geodesic length.
Throughout this subsection G denotes a Coxeter group given by a fixed $n \times n$ matrix M as above. Recall that $\alpha(x)$ denotes the alphabet of the word x. One can show that if u and v are geodesics with $u = v$ in G then $\alpha(u) = \alpha(v)$ [**3**, Cor. 1.4.8]. We will

---

[3] An *involution* on a set Γ is a permutation $a \mapsto \bar{a}$ such that $\bar{\bar{a}} = a$. An involution of a monoid satisfies in addition $\overline{xy} = \bar{y}\,\bar{x}$.

show how to compute this alphabet in logspace. Moreover, we will show that also the geodesic length $|w|$ for a given $w \in G$ can be computed in logspace.

Let $\mathbb{R}^\Sigma$ be the n dimensional real vector space where the letter a is identified with the a-th unit vector. Thus, vectors will be written as formal sums $\sum_{b \in \Sigma} \lambda_b b$ with real coefficients $\lambda_b \in \mathbb{R}$. We fix the standard geometric representation $\sigma : G \to \mathrm{GL}(n, \mathbb{R})$, where we write σ_w for the mapping $\sigma(w)$, see e.g. [3, Sect. 4.2]:

$$(3.1) \qquad \sigma_{a_i}(a_j) = \begin{cases} a_j + 2\cos(\pi/m_{i,j}) \cdot a_i & \text{if } m_{i,j} \neq 0 \\ a_j + 2 \cdot a_i & \text{if } m_{i,j} = 0 \end{cases}$$

Note that for $a \in \Sigma$, $\sigma_w(a)$ cannot be the zero vector, since σ_w is invertible. We write $\sum_{b \in \Sigma} \lambda_b b \geq 0$ if $\lambda_b \geq 0$ for all $b \in \Sigma$. The following fundamental lemma can be found in [3, Prop. 4.2.5]:

LEMMA 3.1. *Let $w \in G$ and $a \in \Sigma$. We have*

$$\|wa\| = \begin{cases} \|w\| + 1 & \text{if } \sigma_w(a) \geq 0 \\ \|w\| - 1 & \text{if } \sigma_w(a) \leq 0 \end{cases}$$

LEMMA 3.2. *For a given $w \in G$ and $a, b \in \Sigma$, one can check in logspace, whether $\lambda_b \geq 0$, where $\sum_{b \in \Sigma} \lambda_b b = \sigma_w(a)$.*

In order to prove Lemma 3.2, we need several tools. Let p_i denote the ith prime number. It is well-known from number theory that the ith prime requires $O(\log(i))$ bits in its binary representation. For a number $0 \leq M < \prod_{i=1}^m p_i$ we define the *Chinese remainder representation* $\mathsf{CRR}_m(M)$ as the m-tuple

$$\mathsf{CRR}_m(M) = (M \bmod p_i)_{1 \leq i \leq m}.$$

By the Chinese remainder theorem, the mapping $M \mapsto \mathsf{CRR}_m(M)$ is a bijection from the interval $[0, \prod_{i=1}^m p_i - 1]$ to $\prod_{i=1}^m [0, p_i - 1]$. By the following theorem, we can transform a CRR-representation very efficiently into binary representation.

THEOREM 3.3 ([8, Thm. 3.3]). *For a given tuple $(r_1, \ldots, r_m) \in \prod_{i=1}^m [0, p_i - 1]$, we can compute in logspace the binary representation of the unique number $M \in [0, \prod_{i=1}^m p_i - 1]$ such that $\mathsf{CRR}_m(M) = (r_1, \ldots, r_m)$.*

By [24], the transformation from the CRR-representation to the binary representation can be even computed by $\mathsf{DLOGTIME}$-uniform TC^0-circuits. Our second tool is a gap theorem for values $p(\zeta)$, where $p(x) \in \mathbb{Z}[x]$ and ζ is a root of unity. For a polynomial $p(x) = \sum_{i=0}^n a_i x^i$ with integer coefficients a_i let $|p(x)| = \sum_{i=0}^n |a_i|$.

THEOREM 3.4 ([30, Thm. 3]). *Let $p(x) \in \mathbb{Z}[x]$ and let ζ be a dth root of unity such that $p(\zeta) \neq 0$. Then $|p(\zeta)| > |p(x)|^{-d}$.*

Finally, our third tool for the proof of Lemma 3.2 is the following result, which was recently shown (independently) in [11, 26].

THEOREM 3.5 ([11, Thm .2],[26, Cor. 4.6]). *For every fixed algebraic number $\alpha \in \mathbb{R}$ the following problem can be computed in logspace:*
INPUT: *A unary coded number n.*
OUTPUT: *The binary representation of the integer $\lfloor 2^n \alpha \rfloor$.*

REMARK 3.6. The result of [26] is actually stronger showing that the output in Theorem 3.5 can be computed in uniform TC^0.

Proof of Lemma 3.2. We decompose the logspace algorithm into several logspace computations. The linear mapping σ_w can be written as a product of matrices $A_1 A_2 \cdots A_{|w|}$, where every A_i is an $(n \times n)$-matrix with entries from $\{0, 1, 2\} \cup \{2\cos(\pi/m_{i,j}) \mid m_{i,j} \neq 0\}$ (which is the set of coefficients that appear in (3.1)). Then, we have to check whether this matrix product applied to the unit vector a has a non-negative value in the b-coordinate. This value is the entry $(A_1 A_2 \cdots A_{|w|})_{a,b}$ of the product matrix $A_1 A_2 \cdots A_{|w|}$.

Let m be the least common multiple of all $m_{i,j} \neq 0$; this is still a constant. Let $\zeta = e^{\pi i/m}$, which is a primitive $(2m)^{th}$ root of unity. If $m = m_{i,j} \cdot k$, we have

$$2 \cdot \cos\left(\frac{\pi}{m_{i,j}}\right) = \zeta^k + \zeta^{2m-k}.$$

Hence, we can assume that every A_i is an $(n \times n)$-matrix over $\mathbb{Z}[\zeta]$. We now replace ζ by a variable X in all matrices $A_1, \ldots, A_{|w|}$; let us denote the resulting matrices over the ring $\mathbb{Z}[X]$ with $B_1, \ldots, B_{|w|}$. Each entry in one of these matrices is a polynomial of degree $< 2m$ with coefficients bounded by 2. More precisely, for every entry $p(X)$ of a matrix B_i we have $|p(X)| \leq 2$. Let $|B_i|$ be the sum of all $|p(X)|$ taken over all entries of the matrix B_i. Hence, $|B_i| \leq 2n^2$.

Step 1. In a first step, we show that the product $B_1 \cdots B_{|w|}$ can be computed in logspace in the ring $\mathbb{Z}[X]/(X^{2m} - 1)$ (keeping in mind that $\zeta^{2m} = 1$). Every entry in the product $B_1 \cdots B_{|w|}$ is a polynomial of degree $< 2m$ with coefficients bounded in absolute value by $|B_1| \cdots |B_{|w|}| \leq (2n^2)^{|w|}$. Here n is a fixed constant. Hence, every coefficient in the matrix $B_1 \cdots B_{|w|}$ can be represented with $O(|w|)$ bits. In logspace, one can compute a list of the first k prime numbers $p_1, p_2, \ldots, p_k$, where $k \in O(|w|)$ is chosen such that $\prod_{i=1}^{k} p_i > (2n^2)^{|w|}$ [8]. Each p_i is bounded by $|w|^{O(1)}$.

For every $1 \leq i \leq k$, we can compute in logspace the matrix product $B_1 \cdots B_{|w|}$ in $\mathbb{F}_{p_i}[X]/(X^{2m}-1)$, i.e., we compute the coefficient of each polynomial in $B_1 \cdots B_{|w|}$ modulo p_i. In the language of [8]: For each coefficient of a polynomial in $B_1 \cdots B_{|w|}$, we compute its Chines remainder representation. From this representation, we can compute in logspace by Theorem 3.3 the binary representation of the coefficient. This shows that the product $B = B_1 \cdots B_{|w|}$ can be computed in the ring $\mathbb{Z}[X]/(X^{2m} - 1)$.

Step 2. We know that if X is substitued by ζ in the matrix B, then we obtain the product $A = A_1 \cdots A_{|w|}$ (the matrix we are actully interested in), which is a matrix over $\mathbb{R}$. Every entry of the matrix A is of the form $\sum_{j=0}^{2m-1} a_j \zeta^j$, where a_j is a number with $O(|w|)$ bits that we have computed in Step 1. If $\sum_{j=0}^{2m-1} a_j \zeta^j \neq 0$, then by Theorem 3.4, we have

$$\left| \sum_{j=0}^{2m-1} a_j \zeta^j \right| > \left(\sum_{j=0}^{2m-1} |a_j| \right)^{-2m}.$$

Since m is a constant, and $|a_j| \leq 2^{O(|w|)}$, we have

$$\sum_{j=0}^{2m-1} a_j \zeta^j = 0 \quad \text{or} \quad \left| \sum_{j=0}^{2m-1} a_j \zeta^j \right| > 2^{-c|w|}$$

for a constant c. Therefore, to check whether $\sum_{j=0}^{2m-1} a_j \zeta^j \geq 0$ or $\sum_{j=0}^{2m-1} a_j \zeta^j \leq 0$, it suffices to approximate this sum up to $c|w|$ many fractional bits. This is the goal of the second step.

Since we are sure that $\sum_{j=0}^{2m-1} a_j \zeta^j \in \mathbb{R}$, we can replace the sum symbolically by its real part, which is $\sum_{j=0}^{2m-1} a_j \cos(j\pi/m)$. In order to approximate this sum up to $c|w|$ many fractional bits, it suffices to approximate each $\cos(j\pi/m)$ up to $d|w|$ many fractional bits (recall that $a_j \in 2^{O(|w|)}$), where the constant d is large enough.

Every number $\cos(q \cdot \pi)$ for $q \in \mathbb{Q}$ is algebraic.[4] Theoreore, by Theorem 3.5, every number $\cos(j\pi/m)$ ($0 \leq j \leq 2m-1$) can be approximated in logspace up to $d|w|$ many fractional bits. This concludes the proof. □

Lemma 3.2 can be used in order to compute in logspace the geodesic length $\|w\|$ for a given group element $w \in G$:

THEOREM 3.7. *For a given word $w \in \Sigma^*$, the geodesic length $\|w\|$ can be computed in logspace.*

PROOF. By Lemma 3.1, the following algorithm correctly computes $\|w\|$ for $w = a_1 \cdots a_k$.

$\ell := 0$;
for $i = 1$ **to** k **do**
   if $\sigma_{a_1 \cdots a_{i-1}}(a_i) \geq 0$ **then**
      $\ell := \ell + 1$
   else
      $\ell := \ell - 1$
   endif
endfor
return ℓ

By Lemma 3.2 it can be implemented in logspace. □

We finally apply Lemma 3.2 in order to compute in logspace the set of all letters that occur in a geodesic for a given group element $w \in G$. As remarked before, this alphabet is independent of the particular geodesic for w.

Introduce a new letter $x \notin \Sigma$ with $x^2 = 1$, but no other new defining relation. This yields the Coxeter group $G' = G * (\mathbb{Z}/2\mathbb{Z})$ generated by $\Sigma' = \Sigma \cup \{x\}$. Thus, ax is of infinite order in G' for all $a \in \Sigma$. Clearly, $\|wx\| > \|w\|$ for all $w \in G$. Hence, $\sigma_w(x) \geq 0$ for all $w \in G$ by Lemma 3.1.

LEMMA 3.8. *Let $w \in G$ and $\sigma_w(x) = \sum_{b \in \Sigma'} \lambda_b b$. Then for all $b \in \Sigma$ we have $\lambda_b \neq 0$ if and only if the letter b appears in the shortlex normal form of w.*

PROOF. We may assume that w is a geodesic in G. We prove the result by induction on $\|w\| = |w|$. If $w = 1$, then the assertion is trivial. If $b \in \Sigma$ does not occur as a letter in w, then it is clear that $\lambda_b = 0$. Thus, we may assume that $b \in \alpha(w)$ and we have to show that $\lambda_b \neq 0$. By induction, we may write $w = ua$ with $\|uax\| > \|ua\| > \|u\|$. We have $\sigma_w(x) = \sigma_u \sigma_a(x) = \sigma_u(x + 2a) = \sigma_u(x) + 2\sigma_u(a)$.

---

[4]This seems to be a folklore fact, for which we could not find a reference. If $q \cdot \pi = \frac{2a\pi}{b}$ for $a \in \mathbb{Z}, b \in \mathbb{N}$, then by DeMoivre's formula $((\cos\theta + i\sin\theta)^n = \cos(n\theta) + i\sin(n\theta))$ we have $(\cos(q \cdot \pi) + i\sin(q \cdot \pi))^b = \cos(2a\pi) + i\sin(2a\pi) = 1$. Hence, $z := \cos(q \cdot \pi) + i\sin(q \cdot \pi)$ is algebraic, which implies that $\cos(q \cdot \pi) = \frac{1}{2}(z + \overline{z})$ is algebraic too.

The standard geometric representation yields moreover $\sigma_w(x) = x + \sum_{c \in \Sigma} \lambda_c c$, where $\lambda_c \geq 0$ for all $c \in \Sigma$ by Lemma 3.1. As $\|ua\| > \|u\|$ we get $\sigma_u(a) \geq 0$ by Lemma 3.1. Moreover, by induction (and the fact $\|ux\| > \|u\|$), we know that for all letters $c \in \alpha(u)$ the corresponding coefficient in $\sigma_u(x)$ is strictly positive. Thus, we are done if $b \in \alpha(u)$. So, the remaining case is that $b = a \notin \alpha(u)$. However, in this case $\sigma_u(a) = a + \sum_{c \in \Sigma \setminus \{a\}} \mu_c c$. Hence $\lambda_a \geq 2$. □

THEOREM 3.9. *There is a logspace transducer which on input $w \in \Sigma^*$ computes the set of letters occurring in the shortlex normal form of w.*

PROOF. By Lemma 3.8, we have to check for every letter $b \in \Sigma$, whether $\lambda_b = 0$, where $\sum_{b \in \Sigma'} \lambda_b b = \sigma_w(x)$. By Lemma 3.2 (applied to the Coxeter group G') this is possible in logspace. □

Let us remark that the use of Lemma 3.2 in the proof of Theorem 3.9 can be avoided, using the technique from [29] and Lemma 3.8. Every λ_b belongs to the ring $\mathbb{Z}[\zeta] \cong \mathbb{Z}[X]/\Phi(X)$, where ζ is a primitive $(2m)^{th}$ root of unity, $\Phi(X)$ is the $(2m)^{th}$ cyclotomic polynomial, and m is the least common multiple of all $m_{i,j} \neq 0$. In order to check whether $\lambda_b = 0$, we can check whether the value is zero mod r with respect to all r up to a polynomial threshold.

3.2. Computing the geodesic Parikh-image in even Coxeter groups.

In this section we assume that G is an even Coxeter group. Thus, the entries $m_{i,j}$ of the matrix M are even for all $i \neq j$.

Let $a \in \Sigma$ be a letter and $w \in \Sigma^*$. By $|w|_a$ we denote the number of a's in a word $w \in \Gamma^*$. The *Parikh-image* of w is the vector $[|w|_a]_{a \in \Sigma} \in \mathbb{N}^\Sigma$. In other words, the Parikh-image of w is the image of w under the canonical homomorphism from the free monoid Σ^* to the free commutative monoid $\mathbb{N}^\Sigma$.

We show that for even Coxeter groups, the Parikh-image of geodesics can be computed in logspace. Actually, all geodesics for a given group element of an even Coxeter group have the same Parikh-image:

LEMMA 3.10. *Let G be an even Coxeter group and let $u, v \in \Sigma^*$ be geodesics with $u = v$ in G. Then we have $|u|_a = |v|_a$ for all $a \in \Sigma$.*

PROOF. Let $1 \leq i, j \leq n$ with $i \neq j$. Since the entry $m_{i,j}$ of the matrix M is even and $(a_i a_j)^{m_{i,j}} = 1$ in G, we get the relation $(a_i a_j)^{m_{i,j}/2} = (a_j a_i)^{m_{i,j}/2}$ which does not effect the Parikh-image. Let the string rewriting system T consist of all rules of the form $(a_i a_j)^{m_{i,j}/2} \to (a_j a_i)^{m_{i,j}/2}$ for $1 \leq i, j \leq n$ with $i \neq j$. Now, it follows from a well-known result of Tits (c.f. [3]) that if u and v are geodesic, then u can be transformed into v using applications of rules from T. Consequently, $|u|_a = |v|_a$ for all $a \in \Sigma$. □

LEMMA 3.11. *Let G be an even Coxeter group, $a \in \Sigma$, and let u, w be geodesics such that $wa = u$ in G. Then there exists $\varepsilon \in \{1, -1\}$ such that $|u| = |w| + \varepsilon$ and $|u|_a = |w|_a + \varepsilon$. For all $b \in \Sigma \setminus \{a\}$ we have $|u|_b = |w|_b$.*

PROOF. By Lemma 3.1 there exists $\varepsilon \in \{1, -1\}$ with $|u| = |w| + \varepsilon$. Moreover, since $a^2 = 1$, we have $ua = w$ and $wa = u$ in G. Hence, if $|w| = |u| + 1$ (resp., $|u| = |w| + 1$), then ua and w (resp., wa and u) are geodesics defining the same group element in G. Lemma 3.10 implies that $|ua|_c = |w|_c$ (resp., $|wa|_c = |u|_c$) for all $c \in \Sigma$. This implies the conclusion of the lemma. □

THEOREM 3.12. *Let G be an even Coxeter group. For a given word $w \in \Sigma^*$, the Parikh-image of the shortlex normal form for w can be computed in logspace.*

PROOF. Lemma 3.11 shows that the following straightforward modification of the logspace algorithm in (the proof of) Theorem 3.7 computes the Parikh-image of the shortlex normal form for w correctly. Let $w = a_1 \cdots a_k$ be the input word.

for all $a \in \Gamma$ **do**
    $\ell_a := 0$
endfor
for $i = 1$ **to** k **do**
    if $\sigma_{a_1 \cdots a_{i-1}}(a_i) \geq 0$ **then**
        $\ell_{a_i} := \ell_{a_i} + 1$
    else
        $\ell_{a_i} := \ell_{a_i} - 1$
    endif
endfor
return $[\ell_a]_{a \in \Gamma}$ □

4. Mazurkiewicz traces and graph groups

In the rest of the paper, we will deal with right-angled Coxeter groups. As explained in Section 3, a right-angled Coxeter group can be specified by a finite undirected graph (Σ, I). The set Σ is the generating set and the relations are $a^2 = 1$ for all $a \in \Sigma$ and $ab = ba$ for all $(a, b) \in I$. Hence, I specifies a partial commutation relation, and elements of a right-angled Coxeter group can be represented by partially commutative words, also known as Mazurkiewicz traces. In this section, we will introduce some basic notions from the theory of Mazurkiewicz traces, see [13, 17] for more details.

An *independence alphabet* is a pair (Σ, I), where Σ is a finite set (or *alphabet*) and $I \subseteq \Sigma \times \Sigma$ is an irreflexive and symmetric relation, called the *independence relation*. Thus, (Σ, I) is a finite undirected graph. The complementary relation $D = (\Sigma \times \Sigma) \setminus I$ is called a *dependence relation*. It is reflexive and symmetric. We extend (Σ, I) to a graph (Γ, I_Γ), where $\Gamma = \Sigma \cup \overline{\Sigma}$ with $\Sigma \cap \overline{\Sigma} = \emptyset$, and I_Γ is the minimal independence relation with $I \subseteq I_\Gamma$ and such that $(a, b) \in I_\Gamma$ implies $(a, \bar{b}) \in I_\Gamma$. The independence alphabet (Σ, I) defines a *free partially commutative monoid* (or *trace monoid*) $M(\Sigma, I)$ and a *free partially commutative group* $G(\Sigma, I)$ by:

$$M(\Sigma, I) = \Sigma^* / \{ab = ba \mid (a, b) \in I\},$$
$$G(\Sigma, I) = F(\Sigma) / \{ab = ba \mid (a, b) \in I\}.$$

Free partially commutative groups are also known as *right-angled Artin groups* or *graph groups*. Elements of $M(\Sigma, I)$ are called *(Mazurkiewicz) traces*. They have a unique description as *dependence graphs*, which are node-labelled acyclic graphs defined as follows. Let $u = a_1 \cdots a_n \in \Sigma^*$ be a word. The vertex set of the dependence graph $\mathrm{DG}(u)$ is $\{1, \ldots, n\}$ and vertex i is labelled with $a_i \in \Sigma$. There is an arc from vertex i to j if and only if $i < j$ and $(a_i, a_j) \in D$. Now, two words define the same trace in $M(\Sigma, I)$ if and only if their dependence graphs are isomorphic. A dependence graph is acyclic, so its transitive closure is a labelled partial order $\prec$, which can be uniquely represented by its *Hasse diagram*: There is an arc from i to j in the Hasse diagram, if $i \prec j$ and there does not exist k with $i \prec k \prec j$.

A trace $u \in M(\Sigma, I)$ is a *factor* of $v \in M(\Sigma, I)$, if $v \in M(\Sigma, I)uM(\Sigma, I)$. The set of letters occurring in a trace u is denoted by $\alpha(u)$. The independence relation I is extended to traces by letting $(u, v) \in I$, if $\alpha(u) \times \alpha(v) \subseteq I$. We also write $I(a) = \{b \in \Sigma \mid (a, b) \in I\}$. A trace u is called a *prime* if $DG(u)$ has exactly one maximal element. Thus, if u is a prime, then we can write u as $u = va$ in $M(\Sigma, I)$, where $a \in \Sigma$ and $v \in M(\Sigma, I)$ are uniquely defined. Moreover, this property characterizes primes. A *prime prefix* of a trace u is a prime trace v such that $u = vx$ in $M(\Sigma, I)$ for some trace x. We will use the following simple fact.

LEMMA 4.1. *Let (Σ, I) be a fixed independence relation. There is a logspace transducer that on input $u \in M(\Sigma, I)$ outputs a list of all prime prefixes of u.*

PROOF. The prime prefixes of u correspond to the downward-closed subsets of the dependence graph $DG(u)$ that have a unique maximal element. Assume that $u = a_1 a_2 \cdots a_n$ with $a_i \in \Sigma$. Our logspace transducer works in n phases. In the i-th phase it outputs the sequence of all symbols a_j ($j \leq i$) such that there exists a path in $DG(u)$ from j to i. Note that there exists a path from j to i in $DG(u)$ if and only if there is such a path of length at most $|\Sigma|$. Since Σ is fixed, the existence of such a path can be checked in logspace by examining all sequences $1 \leq i_1 < i_2 < \cdots < i_k = i$ with $k \leq |\Sigma|$. Such a sequence can be stored in logarithmic space since $|\Sigma|$ is a constant. □

We use standard notations from the theory of rewriting systems, cf [4]. Let $M = M(\Sigma, I)$. A *trace rewriting system* is a finite set of rules $S \subseteq M \times M$. A rule is often written in the form $\ell \longrightarrow r$. The system S defines a one-step rewriting relation $\Longrightarrow_S \subseteq M \times M$ by $x \Longrightarrow_S y$ if there exist $(\ell, r) \in S$ and $u, v \in M$ with $x = u\ell v$ and $y = urv$ in M. By $\stackrel{*}{\Longrightarrow}_S$, we denote the reflexive and transitive closure of $\Longrightarrow_S$. The set $IRR(S)$ denotes the set of traces to which no rule of S applies. If S is confluent and terminating, then for every $u \in M$ there is a unique $\widehat{u} \in IRR(S)$ with $u \stackrel{*}{\Longrightarrow}_S \widehat{u}$, and $IRR(S)$ is a set of normal forms for the quotient monoid M/S. If, in addition, S is length-reducing (i.e., $|\ell| > |r|$ for all $(\ell, r) \in S$), then $\|\pi(u)\| = |\widehat{u}|$ for the canonical homomorphism $\pi : M \to M/S$.

EXAMPLE 4.2. The system $S_G = \{a\bar{a} \longrightarrow 1 \mid a \in \Gamma\}$ is (strongly) confluent and length-reducing over $M(\Gamma, I_\Gamma)$ [13]. The quotient monoid $M(\Gamma, I_\Gamma)/S_G$ is the graph group $G(\Sigma, I)$.

By Example 4.2 elements in graph groups have a unique description as *dependence graphs*, too. A trace belongs to $IRR(S_G)$ if and only if it does not contain a factor $a\bar{a}$ for $a \in \Gamma$. In the dependence graph, this means that the Hasse diagram does not contain any arc from a vertex labeled a to a vertex labeled $\bar{a}$ with $a \in \Gamma$. Moreover, a word $u \in \Gamma^*$ represents a trace from $IRR(S_G)$ if and only if u does not contain a factor of the form $av\bar{a}$ with $a \in \Gamma$ and $\alpha(v) \subseteq I(a)$.

5. Right-angled Coxeter groups

Some of the results on right-angled Coxeter groups in this section are covered by more general statements in Section 3. However, the former section used quite involved tools from computational algebra and an advanced theory of Coxeter groups. In contrast the results we prove here on right-angled Coxeter groups are purely combinatorial. Hence we can give simple and elementary proofs which makes this section fully self-contained. Moreover, in contrast to the general case, for the right-angled case we will succeed in computing shortlex normal forms in logspace.

Recall that a *right-angled Coxeter group* is specified by a finite undirected graph (Σ, I), i.e., an independence alphabet. The set Σ is the generating set and the relations are $a^2 = 1$ for all $a \in \Sigma$ and $ab = ba$ for all $(a,b) \in I$. We denote this right-angled Coxeter group by $C(\Sigma, I)$. Similarly to the graph group $G(\Sigma, I)$, the right-angled Coxeter group $C(\Sigma, I)$ can be defined by a (strongly) confluent and length-reducing trace rewriting system (this time on $M(\Sigma, I)$ instead of $M(\Gamma, I_\Gamma)$). Let
$$S_C = \{a^2 \to 1 \mid a \in \Sigma\}.$$
Then S_C is indeed (strongly) confluent and length-reducing on $M(\Sigma, I)$ and the quotient $M(\Sigma, I)/S_C$ is $C(\Sigma, I)$. Hence we have two closely related (strongly) confluent and length-reducing trace rewriting systems: S_G defines the graph group $G(\Sigma, I)$ and S_C defines the right-angled Coxeter group $C(\Sigma, I)$. Both systems define unique normal forms of geodesic length: $\widehat{u} \in M(\Gamma, I_\Gamma)$ for S_G and $\widehat{u} \in M(\Sigma, I)$ for S_C. Note that there are no explicit commutation rules as they are *built-in* in trace theory. There is a linear time algorithm for computing $\widehat{u}$; see [13] for a more general result of this type.

It is well known that a graph group $G(\Sigma, I)$ can be embedded into a right-angled Coxeter group [25]. For this, one has to duplicate each letter from Σ. Formally, we can take the right-angled Coxeter group $C(\Gamma, I_\Gamma)$ (in which $\bar{a}$ does not denote the inverse of a). Consider the mapping $\varphi(a) = a\bar{a}$ from Γ to Γ^*. Obviously, φ induces a homomorphism from $G(\Sigma, I)$ to the Coxeter group $C(\Gamma, I_\Gamma)$. As $\mathrm{IRR}(S_G) \subseteq M(\Gamma, I_\Gamma)$ is mapped to $\mathrm{IRR}(S_C) \subseteq M(\Gamma, I_\Gamma)$, we recover the well-known fact that φ is injective. Actually we see more. Assume that $\widehat{w}$ is the shortlex normal form of some $\varphi(g)$. Then replacing in $\widehat{w}$ factors $a\bar{a}$ with a and replacing factors $\bar{a}a$ with $\bar{a}$ yields a logspace reduction of the problem of computing shortlex normal forms in graph groups to the problem of computing shortlex normal forms in right-angled Coxeter groups. Thus, for our purposes it is enough to calculate shortlex normal forms for right-angled Coxeter groups of type $C(\Sigma, I)$ in logspace. For the latter, it suffices to compute in logspace on input $u \in \Sigma^*$ some trace (or word) v such that $u = v$ in $C(\Sigma, I)$ and $|v| = \|u\|$. Then, the shortlex normal form for u is the lexicographic normal form of the trace v, which can be easily computed in logspace from u.

A trace in $M(\Sigma, I)$ is called a *Coxeter-trace*, if it does not have any factor a^2 where $a \in \Sigma$. It follows that every element in $C(\Sigma, I)$ has a unique representation as a Coxeter-trace. Let $a \in \Sigma$. A trace u is called a-short, if during the derivation $u \overset{*}{\Longrightarrow}_{S_C} \widehat{u} \in \mathrm{IRR}(S_C)$ the rule $a^2 \longrightarrow 1$ is not applied. Thus, u is a-short if and only if the number of occurrences of the letter a is the same in the trace u and its Coxeter-trace $\widehat{u}$. We are interested in the set of letters which survive the reduction process. By $\widehat{\alpha}(u) = \alpha(\widehat{u})$ we denote the alphabet of the unique Coxeter-trace $\widehat{u}$ with $u = \widehat{u}$ in $C(\Sigma, I)$. Here is a crucial observation:

LEMMA 5.1. *A trace u is a-short if and only if u has no factor ava such that $\widehat{\alpha}(v) \subseteq I(a)$.*

PROOF. If u contains a factor ava such that $\widehat{\alpha}(v) \subseteq I(a)$, then u is clearly not a-short. We prove the other direction by induction on the length of u. Write

$u = a_1 \cdots a_n$ with $a_i \in \Sigma$. We identify u with its dependence graph $\mathrm{DG}(u)$ which has vertex set $\{1, \ldots, n\}$. Assume that u is not a-short. Then, during the derivation $u \overset{*}{\Longrightarrow}_{S_C} \widehat{u}$, for a first time a vertex i with label $a_i = a$ is canceled with vertex j with label $a_j = a$ and $i < j$. It is enough to show that $\widehat{\alpha}(a_{i+1} \cdots a_{j-1}) \subseteq I(a)$. If the cancellation of i and j happens in the first step of the rewriting process, then $\alpha(a_{i+1} \cdots a_{j-1}) \subseteq I(a)$ and we are done. So, let the first step cancel vertices k and ℓ with labels $a_k = a_\ell = b$ and $k < \ell$. Clearly, $\{i, j\} \cap \{k, \ell\} = \emptyset$. The set $\widehat{\alpha}(a_{i+1} \cdots a_{j-1})$ can change, only if either $i < k < j < \ell$ or $k < i < \ell < j$. However in both cases we must have $(b, a) \in I$, and we are done by induction. $\square$

In the right-angled case, the standard geometric representation (see (3.1)) $\sigma : C(\Sigma, I) \to \mathrm{GL}(n, \mathbb{Z})$ (where $n = |\Sigma|$) can be defined as follows, where again we write σ_a for the mapping $\sigma(a)$:

$$\sigma_a(a) = -a,$$
$$\sigma_a(b) = \begin{cases} b & \text{if } (a, b) \in I, \\ b + 2a & \text{if } (a, b) \in D \text{ and } a \neq b. \end{cases}$$

In this definition, a, b are letters. We identify $\mathbb{Z}^n = \mathbb{Z}^\Sigma$ and vectors from $\mathbb{Z}^n$ are written as formal sums $\sum_b \lambda_b b$. One can easily verify that $\sigma_{ab}(c) = \sigma_{ba}(c)$ for $(a, b) \in I$ and $\sigma_{aa}(b) = b$. Thus, σ defines indeed a homomorphism from $C(\Sigma, I)$ to $\mathrm{GL}(n, \mathbb{Z})$ (as well as homomorphisms from Σ^* and from $M(\Sigma, I)$ to $\mathrm{GL}(n, \mathbb{Z})$). Note that if $w = uv$ is a trace and $(b, v) \in I$ for a symbol b, then $\sigma_w(b) = \sigma_u(b)$. The following proposition is fundamental for understanding how the internal structure of w is reflected by letting σ_w act on letters (called *simple roots* in the literature). For lack of a reference for this variant (of a well-known general fact) and since the proof is rather easy in the right-angled case (in contrast to the general case), we give a proof, which is purely combinatorial.

PROPOSITION 5.2. *Let wd be a Coxeter-trace, $\sigma_w(d) = \sum_b \lambda_b b$ and $wd = udv$ where ud is prime and $(d, v) \in I$. Then it holds:*

(1) $\lambda_b \neq 0$ *if and only if $b \in \alpha(ud)$. Moreover, $\lambda_b > 0$ for all $b \in \alpha(ud)$.*
(2) *Let $b, c \in \alpha(ud)$, $b \neq c$, and assume that the first b in $\mathrm{DG}(ud)$ appears before the first c in $\mathrm{DG}(ud)$. Then we have $\lambda_b > \lambda_c > 0$.*

PROOF. We prove both statements of the lemma by induction on $|u|$. For $|u| = 0$ both statements are clear. Hence, let $u = au'$ and $\sigma_{u'}(d) = \sum_b \mu_b b$. Thus,

$$\sigma_u(d) = \sum_b \lambda_b b = \sigma_a(\sum_b \mu_b b) = \sum_b \mu_b \sigma_a(b).$$

Note that $\mu_b = \lambda_b$ for all $b \neq a$. Hence, by induction $\lambda_b = 0$ for all $b \notin \alpha(ud)$ and $\lambda_b > 0$ for all $b \in \alpha(ud) \setminus \{a\}$.

Let us now prove (2) for the trace u (it implies $\lambda_a > 0$ and hence (1)). Consider $b, c \in \alpha(ud)$, $b \neq c$, such that the first b in $\mathrm{DG}(ud)$ appears before the first c in $\mathrm{DG}(ud)$. Clearly, this implies $c \neq a$. For $b \neq a$ we obtain that the first b in $\mathrm{DG}(u'd)$ appears before the first c in $\mathrm{DG}(u'd)$. Hence, by induction we get $\mu_b > \mu_c > 0$. Claim (2) follows since $b \neq a \neq c$ implies $\mu_b = \lambda_b$ and $\mu_c = \lambda_c$.

Thus, let $a = b$. As there is path from the first a to every c in $\mathrm{DG}(ud)$ we may replace c by the first letter we meet on such a path. Hence we may assume that a and c are dependent. Note that $a \neq c$ because u is a Coxeter-trace. Hence,

$\lambda_c = \mu_c > 0$ and it is enough to show $\lambda_a > \mu_c$. But $\lambda_a \geq 2\mu_c - \mu_a$ by the definition of σ_a. If $\mu_a = 0$, then $\lambda_a \geq 2\mu_c$, which implies $\lambda_a > \mu_c$, since $\mu_c > 0$. Thus, we may assume $\mu_a > 0$. By induction, we get $a \in \alpha(u'd)$. Here comes the crucial point: the first c in $\mathrm{DG}(u'd)$ must appear before the first a in $u'd$. Thus, $\mu_c > \mu_a$ by induction, which finally implies $\lambda_a \geq 2\mu_c - \mu_a = \mu_c + (\mu_c - \mu_a) > \mu_c$. □

COROLLARY 5.3. *Let $C(\Sigma, I)$ be a fixed right-angled Coxeter group. Then on input $w \in \Sigma^*$ we can calculate in logspace the alphabet $\widehat{\alpha}(w)$ of the corresponding Coxeter-trace $\widehat{w}$.*

PROOF. Introduce a new letter x which depends on all other letters from Σ. We have $\sigma_w(x) = \sigma_{\widehat{w}}(x) = \sum_b \lambda_b b$. As $\widehat{w}x$ is a Coxeter-trace and prime, we have for all $b \in \Sigma$:
$$b \in \widehat{\alpha}(w) \iff b \in \alpha(\widehat{w}x) \iff \lambda_b \neq 0,$$
where the last equivalence follows from Proposition 5.2. Whether $\lambda_b \neq 0$ can be checked in logspace, by computing $\lambda_b \bmod m$ for all numbers $m \leq |w|$, since the least common multiple of the first n numbers is larger than 2^n (if $n \geq 7$) and the λ_b are integers with $|\lambda_b| \leq 2^{|w|}$. See also [29] for an analogous statement in the general context of linear groups. □

The hypothesis in Corollary 5.3 being a right-angled Coxeter group is actually not necessary as we have seen in Theorem 3.9. It remains open whether this hypothesis can be removed in the following theorem.

THEOREM 5.4. *Let G be a fixed graph group or a fixed right-angled Coxeter group. Then we can calculate in logspace shortlex normal forms in G.*

PROOF. As remarked earlier, it is enough to consider a right-angled Coxeter group $G = C(\Sigma, I)$. Fix a letter $a \in \Sigma$. We first construct a logspace transducer, which computes for an input trace $w \in M(\Sigma, I)$ a trace $u \in M(\Sigma, I)$ with the following properties: (i) $u = w$ in $C(\Sigma, I)$, (ii) u is a-short, and (iii) for all $b \in \Sigma$, if w is b-short, then also u is b-short. Having such a logspace transducer for every $a \in \Sigma$, we can compose all of them in an arbitrary order (note that $|\Sigma|$ is a constant) to obtain a logspace transducer which computes for a given input trace $w \in M(\Sigma, I)$ a trace u such that $w = u$ in $C(\Sigma, I)$ and u is a-short for all $a \in \Sigma$, i.e., $u \in \mathrm{IRR}(S_C)$. Thus $u = \widehat{w}$. From u we can compute easily in logspace the Hasse diagram of $\mathrm{DG}(u)$ and then the shortlex normal form.

So, let us fix a letter $a \in \Sigma$ and an input trace $w = a_1 \cdots a_n$, where $a_1, \ldots, a_n \in \Sigma$. We remove from left to right positions (or vertices) labeled by the letter a which cancel and which therefore do not appear in $\widehat{w}$. We read $a_1 \cdots a_n$ from left to right. In the i-th stage do the following: If $a_i \neq a$ output the letter a_i and switch to the $(i+1)$-st stage. If however $a_i = a$, then compute in logspace (using Corollary 5.3) the maximal index $j > i$ (if it exists) such that $a_j = a$ and $\widehat{\alpha}(a_{i+1} \cdots a_{j-1}) \subseteq I(a)$. If no such index j exists, then append the letter a_i to the output tape and switch to the $(i+1)$-st stage. If j exists, then append the word $a_{i+1} \cdots a_{j-1}$ to the output tape, but omit all a's. After that switch immediately to stage $j+1$. Here is a pseudo code description of the algorithm, where $\pi_{\Sigma \setminus \{a\}} : \Sigma^* \to (\Sigma \setminus \{a\})^*$ denotes the homomorphism that deletes all occurrences of a.

$i := 1;$
$w := 1;$ \hfill (= empty word)

```
while i ≤ n do
    if aᵢ ≠ a then
        w := waᵢ;
        i := i + 1
    else
        j := undefined;
        for k = i + 1 to n do
            if aₖ = a and α̂(aᵢ₊₁ ··· aₖ₋₁) ⊆ I(a) then
                j := k
            endif
        endfor
        if j = undefined then
            w := waᵢ;
            i := i + 1
        else
            w := w π_{Σ\{a}}(aᵢ ··· aⱼ₋₁);
            i := j + 1
        endif
    endif
endwhile
return(w)                    (= the content of the output tape of the transducer)
```

Let w_s be the content of the output tape at the beginning of stage s, i.e., when the algorithm checks the condition of the while-loop and variable i has value s. (hence, $w_1 = 1$ and w_{n+1} is the final output). The invariant of the algorithm is that

- $w_s = a_1 \cdots a_{s-1}$ in $C(\Sigma, I)$,
- w_s is a-short, and
- if $a_1 \cdots a_{s-1}$ is b-short, then also w_s is b-short.

The proof of this fact uses Lemma 5.1. □

6. Free partially commutative inverse monoids

A monoid M is *inverse*, if for every $x \in M$ there is $\bar{x} \in M$ with

(6.1) $$x\bar{x}x = x, \quad \bar{x}x\bar{x} = \bar{x}, \text{ and } \quad x\bar{x}\,y\bar{y} = y\bar{y}\,x\bar{x}.$$

The element $\bar{x}$ is uniquely defined by these properties and it is called the *inverse* of x. Thus, we may also use the notation $\bar{x} = x^{-1}$. It is easy to see that every idempotent element in an inverse monoid has the form xx^{-1}, and all these elements are idempotent. Using equations (6.1) for all $x, y \in \Gamma^*$ as defining relations we obtain the *free inverse monoid* FIM(Σ) which has been widely studied in the literature. More details on inverse monoids can be found in [28].

An *inverse monoid over an independence alphabet* (Σ, I) is an inverse monoid M together with a mapping $\varphi : \Sigma \to M$ such that $\varphi(a)\varphi(b) = \varphi(b)\varphi(a)$ and $\overline{\varphi(a)}\varphi(b) = \varphi(b)\overline{\varphi(a)}$ for all $(a, b) \in I$. We define the *free partially commutative inverse monoid over* (Σ, I) as the quotient monoid

$$\text{FIM}(\Sigma, I) = \text{FIM}(\Sigma)/\{ab = ba, \bar{a}b = b\bar{a} \mid (a,b) \in I\}.$$

It is an inverse monoid over (Σ, I). Da Costa has studied FIM(Σ, I) in his PhD thesis [39]. He proved that FIM(Σ, I) has a decidable word problem, but he did not show any complexity bound. The first upper complexity bound for the word problem is due to [**15**], where it was shown to be solvable in time $O(n \log(n))$ on a RAM. The aim of this section is to show that the space complexity of the word problem of $\mathrm{FIM}(\Sigma, I)$ is very low, too.

THEOREM 6.1. *The word problem of* $\mathrm{FIM}(\Sigma, I)$ *can be solved in logspace.*

PROOF. For a word $u = a_1 \cdots a_n$ $(a_1, \ldots, a_n \in \Gamma)$ let $u_i \in M(\Gamma, I_\Gamma)$ $(1 \leq i \leq n)$ be the trace represented by the prefix $a_1 \cdots a_i$ and define the following subset of the trace monoid $M(\Gamma, I_\Gamma)$.

$$(6.2) \qquad M(u) = \bigcup_{i=1}^{n} \{p \mid p \text{ is a prime prefix of } \widehat{u_i}\} \subseteq M(\Gamma, I_\Gamma).$$

(This set is a partial commutative analogue of the classical notion of *Munn tree* introduced in [**34**].) It is shown in [**15**, Sect. 3] that for all words $u, v \in \Gamma^*$, $u = v$ in $\mathrm{FIM}(\Sigma, I)$ if and only if
(a) $u = v$ in the graph group $G(\Sigma, I)$ and
(b) $M(u) = M(v)$.
Since $G(\Sigma, I)$ is linear, condition (a) can be checked in logspace [**29, 38**]. For (b), it suffices to show that the set $M(u)$ from (6.2) can be computed in logspace from the word u (then $M(u) = M(v)$ can be checked in logspace, since the word problem for the trace monoid $M(\Gamma, I_\Gamma)$ belongs to uniform TC^0 [**1**] and hence to logspace). By Theorem 5.4 we can compute in logspace a list of all normal forms $\widehat{u_i}$ $(1 \leq i \leq n)$, where u_i is the prefix of u of length i. By composing this logspace transducer with a logspace transducer for computing prime prefixes (see Lemma 4.1), we obtain a logspace transducer for computing the set $M(u)$. □

7. Concluding remarks and open problems

We have shown that shortlex normal forms can be computed in logspace for graph groups and right-angled Coxeter groups. For general Coxeter groups, we are able to compute in logspace the length of the shortlex normal form and the set of letters appearing in the shortlex normal form. For even Coxeter groups we can do better and enhance the general result since we can compute the Parikh-image of geodesics. An obvious open problem is, whether for every (even) Coxeter group shortlex normal forms can be computed in logspace. We are tempted to believe that this is indeed the case. A more general question is, whether shortlex normal forms can be computed in logspace for automatic groups. Here, we are more sceptical. It is not known whether the word problem of an arbitrary automatic group can be solved in logspace. In [**31**], an automatic *monoid* with a P-complete word problem was constructed. In fact, it is even open, whether the word problem for a hyperbolic group belongs to logspace. The best current upper bound is LOGCFL [**31**]. So, one might first try to lower this bound e.g. to $\mathsf{LOGDCFL}$ (the class of all languages that are logspace-reducible to a deterministic context-free language).

References

1. Carme Àlvarez and Joaquim Gabarró, *The parallel complexity of two problems on concurrency*, Inform. Process. Lett. **38** (1991), no. 2, 61–70. MR1113338 (92d:68057)
2. Mladen Bestvina and Noel Brady, *Morse theory and finiteness properties of groups*, Inventiones Mathematicae **129** (1997), no. 3, 445–470. MR1465330 (98i:20039)

3. Anders Björner and Francesco Brenti, *Combinatorics of Coxeter groups*, Graduate Texts in Mathematics, vol. 231, Springer, New York, 2005. MR2133266 (2006d:05001)
4. Ron Book and Friedrich Otto, *String-rewriting systems*, Springer-Verlag, 1993. MR1215932 (94f:68108)
5. Brigitte Brink and Robert B. Howlett, *A finiteness property and an automatic structure for Coxeter groups*, Math. Ann. **296** (1993), 179–190. MR1213378 (94d:20045)
6. Jin-Yi Cai, *Parallel computation over hyperbolic groups*, Proc. 24th ACM Symp. on Theory of Computing, STOC 92, ACM-press, 1992, pp. 106–115.
7. William A. Casselman, *Automata to perform basic calculations in Coxeter groups*, Representations of groups (Banff, AB, 1994), CMS Conf. Proc., vol. 16, Amer. Math. Soc., Providence, RI, 1995, pp. 35–58. MR1357194 (96i:20050)
8. Andrew Chiu, George Davida, and Bruce Litow, *Division in logspace-uniform NC^1*, Theoretical Informatics and Applications. Informatique Théorique et Applications **35** (2001), no. 3, 259–275. MR1869217 (2002k:68056)
9. John Crisp, Eddy Godelle, and Bert Wiest, *The conjugacy problem in right-angled Artin groups and their subgroups*, Journal of Topology **2** (2009), no. 3, 442–460. MR2546582 (2011b:20103)
10. John Crisp and Bert Wiest, *Embeddings of graph braid and surface groups in right-angled Artin groups and braid groups*, Algebraic & Geometric Topology **4** (2004), 439–472. MR2077673 (2005e:20052)
11. Samir Datta and Rameshwar Pratap, *Computing bits of algebraic numbers*, Tech. report, arXiv.org, 2011, http://arxiv.org/abs/1112.4295.
12. Michael W. Davis, *The geometry and topology of Coxeter groups*, London Math. Soc. Monographs Series, vol. 32, Princeton University Press, Princeton, NJ, 2008. MR2360474 (2008k:20091)
13. Volker Diekert, *Combinatorics on traces*, Lecture Notes in Computer Science, no. 454, Springer-Verlag, Heidelberg, 1990. MR1075995 (92d:68070)
14. Volker Diekert, Jonathan Kausch, and Markus Lohrey, *Logspace computations in graph groups and Coxeter groups*, Lecture Notes in Computer Science, Springer-Verlag, 2012, To appear in Proc. LATIN'2012, Arequipa, Peru.
15. Volker Diekert, Markus Lohrey, and Alexander Miller, *Partially commutative inverse monoids*, Semigroup Forum **77** (2008), no. 2, 196–226. MR2443435 (2010h:20144)
16. Volker Diekert and Anca Muscholl, *Solvability of equations in free partially commutative groups is decidable*, International Journal of Algebra and Computation **16** (2006), 1047–1070, Journal version of ICALP 2001, 543–554, LNCS 2076. MR2066532
17. Volker Diekert and Grzegorz Rozenberg (eds.), *The book of traces*, World Scientific, Singapore, 1995. MR1478992 (98i:68001)
18. Carl Droms, Jacques Lewin, and Herman Servatius, *The length of elements in free solvable groups*, Proc. Amer. Math. Soc. **119** (1993), 27–33. MR1160298 (93k:20051)
19. Murray Elder, *A linear-time algorithm to compute geodesics in solvable Baumslag-Solitar groups*, Illinois Journal of Mathematics **54** (2010), no. 1, 109–128. MR2776987 (2012e:20091)
20. Murray Elder, Gillian Elston, and Gretchen Ostheimer, *On groups that have normal forms computable in logspace*, Tech. report, arXiv.org, 2012, http://arxiv.org/abs/1201.4363.
21. Murray Elder and Andrew Rechnitzer, *Some geodesic problems in groups*, Groups. Complexity. Cryptology **2** (2010), no. 2, 223–229. MR2747151 (2012a:20059)
22. David B. A. Epstein, James W. Cannon, Derek F. Holt, Silvio V. F. Levy, Michael S. Paterson, and William P. Thurston, *Word processing in groups*, Jones and Bartlett, Boston, 1992. MR1161694 (93i:20036)
23. R. Ghrist and V. Peterson, *The geometry and topology of reconfiguration*, Advances in Applied Mathematics **38** (2007), no. 3, 302–323. MR2301699 (2008f:68104)
24. William Hesse, Eric Allender, and David A. Mix Barrington, *Uniform constant-depth threshold circuits for division and iterated multiplication*, Journal of Computer and System Sciences **65** (2002), 695–716. MR1964650 (2004b:68057)
25. Tim Hsu and Daniel T. Wise, *On linear and residual properties of graph products*, Michigan Mathematical Journal **46** (1999), no. 2, 251–259. MR1704150 (2000k:20056)
26. Emil Jeřábek, *Root finding with threshold circuits*, Tech. report, arXiv.org, 2011, http://arxiv.org/abs/1112.4295.

27. Olga Kharlampovich and Atefeh Mohajeri Moghaddam, *Approximation of geodesics in metabelian groups*, International Journal of Algebra and Computation **22** (2012), no. 2.
28. Mark V. Lawson, *Inverse semigroups: The theory of partial symmetries*, World Scientific, 1999. MR1694900 (2000g:20123)
29. Richard J. Lipton and Yechezkel Zalcstein, *Word problems solvable in logspace*, Journal of the Association for Computing Machinery **24** (1977), no. 3, 522–526. MR0445901 (56:4234)
30. Bruce E. Litow, *On sums of roots of unity*, Proceedings of ICALP 2010, Lecture Notes in Computer Science, vol. 6198, Springer, 2010, pp. 420–425. MR2734602 (2012e:68383)
31. Markus Lohrey, *Decidability and complexity in automatic monoids*, International Journal of Foundations of Computer Science **16** (2005), no. 4, 707–722. MR2159385 (2006h:20088)
32. Markus Lohrey and Nicole Ondrusch, *Inverse monoids: Decidability and complexity of algebraic questions*, Inf. Comput. **205** (2007), 1212–1234. MR2340901 (2009h:03016)
33. Charles F. Miller III, *Decision problems for groups – survey and reflections*, Algorithms and Classification in Combinatorial Group Theory, Springer, 1992, pp. 1–60. MR1230627 (94i:20057)
34. Walter D. Munn, *Free inverse semigroups*, Proc. London Math. Soc. **29** (1974), no. 3, 385–404. MR0360881 (50:13328)
35. Alexei Myasnikov, Vitalii Roman'kov, Alexander Ushakov, and Anatoly Vershik, *The word and geodesic problems in free solvable groups*, Transactions of the American Mathematical Society **362** (2010), 4655–4682. MR2645045 (2011k:20059)
36. Christos Papadimitriou, *Computation complexity*, Addison-Wesley, 1994. MR1251285 (95f:68082)
37. Michael Paterson and Alexander Razborov, *The set of minimal braids is co-NP-complete*, J. Algorithms **12** (1991), 393–408. MR1114918 (92h:68041)
38. Hans-Ulrich Simon, *Word problems for groups and contextfree recognition*, Proceedings of Fundamentals of Computation Theory (FCT'79), Berlin/Wendisch-Rietz (GDR), Akademie-Verlag, 1979, pp. 417–422. MR563704 (81d:20024)
39. António Augusto Veloso da Costa, *γ-produtos de monóides e semigrupos*, Ph.D. thesis, Universidade do Porto, Faculdade de Ciências, 2003.
40. Heribert Vollmer, *Introduction to circuit complexity*, Springer, Berlin, 1999. MR1704235 (2001b:68047)
41. Stephan Waack, *Tape complexity of word problems*, Proceedings of Fundamentals of Computation Theory (FCT'81) (Ferenc Gécseg, ed.), Lecture Notes in Computer Science, vol. 117, Springer, 1981, pp. 467–471. MR653014 (83k:03055)

FMI, UNIVERSITÄT STUTTGART, GERMANY

FMI, UNIVERSITÄT STUTTGART, GERMANY

INSTITUT FÜR INFORMATIK, UNIVERSITÄT LEIPZIG, GERMANY

Collection by Polynomials in Finite p-groups

Bettina Eick

ABSTRACT. We show how the multiplication and inversion in finite p-groups can be described by polynomials over the field with p elements and we exhibit some examples.

1. Introduction

Let G be a polycyclic group. A sequence of elements $(g_1, \ldots, g_n)$ of G is called polycyclic sequence if the series $G_1 \geq \ldots \geq G_n \geq G_{n+1} = \{1\}$ defined via $G_i = \langle g_i, \ldots, g_n \rangle$ is a subnormal series with cyclic factors and satisfies $G = G_1$. The relative orders $(r_1, \ldots, r_n)$ of the sequence are the indices of the subgroup series $r_i = [G_i : G_{i+1}]$ for $1 \leq i \leq n$. In this setting, every element of G can be written uniquely as
$$g_1^{x_1} \cdots g_n^{x_n}, \quad \text{(written symbolically as } g^x\text{),}$$
with $x_i \in \mathbb{Z}_{r_i}$ for $1 \leq i \leq n$, where $\mathbb{Z}_\infty = \mathbb{Z}$ and $\mathbb{Z}_r = \{0, \ldots, r-1\}$ for $r \in \mathbb{N}$. Hence we can identify G with the cartesian product $\mathbb{Z}_G = \mathbb{Z}_{r_1} \times \ldots \times \mathbb{Z}_{r_n}$ via
$$\mathbb{Z}_G \to G : x \mapsto g^x.$$

The group operations in G then translate via this identification to a multiplication function $m : \mathbb{Z}_G \times \mathbb{Z}_G \to \mathbb{Z}_G$ and an inversion function $i : \mathbb{Z}_G \to \mathbb{Z}_G$ of the form
$$g^x \cdot g^y = g^{m(x,y)} \quad \text{and} \quad (g^x)^{-1} = g^{i(x)}$$
for every $x, y \in \mathbb{Z}_G$. Additionally, one can also consider the powering function $p : \mathbb{Z}_G \times \mathbb{Z} \to \mathbb{Z}_G$ defined by
$$(g^x)^t = g^{p(x,t)}$$
for every $x \in \mathbb{Z}_G$ and every $t \in \mathbb{Z}$. There are various connections between m, i and p. In particular, the inversion i is a special case of the powering function p via $i(x) = p(x, -1)$ for every $x \in \mathbb{Z}_G$. The two functions m and i fully define the group G. Hence the following question seems of central interest.

QUESTION 1.1. What type of functions are m and i?

A famous result by P. Hall [5] asserts that m and p (and thus also i) can be described by integral polynomials if G is torsion-free nilpotent and the subnormal series defined by the polycyclic sequence is a central series with infinite cyclic factors. Sims [9, page 441ff] and Leedham-Green & Soicher [8] describe practical

---

2010 *Mathematics Subject Classification*. Primary 20D15; Secondary 20-04.

©2012 Bettina Eick

methods to determine the polynomials for the multiplication function m from a consistent polycyclic presentation of G.

Hall's result has been extended by M. du Sautoy [3] who proved that every polycyclic group has a subgroup of finite index so that the functions m and p (and thus also i) for the subgroup can be described by polynomials based on an algebraic number field. The subgroup used for this purpose is torsion-free and nilpotent-by-abelian. Assmann [1] implemented a practical method to determine the polynomials for m.

However, Question 1.1 is open in general; In particular, it is open for finite polycyclic groups. If Question 1.1 could be solved for finite polycyclic groups, then combining this solution with the result by du Sautoy [3] might lead the way towards a general solution.

In this note we consider Question 1.1 for the special case of finite p-groups. It is not difficult to show that the multiplication and the inversion can be described by polynomials over the field with p elements in this case. We exhibit a method to determine these polynomials from a consistent polycyclic presentation of a considered group. Based on this, we investigate some examples.

In general, if for some polycyclic group G the multiplication m and the inversion i can be described by polynomials, then the conjugacy and the commutator functions can as well:

$$(g^x)^{g^y} = g^{m(i(y), m(x,y))} \quad \text{and} \quad [g^x, g^y] = g^{m(m(i(x), i(y)), m(x,y))}.$$

Thus we obtain a generic description of the conjugacy classes and the commutators in a finite p-group. Similarly, powering with powers in $\{0, \ldots, p-1\}$ or taking pth powers in a finite p-group can be described by polynomials.

2. Existence of polynomials

Let $m \in \mathbb{N}$ and let $X_1, \ldots, X_m$ denote m different commuting indeterminates. Write X^t for the monomial $X_1^{t_1} \cdots X_m^{t_m}$ with $t = (t_1, \ldots, t_m) \in \mathbb{Z}^m$. We define the componentwise degree $mdeg$ of a monomial via $mdeg(X^t) = \max\{t_1, \ldots, t_m\}$ and the total degree as $deg(X^t) = t_1 + \ldots + t_m$. Both definitions are extended from monomials to polynomials as usual.

The set $\mathbb{Z}_p$ with addition and multiplication modulo p is a field. We define W_m as the $\mathbb{Z}_p$-subspace of $\mathbb{Z}_p[X_1, \ldots, X_m]$ generated by all monomials of componentwise degree at most $p-1$; that is,

$$W_m = \{ \sum_{t \in \mathbb{Z}_p^m} c_t X^t \mid c_t \in \mathbb{Z}_p \}.$$

Clearly, the set $\{X^t \mid t \in \mathbb{Z}_p^m\}$ is a basis of W_m called the *monomial basis*.

For $t_i \in \mathbb{Z}_p$ we denote

$$\binom{X_i}{t_i} = \frac{X_i(X_i - 1) \cdots (X_i - (t_i - 1))}{t_i(t_i - 1) \cdots 1} \text{ if } t_i \neq 0 \quad \text{and} \quad \binom{X_i}{0} = 1.$$

Further, we define for $t = (t_1, \ldots, t_m) \in \mathbb{Z}_p^m$

$$\binom{X}{t} = \binom{X_1}{t_1} \cdots \binom{X_m}{t_m}, \text{ and we also write } (X \mid t) = \binom{X}{t}.$$

as a more compact notation. The degree of $(X \mid t)$ as a univariate polynomial in X_i is t_i and the total degree of $(X \mid t)$ is $t_1 + \ldots + t_m$. The set $\{(X \mid t) \mid t \in \mathbb{Z}_p^m\}$ is also a basis for W_m called the *binomial basis*.

Every polynomial F in W_m defines a function $\hat{F} : \mathbb{Z}_p^m \to \mathbb{Z}_p$ via evaluation. We write $(s \mid t)$ for $(X \mid t)$ evaluated at $s \in \mathbb{Z}_p^m$.

THEOREM 2.1. *Let $m \in \mathbb{N}$ and p an arbitrary prime. For every function $f : \mathbb{Z}_p^m \to \mathbb{Z}_p$ there exists a unique polynomial $F \in W_m$ with $\hat{F} = f$.*

PROOF. The set V_m of all functions $f : \mathbb{Z}_p^m \to \mathbb{Z}_p$ forms a vector space over $\mathbb{Z}_p$ with basis $\{f_x \mid x \in \mathbb{Z}_p^m\}$, where $f_x(y) = \delta_{x,y}$. Hence V_m has dimension p^m over $\mathbb{Z}_p$. The set W_m also forms a vector space over $\mathbb{Z}_p$ of dimension p^m over $\mathbb{Z}_p$ and the map $\psi : W_m \to V_m : F \mapsto \hat{F}$ is $\mathbb{Z}_p$-linear. It thus remains to show that ψ is injective.

Let $(l_1, l_2, \ldots)$ denote the sequence of all elements of $\mathbb{Z}_p^m$ sorted lexicographically. Define the matrix $A_m \in \mathbb{Z}_p^{p^m \times p^m}$ whose entry at position (i, j) is $(l_i \mid l_j)$. Note that $(l_i \mid l_i) = 1$ and $(l_i \mid l_j) = 0$ for $j < i$. Thus A_m is upper unitriangular and hence invertible.

Let $F = \sum_{t \in \mathbb{Z}_p^m} c_t (X \mid t) \in W_m$ and $b_s = \hat{F}(s)$ be the value of F at $s \in \mathbb{Z}_p^m$. Denote $c = (c_{l_1}, c_{l_2}, \cdots)$ and $b = (b_{l_1}, b_{l_2}, \ldots)$. Then

$$cA_m = b.$$

Now if $F \in \ker(\psi)$, then $b = 0$ and hence $c = bA_m^{-1} = 0$. This implies that $F = 0$. Therefore, ψ is injective. □

Given a finite p-group G, we choose a polycyclic sequence $(g_1, \ldots, g_n)$ for G with relative orders $r_i = p$ for $1 \leq i \leq n$. Then $\mathbb{Z}_G = \mathbb{Z}_p^n$ and the multiplication function has the form

$$m : \mathbb{Z}_G \times \mathbb{Z}_G \to \mathbb{Z}_G \quad \text{or} \quad m : \mathbb{Z}_p^n \times \mathbb{Z}_p^n \to \mathbb{Z}_p^n$$

Thus for $1 \leq i \leq n$ the ith component m_i of m has the form $m_i : \mathbb{Z}_p^{2n} \to \mathbb{Z}_p$ and thus is of the form considered in Theorem 2.1. Hence m_i can be described by a (unique) polynomial in W_{2n}. Similar arguments imply the following.

COROLLARY 2.2. *Let G be a finite p-group with a polycyclic sequence whose relative orders are all equal to p. Then the multiplication function m, the inversion function i, the powering function p with powers in $\{0, \ldots, p-1\}$, the conjugacy function, the commutator function and the pth powering function can all be described by polynomials over the field with p elements.*

3. Computing the polynomials

The proof of Theorem 2.1 suggests a straightforward approach towards computing the polynomial corresponding to any function $f : \mathbb{Z}_p^m \to \mathbb{Z}_p$. We investigate this approach in more detail here.

3.1. Constructing the matrix A_m. First, we consider the matrix A_m whose entries consist of generalized binomial coefficients $(s \mid t)$ for $s, t \in \mathbb{Z}_p^m$. Note that this matrix is independent of any considered function; It depends on p and m only. The structure of A_m and its inverse is exhibited in the following lemma. This allows to determine A_m^{-1} or just single entries in this matrix readily.

LEMMA 3.1.

(a) A_m is the m-fold Kronecker product $A_1 \otimes \ldots \otimes A_1$ and A_1 is the unitriangular $p \times p$-matrix over $\mathbb{Z}_p$ whose entry at position (i,j) is the binomial coefficient $\binom{j-1}{i-1}$ mod p.

(b) A_m^{-1} is a mirror image of A_m: the entry at position (i,j) in A_m^{-1} is the entry at position $(p^m + 1 - j, p^m + 1 - i)$ in A_m and vice versa.

PROOF. (a) The structure of A_1 follows directly from its definition. Further, for $s, t \in \mathbb{Z}_p^m$ with $m > 1$ we observe that $(s \mid t) = \prod_i (s_i \mid t_i)$ and thus A_m is the m-fold Kronecker product of A_1.

(b) Let C_m denote the mirror image of A_m as defined in part (b) of the lemma. Our aim is to show that $C_m A_m = 1$. As both C_m and A_m are upper unitriangular, it follows that $C_m A_m$ is upper unitriangular. It thus remains to show that the entries above the diagonal in this product are all zero. First we consider the case $m = 1$. Using the Vandermonde identity, we determine the entry at position (i,j) in $C_1 A_1$ via

$$\sum_k (C_1)_{(i,k)}(A_1)_{(k,j)} = \sum_k (A_1)_{(p+1-k, p+1-i)} \binom{j-1}{k-1}$$
$$= \sum_k \binom{p-i}{p-k}\binom{j-1}{k-1}$$
$$= \binom{(p-1)-i+j}{p-1}.$$

If $j > i$, then this entry is equivalent to 0 mod p. Hence $C_1 = A_1^{-1}$ follows. For the case $m > 1$ we use induction and determine $A_m^{-1} = (A_1 \otimes A_{m-1})^{-1} = A_{m-1}^{-1} \otimes A_1^{-1} = C_{m-1} \otimes C_1 = C_m$. $\square$

3.2. Computing coefficients. Suppose that a function $f : \mathbb{Z}_p^m \to \mathbb{Z}_p$ is given and we wish to determine a polynomial in W_m describing this function. Then we can follow the proof of Theorem 2.1. For this purpose, we enumerate the elements of $\mathbb{Z}_p^m$ as $(l_1, l_2, \ldots)$ sorted lexicographically and determine the vector $b = (f(l_1), f(l_2), \ldots)$ of values of f. Then the coefficients c of f with respect to the binomial basis of W_m are the entries of the vector $c = b A_m^{-1}$. For $s, t \in \mathbb{Z}_p^m$ write $s \succ t$ if $s_i \geq t_i$ for all $1 \leq i \leq m$ and $s \neq t$. Then an explicit formula for the coefficients c_t of f is

$$c_t = f(t) + \sum_{s \succ t} f(s) \cdot \binom{p-1-t_1}{p-1-s_1} \cdots \binom{p-1-t_m}{p-1-s_m}.$$

If every c_t for $t \in \mathbb{Z}_p^m$ is determined by this formula, then this requires the explicit computation of every value $f(s)$ for $s \in \mathbb{Z}_p^m$. Obviously, if either c_t or $f(s)$ are known to be zero *a priori*, then this reduces the amount of work that needs to be done. The following lemma exploits this further.

LEMMA 3.2. *Suppose that $c_t = 0$ for all $t \in T \subseteq \mathbb{Z}_p^m$. Let $S = \mathbb{Z}_p^m \setminus T$ and order the elements of S lexicographically. Determine the matrix A consisting of the generalized binomial coefficients $(s \mid t)$ with $s, t \in S$. Then A is invertible and $(c_t \mid t \in S) = (f(s) \mid s \in S) A^{-1}$.*

PROOF. The matrix A is upper unitriangular and hence invertible. As $f = \sum_{t \in S} c_t(X \mid t)$, it follows that $f(s) = \sum_{t \in S} c_t(s \mid t)$. Hence

$$(c_t \mid t \in S)A = (f(s) \mid s \in S)$$

and the result follows. □

Hence if there is a large set T known *a priori* on which the coefficients of f are zero, then Lemma 3.2 can be used to reduce the amount of work necessary to compute the non-zero coefficients of f.

3.3. Multiplication (and inversion). Let G be a group of order p^n and let $(g_1, \ldots, g_n)$ be a polycyclic sequence for G with relative orders $r_i = p$ for $1 \leq i \leq n$. We assume further that the subnormal series defined by the polycyclic sequence refines the lower p-central series. We assign weights $(w_1, \ldots, w_n)$ so that $g_i \in \lambda_{w_i}(G)$ the w_ith term of the lower p-central series.

We consider the multiplication function $m = (m_1, \ldots, m_n)$ with $m_j : \mathbb{Z}_p^n \times \mathbb{Z}_p^n \to \mathbb{Z}_p$. Let $X = (X_1, \ldots, X_n)$ and $Y = (Y_1, \ldots, Y_n)$ and let M_j be the polynomial describing m_j so that M_j is a polynomial in X und Y.

REMARK 3.3. Let $j \in \{1, \ldots, n\}$.
- If $w_j = 1$, then $M_j(X, Y) = X_j + Y_j$.
- If $w_j > 1$, then $M_j(X, Y) = X_j + Y_j + L_j(X, Y)$ and L_j is a polynomial in $(X_1, \ldots, X_k)$ and $(Y_1, \ldots, Y_k)$, where $k \in \{1, \ldots, j-1\}$ is maximal with $w_k = w_j - 1$.

It remains to determine the polynomial L_j for all j with $w_j > 1$. This is a polynomial in (at most) $2k$ indeterminates. We use the binomial basis of W_m and write this polynomial as

$$L_j = \sum_{t \in \mathbb{Z}_p^{2k}} c_t(XY \mid t),$$

where $(XY \mid t) = (X_1 \mid t_1) \cdots (X_k \mid t_k)(Y_1 \mid t_{k+1}) \cdots (Y_k \mid t_{2k})$. In principle, L_j can be determined with the methods of this section. Lemma 3.2 can be used to reduce the computation of the coefficients of L_j. The following lemma exhibits an example.

LEMMA 3.4. *Consider $t \in \mathbb{Z}_p^{2k}$ and suppose that there exists h so that $t_h = \ldots = t_k = t_{k+1} = \ldots = t_{k+h-1} = 0$. Then $c_t = 0$.*

PROOF. Let $s \in \mathbb{Z}_p^{2k}$ with $s_h = \ldots = s_k = s_{k+1} = \ldots = s_{k+h-1} = 0$ for some h. Then the corresponding multiplication of elements reads

$$(g_1^{s_1} \cdots g_k^{s_k})(g_1^{s_{k+1}} \cdots g_k^{s_{2k}}) = g_1^{s_1 + s_{k+1}} \cdots g_k^{s_k + s_{2k}}.$$

Hence L_j evaluated at s is 0 for every such s. This implies that $c_t = 0$. □

Similarly, we can consider the inversion function $i = (i_1, \ldots, i_n)$ with $i_j : \mathbb{Z}_p^m \to \mathbb{Z}_p$ and the polynomial $I_j(X)$ describing it. Similar remarks as for multiplication apply to this polynomial.

4. Examples

In this section we consider various examples of finite p-groups. As a preliminary step, we note that if G is a direct product of two proper subgroups, then the multiplication and inversion can be performed in the subgroups and thus the corresponding polynomials for G can be read off from those for the subgroups.

4.1. Groups of exponent p.
Suppose that G has a polycyclic sequence $(g_1, \ldots, g_n)$ with $g_i^p = 1$ for $1 \leq i \leq n$. For example, if $exp(G) = p$, then every polycyclic sequence of G has this property. If G has such a polycyclic sequence, then the multiplication polynomials can also be computed using the Deep-Thought-method by Leedham-Green & Soicher [8]. We note that in the special case of groups of prime exponent, the method of Section 3 can be improved using the following remark.

REMARK 4.1. Let $exp(G) = p$ and consider $M_j(X, Y) = X_j + Y_j + L_j(X, Y)$. Then the total degree of L_j is bounded by j.

In the following we give a list of explicit examples. We describe the example groups via polycyclic presentations on generators $(g_1, \ldots, g_n)$. As $g_i^p = 1$ for $1 \leq i \leq n$ by assumption, we list the non-trivial commutators for each group only. Let $p > 3$ be a prime.

- Let $|G| = p^3$ with $[g_2, g_1] = g_3$. Then
  $M_3(X, Y) = X_3 + Y_3 + X_2 Y_1$.
- Let $|G| = p^4$ with $[g_2, g_1] = g_3$, $[g_3, g_1] = g_4$. Then
  $M_4(X, Y) = X_4 + Y_4 + X_3 Y_1 + X_2 (Y_1 \mid 2)$.
- Let $|G| = p^5$ with $[g_2, g_1] = g_3$, $[g_3, g_1] = g_4$, $[g_3, g_2] = g_5$. Then
  $M_5(X, Y) = X_5 + Y_5 + X_3 Y_2 + X_2 Y_1 Y_2 + (X_2 \mid 2) Y_1$.
- Let $|G| = p^5$ with $[g_2, g_1] = g_3$, $[g_3, g_1] = g_4$, $[g_4, g_1] = g_5$. Then
  $M_5(X, Y) = X_5 + Y_5 + X_4 Y_1 + X_3 (Y_1 \mid 2) + X_2 (Y_1 \mid 3)$.
- Let $|G| = p^5$ with $[g_2, g_1] = g_3$, $[g_3, g_1] = g_4$, $[g_4, g_1] = g_5$, $[g_3, g_2] = g_5$. Then
  $M_5(X, Y) = X_5 + Y_5 + X_2 Y_1 Y_2 + X_3 Y_2 + (X_2 \mid 2) Y_1 + X_4 Y_1 + X_3 (Y_1 \mid 2) + X_2 (Y_1 \mid 3)$.
- Let $|G| = p^5$ with $[g_2, g_1] = g_4$, $[g_3, g_1] = g_5$. Then
  $M_5(X, Y) = X_5 + Y_5 + X_3 Y_1$.
- Let $|G| = p^5$ with $[g_2, g_1] = g_4$, $[g_3, g_1] = g_5$, $[g_4, g_2] = g_5$. Then
  $M_5(X, Y) = X_5 + Y_5 + X_3 Y_1 + X_4 Y_2 + X_2 Y_1 Y_2 + (X_2 \mid 2) Y_1$.
- Let $|G| = p^5$ with $[g_2, g_1] = g_5$, $[g_4, g_3] = g_5$. Then
  $M_5(X, Y) = X_5 + Y_5 + X_4 Y_3 + X_2 Y_1$.

4.2. Abelian groups.
Let G be the cyclic group of order p^n. Let $(g_1, \ldots, g_n)$ be a polycyclic sequence for G with $g_i^p = g_{i+1}$. Then $L_n = \sum_{i=1}^{n-1} L_{n,i}$, where each $L_{n,i}$ is a sum of $(p-1)p^{i-1}$ monomials of degree $(i-1)(p-1) + 1$. For example,

$$L_2(X, Y) = \sum_{t=1}^{p-1} \binom{X_1}{t} \binom{Y_1}{p-t}.$$

Hence for $n > 1$ the multiplication polynomials for the cyclic group depend heavily on the prime p.

4.3. The groups of order at most p^3.
Using direct products and the groups already listed in this section, we can determine multiplication polynomials for all groups of order at most p^3, except for the group with the following presentation $\langle g_1, g_2, g_3 \mid g_1^p = g_3, g_2^p = g_3^p = 1, [g_2, g_1] = g_3 \rangle$. For this group we find that

$$L_3(X,Y) = X_2 Y_1 + \sum_{t=1}^{p-1} \binom{X_1}{t}\binom{Y_1}{p-t}.$$

4.4. One more example.
As a slightly larger and more interesting example, we consider the group number 1000 in the SmallGroups library of groups of order 2^9. This is a group of exponent 16 and class 5. It has the following multiplication and inversion functions.

$$\begin{aligned}
M_1(X,Y) &= X_1 + Y_1, \\
M_2(X,Y) &= X_2 + Y_2, \\
M_3(X,Y) &= X_2 Y_1 + X_3 + Y_3, \\
M_4(X,Y) &= X_1 Y_1 + X_4 + Y_4, \\
M_5(X,Y) &= X_1 X_2 Y_1 + X_1 Y_1 Y_2 + X_3 Y_1 + X_4 Y_2 + X_5 + Y_5, \\
M_6(X,Y) &= X_1 X_2 Y_1 + X_1 Y_1 Y_2 + X_2 Y_1 Y_2 + X_3 Y_2 + X_4 Y_2 + X_6 + Y_6, \\
M_7(X,Y) &= X_1 X_2 Y_1 + X_1 X_3 Y_1 + X_1 Y_1 Y_3 + X_2 Y_2 + X_4 Y_3 + X_5 Y_1 + X_7 + Y_7, \\
M_8(X,Y) &= X_1 X_3 Y_1 + X_1 Y_1 Y_2 + X_1 Y_1 Y_3 + X_2 X_3 Y_1 + X_2 Y_1 Y_3 + X_3 Y_1 Y_2 + X_2 Y_2 \\
&\quad + X_3 Y_3 + X_4 Y_2 + X_4 Y_3 + X_5 Y_2 + X_6 Y_1 + X_6 Y_2 + X_8 + Y_8, \\
M_9(X,Y) &= X_1 X_2 X_3 Y_1 + X_1 X_2 X_4 Y_1 + X_1 X_2 Y_1 Y_3 + X_1 X_2 Y_1 Y_4 + X_1 X_3 Y_1 Y_2 \\
&\quad + X_1 Y_1 Y_2 Y_4 + X_2 X_3 Y_1 Y_2 + X_2 X_4 Y_1 Y_2 + X_2 Y_1 Y_2 Y_3 + X_1 X_2 Y_1 + X_1 Y_1 Y_2 \\
&\quad + X_3 X_4 Y_1 + X_3 X_4 Y_2 + X_3 Y_1 Y_4 + X_3 Y_2 Y_3 + X_4 Y_2 Y_4 + X_4 Y_2 + X_5 Y_4 + X_6 Y_2 \\
&\quad + X_7 Y_1 + X_9 + Y_9,
\end{aligned}$$

$$\begin{aligned}
I_1(X) &= X_1, \\
I_2(X) &= X_2, \\
I_3(X) &= X_1 X_2 + X_3, \\
I_4(X) &= X_1 + X_4, \\
I_5(X) &= X_1 X_3 + X_2 X_4 + X_5, \\
I_6(X) &= X_1 X_2 X_3 + X_2 X_3 + X_2 X_4 + X_3 + X_6, \\
I_7(X) &= X_1 X_2 X_4 + X_1 X_5 + X_3 X_4 + X_2 + X_7, \\
I_8(X) &= X_1 X_2 X_3 X_6 + X_1 X_2 X_4 + X_1 X_3 X_5 + X_2 X_3 X_6 + X_2 X_4 X_5 + X_2 X_4 X_6 + X_1 X_3 \\
&\quad + X_1 X_6 + X_2 X_4 + X_2 X_5 + X_2 X_6 + X_3 X_4 + X_3 X_6 + X_2 + X_5 + X_6 + X_8, \\
I_9(X) &= X_1 X_2 X_3 X_4 + X_1 X_2 X_3 X_6 + X_1 X_2 X_4 + X_2 X_3 X_6 + X_2 X_4 X_6 + X_1 X_2 + X_1 X_3 \\
&\quad + X_1 X_5 + X_1 X_7 + X_2 X_4 + X_2 X_6 + X_3 X_6 + X_4 X_5 + X_3 + X_6 + X_9.
\end{aligned}$$

This allows to read off 2nd powers and conjugacy:

$$\begin{aligned}
P_1(X) &= 0, \\
P_2(X) &= 0, \\
P_3(X) &= X_1 X_2, \\
P_4(X) &= X_1, \\
P_5(X) &= X_1 X_3 + X_2 X_4, \\
P_6(X) &= X_1 X_2 + X_2 X_3 + X_2 X_4, \\
P_7(X) &= X_1 X_2 + X_1 X_5 + X_3 X_4 + X_2, \\
P_8(X) &= X_1 X_2 X_3 + X_1 X_2 + X_1 X_6 + X_2 X_4 + X_2 X_5 + X_2 X_6 + X_3 X_4 + X_2 + X_3, \\
P_9(X) &= X_1 X_2 X_3 + X_2 X_3 X_4 + X_1 X_7 + X_2 X_3 + X_2 X_6 + X_4 X_5,
\end{aligned}$$

$C_1(X,Y) = X_1,$
$C_2(X,Y) = X_2,$
$C_3(X,Y) = X_1Y_2 + X_2Y_1 + X_3,$
$C_4(X,Y) = X_4,$
$C_5(X,Y) = X_1Y_3 + X_2Y_4 + X_3Y_1 + X_4Y_2 + X_5,$
$C_6(X,Y) = X_1X_2Y_2 + X_1Y_1Y_2 + X_2Y_1Y_2 + Y_1Y_2Y_3 + X_1Y_2 + X_2Y_3 + X_2Y_4 + X_3Y_2$
$\quad + X_4Y_2 + Y_1Y_2 + X_6 + Y_3,$
$C_7(X,Y) = X_1Y_2Y_4 + X_2Y_1Y_4 + X_1Y_5 + X_3Y_4 + X_4Y_3 + X_5Y_1 + X_7,$
$C_8(X,Y) = X_1Y_1Y_2Y_3 + X_2Y_1Y_2Y_3 + Y_1Y_2Y_3Y_6 + X_1X_2Y_3 + X_1X_3Y_2 + X_1Y_1Y_2$
$\quad + X_1Y_1Y_3 + X_1Y_2Y_4 + X_2X_3Y_1 + X_2Y_1Y_2 + X_2Y_1Y_4 + X_2Y_2Y_3 + X_3Y_1Y_2$
$\quad + Y_1Y_3Y_5 + Y_2Y_3Y_6 + Y_2Y_4Y_5 + Y_2Y_4Y_6 + X_1Y_3 + X_1Y_6 + X_2Y_3 + X_2Y_4$
$\quad + X_2Y_5 + X_2Y_6 + X_3Y_4 + X_4Y_2 + X_4Y_3 + X_5Y_2 + X_6Y_1 + X_6Y_2 + Y_1Y_2$
$\quad + Y_1Y_3 + Y_3Y_6 + X_8 + Y_3 + Y_5 + Y_6,$
$C_9(X,Y) = X_1X_2X_3Y_1 + X_1X_2X_3Y_2 + X_1X_2Y_1Y_4 + X_1X_2Y_2Y_4 + X_1X_4Y_1Y_2 + X_1Y_1Y_2Y_3$
$\quad + X_1Y_1Y_2Y_4 + X_2X_3Y_1Y_2 + X_2X_4Y_1Y_2 + Y_1Y_2Y_3Y_4 + Y_1Y_2Y_3Y_6 + X_1X_2Y_1$
$\quad + X_1X_3Y_2 + X_1X_4Y_3 + X_1Y_2Y_4 + X_1Y_3Y_4 + X_2X_3Y_3 + X_2X_4Y_4 + X_2Y_1Y_2$
$\quad + X_2Y_1Y_3 + X_2Y_2Y_3 + X_2Y_2Y_4 + X_2Y_3Y_4 + X_3X_4Y_1 + X_3X_4Y_2 + X_3Y_1Y_2$
$\quad + X_3Y_1Y_4 + X_4Y_2Y_4 + Y_2Y_3Y_4 + Y_2Y_3Y_6 + Y_2Y_4Y_6 + X_1Y_2 + X_1Y_7 + X_2Y_6$
$\quad + X_4Y_2 + X_4Y_5 + X_5Y_4 + X_6Y_2 + X_7Y_1 + Y_2Y_3 + Y_2Y_4 + Y_3Y_6 + X_9 + Y_3 + Y_6.$

5. The Complexity of multiplication

The multiplication in a finite p-group G given by a consistent polycyclic presentation is usually performed by an algorithm called *collection*, see [6, Chapter 8]. This is an interative method which splits the multiplication of two elements into various small steps. The complexity and the effectivity of this algorithm is considered in [7], [4] and [2]. As a result, it is known that this algorithm is reasonably effective in practice, but no explicit complexity analysis is known for it.

5.1. Multiplication by polynomials. If polynomials for the multiplication function m of G are given, then the multiplication reduces to the evaluation of polynomials. The complexity of this evaluation is bounded in the following lemma.

LEMMA 5.1. *Let F be a polynomial in W_{2n}.*

(a) *The evaluation of F costs $O(l(log_2(n) + p))$ arithmetic operations (additions and multiplications) in $\mathbb{Z}_p$, where l is the number of non-zero coefficients of F with respect to the natural basis of W_{2n}.*

(b) *The evaluation of F costs $O(l(log_2(n) + p^2))$ arithmetic operations (additions and multiplications) in $\mathbb{Z}_p$, where l is the number of non-zero coefficients of F with respect to the binomial basis of W_{2n}.*

PROOF. (a) We precompute the values a^i for $a \in \mathbb{Z}_p$ and $i \in \mathbb{Z}_p$. This costs $O(p)$ multiplications in $\mathbb{Z}_p$, as the multiplicative group of $\mathbb{Z}_p$ is cyclic. Then the evaluation of a monomial in $2n$ variables costs $O(log_2(n))$ multiplications in $\mathbb{Z}_p$ using a divide-and-conquer approach. Hence the full evaluation of F costs $O(l(log_2(n) + p))$ arithmetic operations in $\mathbb{Z}_p$.

(b) In this case we precompute the values $\binom{a}{b}$ for $a,b \in \mathbb{Z}_p$. This costs $O(p^2)$ multiplications in $\mathbb{Z}_p$. The remaining proof is as in (a). $\square$

Hence the number of non-zero coefficients in the polynomials $M_1, \ldots, M_n$ plays a central role for the efficiency of the multiplication in G. Note that this number is usually smaller if the binomial basis is used. Further, the examples in Section 4 indicate that the exponent of the considered group has a significant impact on this number.

5.2. Hybrid methods. Leedham-Green & Soicher [8] suggest an alternative approach to perform multiplication in finite p-groups. If a finite p-group G is given by a consistent polycyclic presentation, then they eliminate all power relations from this presentation, determine the multiplication polynomials for the reduced presentation and then perform multiplication by first evaluating polynomials and then evaluating powers using ordinary collection.

This hybrid method has the advantage that the multiplication polynomials have comparatively few non-zero coefficients and thus are effective to compute and effective to evaluate. However, this hybrid approach has the disadvantage that a second step of evaluating powers is necessary in general. Hence this method does not allow further applications such as the generic computation of conjugacy classes.

References

1. B. Assmann. Symbolic collection in polycyclic groups. Preprint, 2006.
2. B. Assmann and S. Linton. Using the Malcev correspondence for collection in polycyclic groups. *J. Algebra*, 316(2):828–848, 2007. MR2358616 (2008g:20080)
3. M. du Sautoy. Polycyclic groups, analytic groups and algebraic groups. *Proc. London Math. Soc. (3)*, 85(1):62–92, 2002. MR1901369 (2003c:20037)
4. V. Gebhardt. Efficient collection in infinite polycyclic groups. *J. Symb. Comput.*, 34:213 – 228, 2002. MR1935079 (2003g:20001)
5. P. Hall. *The collected works of Philip Hall*. Oxford Science Publications. The Clarendon Press Oxford University Press, New York, 1988. Compiled and with a preface by K. W. Gruenberg and J. E. Roseblade, With an obituary by Roseblade. MR986732 (90b:01108)
6. D. F. Holt, B. Eick, and E. A. O'Brien. *Handbook of computational group theory*. Discrete Mathematics and its Applications (Boca Raton). Chapman & Hall/CRC, Boca Raton, FL, 2005. MR2129747 (2006f:20001)
7. C. R. Leedham-Green and L. H. Soicher. Collection from the left and other strategies. *J. Symb. Comput.*, 9:665 – 675, 1990. MR1075430 (92b:20021)
8. C. R. Leedham-Green and L. H. Soicher. Symbolic collection using Deep Thought. *LMS J. Comput. Math.*, 1:9 – 24, 1998. MR1635719 (99f:20002)
9. C. C. Sims. *Computation with finitely presented groups*. Cambridge University Press, Cambridge, 1994. MR1267733 (95f:20053)

INSTITUT COMPUTATIONAL MATHEMATICS, TECHNICAL UNIVERSITY BRAUNSCHWEIG, GERMANY

E-mail address: `beick@tu-bs.de`

All Finite Generalized Tetrahedon Groups II

Benjamin Fine, Alexander Hulpke, and Gerhard Rosenberger

ABSTRACT. We correct a technical error in the proof of the main theorem in [1].

The purpose of this note is to point out, and to correct, a technical error in [1] that arose in the application of Theorem 2.7. This error is a consequence of an input error in the computer calculations. It affects only the proof of Theorem 3.1, but not the actual theorem itself. Hence, we have to correct the proof of the classification Theorem 3.1 for these groups where Theorem 2.7, together with the erroneous input, was used. Nevertheless, a combination of a corrected program and calculations in GAP presents easily a proof for the corresponding groups.

(1) Let first
$$G = \langle x, y, z \mid x^3 = y^5 = z^3 = W_1^2(x,y) = (zy^\gamma)^2 = \\ = (xy)^2 = 1 \rangle$$

with $\gamma = 1, 2$ or 3 and $W_1(x,y) = x^{-1}yxy^{-1}xy^{-1}$, $W_1(x,y) = x^{-1}yxyx^{-1}y^2xy^{-1}$ or $W_1(x,y) = x^{-1}yx^{-1}y^2xyxy^{-1}$. If $\gamma = 1$ then we may choose indeed $X, Y, Z \in PSL(2, \mathbb{C})$ with $tr\ X = tr\ Z = 1$, $tr\ Y = \lambda = 2\cos\frac{\pi}{5}$, $tr\ XY = \lambda$ and $tr\ YZ = tr\ XZ = 0$ such that $x \mapsto X$, $y \mapsto Y$, $z \mapsto Z$ defines an essential representation $\rho: G \to PSL(2, \mathbb{C})$ with non-real trace $tr\ XYZ$. If $\gamma = 2$ or 3 then in all cases G has a subgroup H of index 12 for which the abelianization H/H' is infinite.

(2) Let now
$$G = \langle x, y, z \mid x^3 = y^5 = z^2 = (x^{-1}yxyx^{-1}y^2xy^{-1})^2 = \\ = (yz)^2 = (xz)^2 = 1 \rangle.$$

The subgroup H generated by y, $xy^{-1}x$ and $xyx^{-1}yxy^{-1}x^{-1}$ has index 60 in G, and the abelianization H/H' is infinite.

(3) Let now
$$G = \langle x, y, z \mid x^3 = y^5 = z^2 = (x^{-1}yx^{-1}y^2xyxy^{-1})^2 = \\ = (yz)^2 = (xz)^2 = 1 \rangle.$$

G has as an epimorphic image the direct product $A_5 \times A_5$, given by $x \mapsto (2,3,4)(7,10,8)$, $y \mapsto (1,5,2,3,4)(6,10,7,8,9)$ and $z \mapsto (1,5)(2,4)(6,9)(8,10)$. Let U be the preimage in G of the subgroup of $A_5 \times A_5$ generated

2010 *Mathematics Subject Classification.* Primary 20F05.

by $(2,3,4)(7,8,10)$ and $(1,4)(3,5)(6,7)(8,9)$ and $H = U'$ be the derived subgroup of U. The abelianization H/H' is infinite.

(4) Let now
$$G = \langle x, y, z \mid x^3 = y^3 = z^5 = W_1^2(x,y) =$$
$$= (yz^\gamma)^2 = (xz)^2 = 1\rangle$$
with $\gamma = 3$ and $W_1(x,y) = xy^{-1}x^{-1}y^{-1}xyx^{-1}y$ or $\gamma = 2$ and $W_1(x,y) = xyxyx^{-1}y^{-1}$. In both cases G has a subgroup H of index 12 for which the abelianization H/H' is infinite.

(5) Let now
$$G = \langle x, y, z \mid x^3 = y^3 = z^3 = (xyxyx^{-1}y^{-1})^2 =$$
$$= (yz^2)^2 = (xz)^2 = 1\rangle.$$

Here G has a subgroup H of index 864 for which the abelianization H/H' is infinite. We get H as the preimage of a point stabilizer of index 36 in a $P\Gamma L(2,8)$-quotient.

(6) Finally, let
$$G = \langle x, y, z \mid x^3 = y^3 = z^5 = (xyx^2y^2)^2 =$$
$$= (yz^3)^2 = (xz)^2 = 1\rangle$$
or
$$G = \langle x, y, z \mid x^3 = y^5 = z^3 = (xyx^2y^2)^2 =$$
$$= (y^2z)^2 = (xz)^2 = 1\rangle.$$

Here G has a subgroup of index 6 or 21, respectively, which can be written as a non-trivial free product with amalgamation and contains a free subgroup of rank two. For more details see [2].

Hence, altogether Theorem 3.1 remains correct as stated.

References

1. Benjamin Fine, Miriam Hahn, Alexander Hulpke, Volkmar große Rebel, Gerhard Rosenberger, and Martin Scheer, *All finite generalized tetrahedron groups*, Algebra Colloq. **15** (2008), no. 4, 555–580. MR2451990 (2009k:20071)
2. Benjamin Fine, Alexander Hulpke, Volkmar große Rebel, Gerhard Rosenberger, and Stefanie Schauerte, *The Tits alternative for short generalized tetrahedron groups*, Scientia Series A: Math.Sciences **21** (2011), 1–15.

DEPARTMENT OF MATHEMATICS, FAIRFIELD UNIVERSITY, FAIRFIELD, CONNECTICUT 06430
E-mail address: fine@mail.fairfield.edu

DEPARTMENT OF MATHEMATICS, COLORADO STATE UNIVERSITY, FORT COLLINS, COLORADO 80528
E-mail address: hulpke@math.colostate.edu

UNI HAMBURG, DEPARTMENT OF MATHEMATICS, BUNDESSTR. 55, 20146 HAMBURG, GERMANY
E-mail address: gerhard.rosenberger@math.uni-hamburg.de

The Classification of One Relator Limit Groups and the Surface Group Conjecture

Benjamin Fine and Gerhard Rosenberger

Dedicated to Gilbert Baumslag on his retirement from City University

ABSTRACT. Suppose that G is a non-free non-cyclic one- relator group such that each subgroup of finite index is again a one-relator group and each subgroup of infinite index is a free group then G is a surface group. This is known as the **surface group conjecture**. Recent progress has been made on this by placing it within the general classification of one-relator fully residually free groups. In this paper we explain in detail the surface group conjectures and survey the results obtained by assuming the group is fully residually free and also explain several other recent results related to this classification. These include work on a conjecture of Gromov as related to hyperbolic Baumslag doubles, the preservation of quadratic Lyndon properties under certain group amalgams and within one-relator groups with torsion, the Baumslag free-by-cyclic conjectures and the classification of one-relator CT and CSA groups.

TABLE OF CONTENTS
(1) Introduction
(2) Surface Groups and The Surface Group Conjecture
(3) Cyclically Pinched and Conjugacy Pinched One Relator Groups
(4) Fully Residually Free Groups
(5) Property IF and Results on the Surface Group Conjecture
(6) The Gromov Conjecture and Baumslag Doubles
(7) The Free-by-Cyclic Conjecture
(8) Approaches to the Classification of One Relator Limit Groups
(9) The Lyndon Properties
(10) The Classification of One Relator CSA Groups

1. Introduction

A **surface group** is the fundamental group of a compact orientable or non-orientable surface. If the genus of the surface is g then we say that the corresponding

---

2010 *Mathematics Subject Classification.* Primary 20F05, 20F18; Secondary 20F99, 20E07, 20F67, 20E26.

Key words and phrases. Surface group, surface group conjecture, fully residually free group, cyclically pinched one-relator group, hyperbolic group.

©2012 American Mathematical Society

surface group also has genus g. It is an **orientable surface group** if the surface is orientable and a **nonorientable surface group** otherwise.

An orientable surface group S_g of genus $g \geq 2$ has a one-relator presentation of the form
$$S_g = <a_1, b_1, ..., a_g, b_g; [a_1, b_1]...[a_g, b_g] = 1>$$
while a non-orientable surface group T_g of genus $g \geq 2$ also has a one-relator presentation - now of the form
$$T_g = <a_1, a_2, ..., a_g; a_1^2 a_2^2 ... a_g^2 = 1>.$$

An orientable surface group S_g is hyperbolic if $g \geq 2$ whereas a nonorientable surface group T_g is hyperbolic if $g \geq 3$. Here and for the remainder of the paper we use $[x, y] = xyx^{-1}y^{-1}$ for the commutator of x and y.

Much of combinatorial group theory arose originally out of the theory of one-relator groups and the concepts and ideas surrounding the Freiheitssatz or Independence Theorem of Magnus (see section 2). Going backwards the ideas of the Freiheitssatz were motivated by the topological properties of surface groups.

From covering space theory it follows that any subgroup of finite index in a orientable surface group of genus $g \geq 2$ is again a surface group of equal or higher genus while any subgroup of infinite index must be a free group. These results, although known since the early 1900's were proved purely algebraically using the Reidemeister-Schreier rewriting process by Hoare, Karrass and Solitar in 1971 [**HKS 1**], [**HKS 2**]. It is well known (see [**FR**]) that an orientable surface group can be faithfully represented as a discrete subgroup of $PSL_2(\mathbb{C})$ and hence each such group is linear. It follows that surface groups are residually finite. G.Baumslag [**GB 1**] showed that any orientable surface group of genus ≥ 2 must be residually free and 2-free from which it can be deduced using results of Remeslennikov [**Re**] and Gaglione and Spellman [**GS**] that they are fully residually free (see section 3). The article [**AFR**] surveys most of the properties of surface groups and shows how they are the primary motivating examples for much of combinatorial group theory.

In this paper we consider the **surface group conjecture**. In the Kourovka notebook Melnikov asked the following question.

QUESTION 1.1. *Suppose that G is a residually finite non-free, non-cyclic one-relator group such that every subgroup of finite index is again a one-relator group. Then G is a surface group.*

As asked by Melnikov the answer is no. Recall that the Baumslag-Solitar groups $BS(m, n)$ are the groups
$$BS(m, n) = <a, b; a^{-1}b^m a = b^n>.$$
If $|m| = |n|$ or either $|m| = 1$ or $|n| = 1$ these groups are residually finite. In all other cases these groups are not residually finite. Further $BS(m, n)$ is Hopfian if and only if $m = \pm 1$ or $n = \pm 1$ or m, n have the same set of prime ([**CoLe**]). If either $|m| = 1$ or $|n| = 1$ every subgroup of finite index is again a Baumslag-Solitar group and therefore a one-relator group. This is a direct consequence of results in [**GKM**]. It follows that besides the surface groups, the groups $BS(1, m)$, also satisfy Melnikov's question. We then have the following conjecture.

CONJECTURE 1.1. **Surface Group Conjecture A** *Suppose that G is a residually finite non-free, non-cyclic one-relator group such that every subgroup of finite* *index is again a one-relator group. Then G is either a surface group or a Baumslag-Solitar group $BS(1, m)$ for some integer m.*

We note that the groups $BS(1,1)$ and $BS(-1,-1)$ are orientable surface groups while $BS(-1,1)$ and $BS(1,-1)$ are nonorientable surface groups. In surface groups, subgroups of infinite index must be free groups. To avoid the Baumslag-Solitar groups, $BS)(1,m), |m| \geq 2$, Surface Group Conjecture A, was modified to:

CONJECTURE 1.2. **Surface Group Conjecture B** *Suppose that G is a non-free, non-cyclic one-relator group such that every subgroup of finite index is again a one-relator group, there exists a noncyclic subgroup of infinite index and and every subgroup of infinite index is a free group . Then G is a surface group (of genus $g \geq 2$.*

Although this problem seems very tantalizing very little was done on it (see section 2). However by placing the problem in the wider context of the classification of one-relator fully residually free groups also know as limit groups partial progress has been made (see [**FKMRR**] and section 3).

In this paper we explain in detail the surface group conjectures and survey the results obtained by assuming the group is fully residually free and placing the problem within the classification of one-relator limit groups. We also explain several other recent results related to this classification. These include work on a conjecture of Gromov as related to hyperbolic Baumslag doubles, the preservation of quadratic Lyndon properties under certain group amalgams and within one-relator groups with torsion, the Baumslag free-by-cyclic conjectures and the classification of one-relator CT and CSA groups.

2. Surface Groups and The Surface Group Conjecture

Let G be a hyperbolic surface group. Then as described in the introduction G has a one-relator presentation of the form

$$G =< a_1, b_1, ..., a_g, b_g; [a_1, b_1]...[a_g, b_g] = 1 > \text{ with } g \geq 2$$

in the orientable case or

$$G =< a_1, a_2, ..., a_g; a_1^2 a_2^2...a_g^2 = 1 > \text{ with } g \geq 3$$

in the nonorientable case.

Let G be an orientable surface group of genus $g \geq 2$ and let H be a subgroup of G. From covering space theory H then corresponds to the fundamental group of a finite cover of the surface for which G is the fundamental group. If H has finite index the genus of the cover must be higher or equal. If H has infinite index then homotopically H must be the fundamental group of a wedge of circles and hence H is a free group. It follows that a subgroup of finite index in an orientable surface group of genus $g \geq 2$ is again a surface group of higher or equal genus. This implies that a subgroup of finite index in a surface group again has a one-relator presentation. Further a subgroup of infinite index must be a free group. It is well known (see [**FR**]) that an orientable surface group can be faithfully represented as a discrete subgroup of $PSL_2(\mathbb{C})$ and hence each such group is linear. It follows that surface groups are residually finite. G.Baumslag [**GB 1**] showed that any orientable surface group of genus ≥ 2 must be residually free and 2-free from which it can be deduced using results of Remeslennikov [**Re**] and Gaglione and Spellman [**GS**] that they are fully residually free (see section 2).

As described in the introduction, based on a question proposed by Melnikov, we have the following two conjectures that we term the **Surface Group Conjectures**.

CONJECTURE 2.1. *Surface Group Conjecture A Suppose that G is a residually finite non-free, non-cyclic one-relator group such that every subgroup of finite index is again a one-relator group. Then G is either a surface group or a Baumslag-Solitar group $B(1,m)$ for some integer m.*

CONJECTURE 2.2. *Surface Group Conjecture B Suppose that G is a non-free, non-cyclic one-relator group such that every subgroup of finite index is again a one-relator group, there exists a non-cyclic subgroup of infinite index and every subgroup of infinite index is a free group. Then G is a surface group (of genus $g \geq 2$).*

Suppose that G is a one-relator group with the properties of Surface Group Conjecture B. Then G must be torsion-free for otherwise G would contain nonfree subgroups of infinite index. Along these lines Zieschang [**Z**] proved the following. Recall that if $\mathcal{P}$ is a group property then G is **virtually** $\mathcal{P}$ if G contains a subgroup of finite index satisfying $\mathcal{P}$.

THEOREM 2.1. *Let G be a torsion-free hyperbolic virtually surface group. Then G is also a surface group.*

Notice the similarity of this theorem to the Stallings-Swan theorem (see [**FR**]) on free groups which asserts that a torsion-free virtually free group must be free. Zieschang's result began a whole project which eventually evolved into the complete solution by Eckmann and Mueller of the **Nielsen Realization Problem** (see [**Z**] which gives a complete classification of virtually Fuchsian groups).

Note further that an orientable surface group of genus $g \geq 2$ with the presentation
$$G = < a_1, b_1, ..., a_g, b_g; [a_1, b_1]....[a_g, b_g] >$$
also has a presentation
$$G = < x_1, ..., x_n; x_1...x_n x_1^{-1}...x_n^{-1} = 1 >$$
with n even. P. M. Curran [**C**] proved the following.

THEOREM 2.2. *Let G be a one-relator group with the presentation*
$$G = < x_1, ..., x_n; x_1^{\nu_1}...x_n^{\nu_n} x_1^{-\nu_1}...x_n^{-\nu_n} = 1 >.$$
Then, if n is odd, there exist normal subgroups of finite index which do not have one-relator presentations. In particular if
$$G = < x_1, ..., x_n; x_1...x_n x_1^{-1}...x_n^{-1} = 1 >$$
then every subgroup of finite index is again a one-relator group if and only if n is even and hence a surface group.

Beyond these very little was done on the Surface Group Conjectures until recently. By placing the problem within the classifcation of one-relator limit groups progress can be made. First we introduce some relevant necessary concept related to the conjectures and the results. In section 5 we will give the results on the conjectures.

3. Cyclically Pinched and Conjugacy Pinched One Relator Groups

The algebraic generalization of the one-relator presentation type of a surface group presentation leads to **cyclically pinched one-relator groups**. These groups have the same general form as a hyperbolic surface group and have proved to be quite amenable to study. In particular:

DEFINITION 3.1. *A **cyclically pinched one-relator group** is a one-relator group with a presentation of the following form*

$$G = <a_1, ..., a_p, a_{p+1}, ..., a_n; U = V>$$

where $1 \neq U = U(a_1, ..., a_p)$ is a cyclically reduced, non-primitive (not part of a free basis) word in the free group F_1 on $a_1, ..., a_p$ and $1 \neq V = V(a_{p+1}, ..., a_n)$ is a cyclically reduced, non-primitive word in the free group F_2 on $a_{p+1}, ..., a_n$.

Clearly such a group is the free product of the free groups on $a_1, ..., a_p$ and $a_{p+1}, ..., a_n$ respectively amalgamated over the cyclic subgroups generated by U and V. Hyperbolic Surface groups are cyclically pinched one-relator groups.

Cyclically pinched one-relator groups have been shown to be extremely similar to surface groups. We summarize many of the most important results.

THEOREM 3.1. *Let G be a cyclically pinched one-relator group. Then*

*(1) G is residually finite (G.Baumslag [**GB 1**])*

*(2) G has a solvable conjugacy problem (S.Lipschutz [**Li**]) and is conjugacy separable (J.Dyer[**Dy**])*

*(3) G is subgroup separable (Brunner,Burns and Solitar[**BBS**])*

*(4) If neither U nor V is a proper power then G has a faithful representation over some commutative field ([**FR**]).*

*(5) If neither U nor V is a proper power then G has a faithful representation in $PSL_2(\mathbb{C})$ (Fine,Rosenberger [**FR**])*

*(6) If either U or V is not a proper power then G is hyperbolic ([**BeF 1**],[**JR**], [**Kh M**])*

*(7) If neither U nor V is in the commutator subgroup of its respective factor then G is free-by-cyclic [**BFMT**]*

Rosenberger [**Ro 2**], using Nielsen cancellation, has given a positive solution to the isomorphism problem for cyclically pinched one-relator groups, that is, he has given an algorithm to determine if an arbitrary one-relator group is isomorphic or not to a given cyclically pinched one-relator group.

THEOREM 3.2. *([**Ro 2**]) The isomorphism problem for any cyclically pinched one-relator group is solvable; given a cyclically pinched one-relator group G there is an algorithm to decide in finitely many steps whether an arbitrary one-relator group is isomorphic or not to G.*

The HNN analogs of cyclically pinched one-relator groups are called **conjugacy pinched one-relator groups** and are also motivated by the structure of orientable surface groups. In particular suppose

$$S_g = <a_1, b_1, ..., a_g, b_g; [a_1, b_1]...[a_g, b_g] = 1> \text{ with } g \geq 2.$$

Let $b_g = t$ then S_g is an HNN group of the form

$$S_g = <a_1, b_1, ..., a_g, t; tUt^{-1} = V>.$$

where $U = a_g$ and $V = [a_1, b_1]...[a_{g-1}, b_{g-1}]a_g$. Generalizing this:

DEFINITION 3.2. *A* **conjugacy pinched one-relator group** *is a one-relator group of the form*

$$G = <a_1,...,a_n,t; tUt^{-1} = V>$$

where $1 \neq U = U(a_1,...,a_n)$ *and* $1 \neq V = V(a_{p+1},...,a_n)$ *are cyclically reduced in the free group F on* $a_1,...,a_n$.

Structurally such a group is an HNN extension of the free group F on $a_1,...,a_n$ with cyclic associated subgroups generated by U and V and is hence the HNN analog of a cyclically pinched one-relator group.

Groups of this type arise in many different contexts and share many of the general properties of the cyclically pinched case. However many of the proofs become tremendously more complicated in the conjugacy pinched case than the cyclically pinched case. Further in most cases additional conditions on the associated elements U and V are necessary. To illustrate this we state a result ([**FRR**], see [**FR**]) which gives a partial solution to the isomorphism problem for conjugacy pinched one-relator groups.

THEOREM 3.3. *Let* $G = <a_1,...,a_n,t; tUt^{-1} = V>$ *be a conjugacy pinched one-relator group and suppose that neither U nor V is a proper power in the free group on* $a_1,...,a_n$. *Suppose further that there is no Nielsen transformation from* $\{a_1,...,a_n\}$ *to a system* $\{b_1,...,b_n\}$ *with* $U \in \{b_1,...,b_{n-1}\}$ *and that there is no Nielsen transformation from* $\{a_1,...,a_n\}$ *to a system* $\{c_1,...,c_n\}$ *with* $V \in \{c_1,...,c_{n-1}\}$. *Then:*

(1) G has rank $n+1$ and for any minimal generating system for G there is a one-relator presentation.

(2) The isomorphism problem is solvable

More information about both cyclically pinched one-relator groups and conjugacy pinched one-relator groups is in [**FR**] or [**FRS 1**].

Theorems 3.2 and 3.3, which use Nielsen cancellation methods in their proof overlap to a great extent with general results on the isomorphism problem in hyperbolic groups. From the results of Bestvinna and Feighn, Juhasz and Rosenberger and Kharlampovich and Myasnikov a cyclically pinched one-relator group with either U or V not a proper power is hyperbolic. Guildenhuys, Kharlampovich and Myasnikov [**GKM**] proved that conjugacy pinched one-relator groups, with U and V conjugacy separated are also hyperbolic. Sela [**Se 1**]–[**Se 6**] using the JSJ decomposition proved that the isomorphism problem for torsion-free hyperbolic groups is solvable. Dahmiani and Guirardel [**DG**] have extended this to show that the ismorphism problem for the class of all hyperbolic groups, possibly with torsion, is solvable. In another direction, Bumagin, Kharlampovich and Myasnikov [**BKM**] have shown that the isomorphism problem for finitely generated fully residually free groups (see the next section) is solvable. They also use the JSJ decomposition of limit groups. Dahmiani and Groves [**DGr**], building on Sela's work, studied the isomorphism problem for toral relatively hyperbolic groups and were able to recover the results of both Sela and Bumagin, Kharlampovich and Myasnikov.

THEOREM 3.4. *(1)* [**DG**] *The isomorpshim problem for the class of all hyperbolic groups is solvable.*

(2) [**BKM**] *The isomorphism problem for the class of finitely generated fully residually free groups is solvable.*

4. Fully Residually Free Groups

Our results on the surface group conjecture depend on the properties of fully residually free groups or limit groups.

DEFINITION 4.1. *A group G is* **residually free** *if given any nontrivial element $g \in G$ there is a homomorphism $\phi_g : G \to F$ with F a free group such that $\phi_g(g) \neq 1$. G is fully residually free if given any finitely many nontrivial elements $g_1, ..., g_n$ in G there is a homomorphism $\phi : G \to F$, where F is a free group, such that $\phi(g_i) \neq 1$ for all $i = 1, ..., n$.*

The terminology G is **discriminated** by a free group is also used to describe fully residually free groups, hence a group G is discriminated by a free group F given any finitely many nontrivial elements $g_1, ..., g_n$ in G there is a homomorphism $\phi : G \to F$, where F is a free group, such that $\phi(g_i) \neq 1$ for all $i = 1, ..., n$. We note that there are several different concepts of discrimination in group theory (see [**FGS**]). Fully residually free groups have played a pivotal role in the study of equations and first order formulas over free groups and in particular the solution of the Tarski problem by Kharlampovich and Myasnikov and independently by Sela (see [**KhM 1**]–[**KhM 5**] and [**Se 1**]–[**Se 6**]). In Sela's work, finitely generated fully residually free groups are known as **limit groups**. In this guise they were studied by Sela (see [**Se 1**]–[**Se 6**] and [**BeF 2**]) in terms of studying homomorphisms of general groups into free groups.

By the **language of group theory** we mean a first order predicate language with equality, a constant symbol 1, a unary operation symobol $^{-1}$ and a binary operation symbol, $\cdot$ (or just juxtapostion). A *universal sentence* in the language of group theory is a first order sentence using only universal quantifiers (see [**FGMRS**]. The *universal theory* of a group G consists of all universal sentences true in G. It is not hard to see that all free groups share the same universal theory. A group G is called a *universally free group* if it shares the same universal theory as the class of free groups. Analogously an *existential sentence* in the language of group theory is a first order sentence using only existential quantifiers and the *existential theory* of a group G consists of all universal sentences true in G. Since any universal sentence is equivalent to an existential sentence the universally free groups have the same existential theory. Remeslennikov calls the universally free groups ∃-free groups. Remeslennikov [**Re**] and independently Gaglione and Spellman [**GS**] proved the following remarkable theorem.

THEOREM 4.1. *([**Re**],[**GS**]) Suppose G is residually free. Then the following are equivalent:*
  (1) G is fully residually free
  (2) G is commutative transitive
  (3) G is universally free

From the fact that they are linear it is easy to see that orientable surface groups are commutative transitive. A result of G. Baumslag [**GB 1**] shows that they are residually free and hence we have

THEOREM 4.2. *An orientable surface group is fully residually free.*

The class of nonabelian finitely generated fully residually free groups coincides with the class of nonabelian finitely generated universally free groups From the solution of the Tarski problem all nonabelian free groups share the same elementary or first-order theory. An *elementary free group* is a group G which shares the same elementary theory as the class of nonabelian free groups. As an outgrowth of the solution to the Tarski problem Kharlampovich and Myasnikov and independently Sela have completely characterized the elementary free groups. In particular, in the language of Kharlampovich and Myasnikov, they are precisely the NTQ-groups - the coordinate groups of regular NTQ-systems of equations over free groups. What is important here is that an elementary free group must be universally free and hence fully residually free. Therefore results proved about fully residually free groups apply to elementary free groups.

Kharlampovich, Myasnikov, Remeslennikov, Sela and others have done extensive work on describing both the subgroups and the subgroup structure of fully residually free groups. We mention a few results that are relevant to our results on the surface group conjecture. The following is a summary of several results:

THEOREM 4.3. *Let G be a finitely generated fully residually free group. Then*
(1) G can be embedded as a subgroup in the free exponential group $F^{\mathbb{Z}[t]}$
(2) G is finitely presentable
(3) G can be constructed in a systematic way starting with free groups and abelian groups using free products with cyclic amalgamation and extensions of centralizers.

The construction mentioned in the theorem leads to the existence of nontrivial JSJ decompositions. This is crucial to our results.

JSJ-decompositions were introduced by Rips and Sela ([**RiS**]) and have played a fundamental role in the study of both hyperbolic groups and fully residually free groups. Roughly a JSJ-decomposition of a group G is a splitting of G as a graph of groups with abelian edges which is canonical in that it encodes all other such abelian splittings. If each edge is cyclic it is called a *cyclic JSJ-decomposition*. For a formal definition we refer to [**RiS**]. There are also full discussions in [**BeF 2**] and the work of Sela [**Se 1**]–[**Se 6**]. The relevant fact for fully residually free groups is the following.

THEOREM 4.4. *(see [**KhM 1**]–[**KhM 5**]) (a) A finitely generated fully residually free group which is indecomposable relative to JSJ-decompositions is either the fundamental group of a closed surface, a free group or a free abelian group.*

(b) A finitely generated fully residually free group admits a non-trivial cyclic JSJ-decomposition if it is not abelian or a surface group.

5. Property IF and Results on the Surface Group Conjecture

In order to consider the surface group conjecture in conjunction with being fully residually free we concentrate on the property that subgroups of infinite index must be free.

DEFINITION 5.1. *A group G satisfies* **Property IF** *if there there exists a noncyclic subgroup of infinite index and every subgroup of infinite index is a free group.*

This property when combined with the cyclic JSJ decompositions of fully residually free groups allowed us to prove several results [**FKMRR**] on the surface group conjecture. First, it is clear that if a group G has a nontrivial free product decomposition and satisfies Property IF then it must be a free group. The main result below shows that property IF forces the group to be either a cyclically pinched one-relator group or a conjugacy pinched one-relator group. We reconstruct the proof because it is straightforward and a good illustration of the technique.

THEOREM 5.1. ([**FKMRR**]) *Suppose that G is a finitely generated fully residually free group with property IF. Then G is either a free group or a cyclically pinched one relator group or a conjugacy pinched one relator group.*

PROOF. Suppose that G is a finitely generated fully residually free group. Then if it is indecomposable relative to JSJ-decompositions it is either the fundamental group of a closed surface, a free group or a free abelian group. The first two cases are covered by the theorem. If it is free abelian of rank > 2 then it cannot satisfy property IF. Therefore if it is abelian it must be either infinite cyclic and hence free or free abelian of rank 2. In this case G has the presentation

$$G = <x, y; [x,y] = 1> \implies G = <x, y; xyx^{-1} = y>$$

and G can be considered as a conjugacy pinched one-relator group.

If G is not abelian or a surface group then from Theorem 4.4, G admits a non-trivial cyclic JSJ-decomposition. Let e be a a non-trivial edge with edge stabilizer G_e. Collapse everything at one vertex into a single group G_1 and everything at the other vertex to a single group G_2. That is G_1 is the subgroup generated by all vertex groups on one side of the vertex e and G_2 is the subgroup generated by all vertex groups on the other side of e. If both collapse totally, that is both G_1 and G_2 are trivial, then G is cyclic which is a contradiction. Hence we can assume that at least one of G_1 and G_2 is nontrivial.

Suppose first that both G_1 and G_2 are non-trivial. Then G is a free product of G_1 and G_2 with cyclic amalgamation along the edge e and hence along G_e. Since this is a free product with amalgamation, both factors G_1 and G_2 have infinite index. By assumption G satisfies property IF and hence both factors are free groups. Therefore G is a free product with amalgamation of two free groups with a cyclic amalgamated subgroup; i.e a cyclically pinched one-relator group or a free group.

Now suppose that G_2 is trivial. Then G has a tree decomposition with a single edge emanating from G_1. Hence G is an HNN extension of G_1 with cyclic associated subgroups. As before from property IF, G_1 must be a free group, and hence G in this case is a conjugacy pinched one-relator group. □

The theorem asserts essentially that if a fully residually free group has Property IF then each subgroup of finite index is a one-relator group.

COROLLARY 5.1. *Suppose that G is a finitely generated fully residually free group with property IF. Then G is either free or every subgroup of finite index is freely indecomposable and hence a one-relator group.*

Orientable surface groups of genus $g \geq 2$ are hyperbolic. Hence if the surface group conjecture were to be true then the resulting group must be hyperbolic unless the group were free abelian of rank 2. We can consider a free abelian group as a surface group of genus $g = 1$; i.e. $G = <x, y; [x, y] = 1>$. A result of Kharlampovich and Myasnikov [**KhM 3**] shows that every finitely generated fully residually free group is either hyperbolic or conatins a free abelian group of rank two. Combining this with Property IF we get

COROLLARY 5.2. *Let G be a finitely generated fully residually free group with property IF. Then either G is hyperbolic or G is free abelian of rank 2.*

In the preceding results we assumed that G was fully residually free and used the JSJ decomposition. We say that a group has the finite index property if every subgroup of finite index is a one-relator group. Property IF will imply the finite index property if we don't assume the fully residually free property but start with a graph of groups decomposition. In particular we get the following which gives further evidence towards the full surface group conjecture;

THEOREM 5.2. *(**[FKMRR]**) Let G be a nonfree cyclically pinched or conjugacy pinched one-relator group with property IF. Then each subgroup of finite index is again a cyclically pinched or conjugacy pinched one-relator group.*

This theorem uses the subgroup theorem for free products with amalgamation and for HNN groups in the form described by Karrass and Solitar (see [**FR**]).

We mention further that the following is known. It appears in [**KRW**] but not stated exactly in the same way.

THEOREM 5.3. *(see [**KRW**]) Suppose that $G = <a_1,...,a_n; a_1^{\alpha_1} \cdots a_n^{\alpha_n} = 1>$ with $n \geq 2$ and all $\alpha_i \geq 2$. Then G has Property IF if and only if $\alpha_1 = ... = \alpha_n = 2$.*

In light of these results we give a modified version of the surface group conjecture. Notice that in this modified version we omit the assumption that G is a one-relator group. Every subgroup of finite idnex is a one-relator group from Corollary 5.1.

CONJECTURE 5.1. **Surface group Conjecture C** *Suppose that G is a finitely generated freely indecomposable fully residually free group with property IF. Then G is a surface group.*

We note that Surface group Conjecture C is true under either of the following two conditions

(1) The original relator is strictly quadratic, that is each generator appears exactly twice in the realtor.

(2) There is only one flexible (QH) vertex in the JSJ decomposition for G.

We refer to the papers [**KhM 1**]–[**KhM 5**] for a description of a QH-vertex.

6. Baumslag Doubles and a Question of Gromov

A **Baumslag double** is an amalgamated product of the form $G = F \star_{\{W = \overline{W}\}} \overline{F}$ where F is a finitely generated free group, $\overline{F}$ is an isomorphic copy, $W \neq 1$ is a word in F and $\overline{W}$ is its copy in $\overline{F}$. An orientable surface group of genus 2 is a Baumslag double and in fact Baumslag doubles were introduced in [**GB 1**] to prove that surface groups are residually free. If G is a Baumslag double and if the identified word W is not a proper power in F it follows from the combination theorems of Juhasz and Rosenberger [**JR**], Kharlampovich and Myasnikov [**KM**] and Bestvina and Feighn [**BeF 1**] that the group G is hyperbolic. In fact the Baumslag double G is hyperbolic if and only if W is not a proper power in F because W is a proper power in F if and only if $\overline{W}$ is a proper power in $\overline{F}$.

Gromov (see [**GW**],[**KW**]) conjectured that a one-ended word hyperbolic group must contain a subgroup isomorphic to the fundamental group of a closed hyperbolic surface. The general question is still open. Recent work by Gordon and

Wilton [**GW**] and Kim and Wilton [**KW**] provide sufficient conditions for hyperbolic surface groups to be embedded in a Baumslag double G. The work of Gordon and Wilton uses group cohomology and 3-manifold theory while that of Kim and Wilton proceeds by realizing a Baumslag double as the fundamental group of a non-positively curved square complex.

In [**FR 2**] Fine and Rosenberger using Nielsen cancellation methods based on the techniques in [**Ro 2**], proved that a hyperbolic orientable surface group of genus 2 is embedded in a hyperbolic Baumslag double if and only if the amalgamated word W is a commutator, that is $W = [U, V]$ for some elements $U, V \in F$. The main result of that paper is the following.

THEOREM 6.1. *Let* $G = F *_{\{W=\overline{W}\}} \overline{F}$ *be a hyperbolic Baumslag double. Then G contains a hyperbolic orientable surface group of genus 2 if and only W is a commutator, that is $W = [U, V]$ for some elements $U, V \in F$. Further a Baumslag double G contains a nonorientable surface group of genus 4 if and only if $W = X^2 Y^2$ for some $X, Y \in F$.*

Since an orientable surface group of genus 2 contains an orientable surface group of any finite genus as a subgroup we immediately get the following corollary.

COROLLARY 6.1. *Let $G = F *_{\{W=\overline{W}\}} \overline{F}$ be a hyperbolic Baumslag double. Then G contains orientable surface groups of all finite genus if and only W is a commutator.*

7. The Free-by-Cyclic Conjectures

Gilbert Baumslag conjectured that any one-relator group with torsion is residually finite. There have been significant partial results on this conjecture (see [**W 1**],[**W 2**]). There has been a purpoted complete solution by D. Wise but as of this writing it has not appeared yet. In a series of papers, Baumslag, Fine, Miller and Troeger [**BT**], [**BFMT**] have results on a stronger conjecture that would actually imply the Baumslag conjecture..

CONJECTURE 7.1. *Every one–relator group with torsion is virtually free–by–cyclic i.e., contains a subgroup of finite index which is an infinite cyclic extension of a free group.*

Since a finitely generated virtually free–by–cyclic group is residually finite [**GB 3**], the residual finiteness conjecture is a consequence of the above. Further the conjecture may be true for a wider class of torsion-free one-relator groups. Each surface group is free-by-cyclic tying this problem to the surface group conjecture. More generally we conjecture that each Baumslag double is virtually free-by-cyclic.

Free-by-cyclic groups and virtually free-by-cyclic groups arise in many different contexts. In particular the fundamental groups of all orientable surfaces of genus $g \geq 2$ and nonorientable of genus $g \geq 3$ are free-by-cyclic. It follows that all finitely generated Fuchsian groups are virtually free-by-cyclic. In the same spirit a result of J.Howie [**H**] gives sufficient conditions for the fundamental group of a 2-complex to be free-by-cyclic in terms of a Morse function.

In [**BFMT**] several results were proved concerning the free-by-cylic structure of both cyclically pinched one-relator groups and conjugacy pinched one-relator groups.

THEOREM 7.1. *Suppose that $G = A \underset{U=V}{\star} B$ is a cyclically pinched one-relator group. If $U \notin [A, A]$ and $V \notin [B.B]$ then G is free-by-cyclic.*

THEOREM 7.2. *Suppose that F is a finitely generated free group and $G = < F, t; t^{-1}Ut = V$ with $U, V \notin [F, F] >$ is a conjugacy pinched one-relator group. If either $U[F, F] = V[F, F]$ or U, V are linearly independent modulo $[F, F]$ then G is free-by-cyclic.*

These two results deal with elements not in the derived group. When they are in the derived group we can prove results in the case of Baumslag doubles.

THEOREM 7.3. *Baumslag doubles are virtually free-by-abelian. That is $G = F \underset{U=\overline{U}}{\star} \overline{F}$ and $H = < F, t; t^{-1}Ut = U >$ are virtually free-by-abelian.*

There are also technical sufficient conditions so that doubles are actually virtually free-by-cyclic (see [**BFMT**]). The groups G in question are all amalgams and hence the structure of subgroups can be studied via the group's Bass-Serre tree. The proofs of all these results depend upon constructing a homomorphism onto an infinite cyclic group and then analyzing the structure of the kernel in terms of the Bass-Serre tree for the groups G. The freeness comes from the Hanna Neumann theorem that a subgroup that intersects each vertex in a graph of groups decomposition trivially must be a free group.

8. Approaches to the Classification of One Relator Limit Groups

Although one-relator groups are so important and have been so heavily studied there is no systematic approach to the whole field. In [**BFR**] an inductive approach based on the Magnus breakdown was suggested. There is a similar approach to one-relator fully residually free groups. Here we briefly describe both methods.

In the course of proving the Freiheitssatz (see [**LS**]), Magnus developed a technique for handling one-relator groups in general. He proved that a one-relator group can be broken down into amalgamated products of simpler one-relator groups, thereby providing an inductive mechanism for handling them. This procedure is called the **Magnus Breakdown**. An in depth focus on the Magnus approach is a primary tool, though not the only one, in the study of a one-relator group.

DEFINITION 8.1. *Let $G = < a, \ldots ; r >$ be a one-relator group. G is said to be* **plain** *if r is a power of a primitive element in the underlying free group on the given generators $a, \ldots$ of G.*

Magnus's basic idea has been reworked in terms of HNN extensions. In fact Moldavanski [**Mo**] has shown that if

$$G = < a, \ldots ; r >$$

is a one-relator group and if G is not plain, then G can be embedded, in a very simple way, in a closely related one-relator group G_0, which often coincides with G. G_0 is an HNN-extension of a second one-relator group H; the length of the given defining relator of H is smaller than the length of r. This procedure gives rise to a series of one relator groups $G_0, G_1, \ldots$ ending up in a plain one-relator group G_d. The length d of such a series depends on a number of choices that are available at almost every step in the production of the series. We say that that G is of **plainarity** d if it has such a series of length d with d minimal. Notice that the one-relator groups of plainarity 0 are the plain ones. The plain one-relator groups are either free or the free product of a finite cyclic group and a free group. Thus they can be considered as well known. The answers to all of the problems and questions that we consider are known for such plain one-relator groups. The one-relator groups of plainarity 1 are more difficult to investigate than the ones of plainarity 0. In [**BFR**] a specific example was given to illustrate how such groups of plainarity 1 arise and how to deal with them. Further in [**BFR**] several suggestions are given as to how plainarity can be further used.

The second approach is known as **level** and can be applied to one-relator fully residually free groups. This approach is somewhat analogous to the plainarity approach. Myasnikov and Remeslennikov [**KhM 1**] have proved that a finitely generated fully residually free group is a subgroup of a group built up in the following manner.

DEFINITION 8.2. *Let $G \neq \{1\}$ be a commutative transitive group. Let $u \in G \setminus \{1\}$ and let $M = Z_G(u)$, the centralizer of u in G. Let $B \neq 1$ be a torsion free abelian group. Then*

$$G(u, B) = < G, B; \text{ rel}(G), \text{ rel}(B), [B, M] = 1 >$$

is the **B-extension of the centralizer** *M of u in G. If $B = < t; >$ is infinite cyclic, then*

$$G(u, B) = < G, t; \text{ rel}(G), t^{-1}zt = z, \text{ for all } z \in M >$$

is the **free rank one extension of the centralizer** *M of u in G.*

Observe that if $B \neq \{1\}$ is an arbitrary torsion free abelian group, then $G(u, B)$ is a direct union of free rank one extensions of centralizers.

One can prove that if $G \neq \{1\}$ is fully residually free, then so is every free rank one extension of a centralizer of G. This is an extension of the method of Baumslag in [**GB 1**] and uses what is now called the big powers argument. From this it follows that if $B \neq \{1\}$ is any torsion free abelian group and $u \in G \setminus \{1\}$, then $G(u, B)$ is locally fully residually free. In the special case where B is residually $\mathbb{Z}$, in particular if B is free abelian, then $G(u, B)$ is fully residually free.

Kharlampovich and Myasnikov (see [**KhM 1**]–[**KhM 5**]) then prove that a finitely generated fully residually free group must be a subgroup of a group built up from a free group by iterated extensions of centralizers.

THEOREM 8.1. *(See [**KhM 1**]–[**KhM 5**]) A finitely generated fully residually free group is in the class of groups formed by taking iterated extensions of centralizers starting with a free group.*

Fine and Rosenberger, utilizing a theorem of Shalen (see [**FR**] and the references there) showed that a cyclically pinched one-relator group with neither U nor V a proper power, has a faithful representation in $PSL(2, \mathbb{C})$. A modification of that argument applied to free rank one extensions of centralizers and combined with Theorem 8.1 provides a direct proof that any hyperbolic limit group has a faithful representation in $PSL(2, \mathbb{C})$ and shows constructively how to find the representation. More recently using nonstandard free groups the result was extended to all finitely generated fully residually free groups. More generally Breuillard, Gelander, Souto and Storm [**BGSS**] show that surface groups can be embedded as dense subgroups of complex semisimple Lie groups. We thank S.Liriano for pointing out the rersult in [**BGSS**].

THEOREM 8.2. *([**FR 3**],[**FMR**] [**BGSS**]) A finitely generated fully residually free group has a faithful representation in $PSL(2,\mathbb{C})$.*

The concept of a centralizer extension actually has its origins in the paper of Baumslag [**GB 1**] mentioned earlier where he proved that each orientable surface group of genus g, with $g \geq 2$, is residually free. Baumslag observed that each such G, with $g \geq 2$, embeds into the surface group of genus 2. Residual freeness is inherited by subgroups so it suffices to show that a surface group of genus 2 is residually free. Baumslag actually showed more. If F is a nonabelian free group and $u \in F$ with $u \neq 1$ is a nontrivial element which is neither primitive nor a proper power then the Baumslag double K given by

$$K = < F \star \overline{F}; u = \overline{u} >$$

is residually free. To prove this he proceeded by embedding K in the free rank one extension of centralizers

$$H = < F, t; t^{-1}ut = u >$$

by

$$K = < F, t^{-1}Ft > .$$

The group H is then residually free and hence K is residually free. Therefore every Baumslag double is residually free.

A **limit group of level** d is a limit group built up starting with a free group by d extensions of centralizers. Similar to our plainarity approach this provides an inductive treatment of limit groups. This was used by Fine, Gaglione, Myasnikov, Rosenberger and Spellman [**FGMRS**] to classify the fully residually free groups of small rank.

9. The Lyndon Properties

A **quadratic equation** in a free group is an equation $W(x_1, ..., x_n) = 1$ where each variable x_i occurs exactly twice. In [**L**] Lyndon proved that in a free group any solution set $\{x, y, z\}$ of the quadratic equation $x^2y^2z^2 = 1$ generates a cyclic group. Lyndon's theorem launched the whole theory of equations over free groups which led eventually to the solution of the Tarski problems by Kharlampovich and Myasnikov [**KhM 1**]–[**KhM 5**] and by Sela [**Se 1**]–[**Se 6**]. Lyndon's theorem can be phrased as a universal sentence (see Section 3) and the result holds in any fully residually free group if we replace cyclic by abelian (see Section 3). In particular it holds in an orientable surface group of genus $g \geq 2$. From a result of Rosenberger [**Ro 1**] Lyndon's result will also hold in most cyclically pinched one-relator groups. From a result of Fine, Roehl and Rosenberger [**FRR**] there is an analogous result for certain HNN groups. In [**FRR 1**],[**FRR 2**] and [**BCMR**] properties related to Lyndon's theorem were considered.

DEFINITION 9.1. *The following are called* **Lyndon properties**. *Let G be a group. Then G satisfies Property*

    (1) **LZ** *if whenever $x^2y^2z^2 = 1$ for $x, y, z \in G$ then $< x, y, z >$ is cyclic.*

    (2) **LA** *if whenever $x^2y^2z^2 = 1$ for $x, y, z \in G$ then $< x, y, z >$ is abelian.*

    (3) **LPZ** *if whenever $x^py^qz^r = 1$ for $x, y, z \in G$ with $2 \leq p, q, r \in \mathbb{N}$ then $< x, y, z >$ is cyclic.*

    (4) **LPA** *if whenever $x^py^qz^r = 1$ for $x, y, z \in G$ with $2 \leq p, q, r \in \mathbb{N}$ then $< x, y, z >$ is abelian.*

(5) **LCZ** if whenever $[x^p, y^q]z^r = 1$ for $x, y, z \in G$ with $1 \le p, q \in \mathbb{N}$, $2 \le r \in \mathbb{N}$ then $<x, y, z>$ is cyclic.

(6) **LCA** if whenever $[x^p, y^q]z^r = 1$ for $x, y, z \in G$ with $1 \le p, q \in \mathbb{N}$, $2 \le r \in \mathbb{N}$ then $<x, y, z>$ is abelian.

All of these properties hold in free groups and Properties LA, LPA and LCA are given by universal sentences. Hence these hold in any fully residually free group. Note that it is an open question as to when a free product with amalgamation of fully residually free groups is still fully residually free. Recall that a subgroup $H \subset G$ is **malnormal** if $g^{-1}Hg \cap H = \{1\}$ for $g \notin H$. The main results in [**FRR 1**] showed that these properties are preserved under many group amalgams.

THEOREM 9.1. ([**FRR 1**]) *Suppose that H_1 and H_2 are groups with no elements of order 2 and that G is the amalgamated product $G = H_1 \star_A H_2$ with $H_1 \ne A \ne H_2$ and A is malnormal in both H_1 and H_2. Then*

(1) if both H_1 and H_2 satisfy Property LZ then G also satisfies Property LZ.

(2) if both H_1 and H_2 satisfy Property LA then G also satisfies Property LA.

THEOREM 9.2. ([**FRR 1**]) *Suppose that H_1 and H_2 are torsion-free groups and that G is the amalgamated product $G = H_1 \star_A H_2$ with $H_1 \ne A \ne H_2$ and A is malnormal in both H_1 and H_2. Then*

(1) if both H_1 and H_2 satisfy Property LPZ then G also satisfies Property LPZ.

(2) if both H_1 and H_2 satisfy Property LPA then G also satisfies Property LPA.

(3) if both H_1 and H_2 satisfy Property LCZ then G also satisfies Property LCZ.

(4) if both H_1 and H_2 satisfy Property LCA then G also satisfies Property LCA.

In particular if $x, y \in G$ with $[x, y] \ne 1$ then if both H_1, H_2 have property LCZ or LCA then $[x, y]$ is never a proper power.

We note that the malnormality condition in the above theorems is essential. For example in $G = <a, b, c; a^2b^2c^2 = 1>$ we have an equation $x^2y^2z^2 = 1$ such that $<x, y, z>$ is not abelian.

THEOREM 9.3. ([**FRR 1**]) *Suppose that G is a cyclically pinched one-relator group*

$$G = <a_1, ..., a_p, b_1, ..., b_q; WV = 1>$$

where $p \ge 2, q \ge 2$, $1 \ne W = W(a_1, ..., a_p)$ is not a proper power nor a primitive element in the free group on $a_1, ..., a_p$ and $1 \ne V = V(b_1, ..., b_q)$ is not a proper power nor a primitive element in the free group on $b_1, ..., b_q$. Then G has properties LZ, LPZ, LCZ.

These results are all technical applications of Nielsen cancellation methods in free products with amalgamation.

The situation for HNN groups is much more complicated.

THEOREM 9.4. ([**FRR 1**]) *Suppose that G is an HNN extension of the base B so that G has the form*

$$G = <B, t; rel(B), t^{-1}K_1t = K_2>.$$

Suppose further that K_1 and K_2 are both malnormal in B and that B does not contain an element of order 2. Suppose further that B satisfies the basic Lyndon Property LZ. Then if $x^2y^2z^2 = 1$ in G and $\mathcal{U} = \{x, y, z\}$ is regular then $<x, y, z>$ is cyclic.

We refer to [**FRR 1**] for the concept of a regular set which is quite technical.

In [**FRR 2**] the basic Lyndon properties were shown to hold in one-relator groups with odd torsion. The proof of this result turned out to be much more complicated that the corresponding results for amalgams.

THEOREM 9.5. *([**FRR 2**]) Let G be the one-relator group with torsion*

$$G = <a, b, c, ...; R^m>$$

with $m \geq 3$ and m odd and R a cyclically redcued word, not a proper power on the free group on $a, b, c, ...$ Let $w(x_1, x_2, x_3)$ be a regular quadratic word in the free group F on x_1, x_2, x_3 and let $\phi : F \to G$ be a homomorphism from F into G with $\phi(x_i) = u_i$ for $i = 1, 2, 3$. If $w(u_1, u_2, u_3) = 1$ in G then the subgroup $<u_1, u_2, u_3>$ is cyclic.

In particular G satisfies the Lyndon property LZ.

N.Brady, L.Ciabanu, A,Martino and S.O'Rourke [**BCMR**] studied the Lyndon properties in groups acting freely on Λ-trees where Λ is an ordered abelian group. They proved.

THEOREM 9.6. *([**BCMR**]) Let G be a group that acts freely on a Λ-tree, where Λ is an ordered abelian group, and let x, y, z be elements in G. If $x^p y^q = z^r$ with integers p, q, $r \geq 4$, then x, y and z must commute. Hence G satsifies LPA.*

It has been unclear whether the same conclusion holds for p, q, r not all larger than 4, and in particular the proof in [**BCMR**] could not be extended to these smaller cases. In [**CFR**] the behavior for small values of p, q, r was considered in certain HNN extensions. In particular it was proved that there are cases where LPA will not hold.

THEOREM 9.7. *Let F be a finitely generated non-cyclic free group, and let u and v be elements in F which are not proper powers. Let $G = \langle F, t \mid tut^{-1} = v \rangle$ and $r \geq 2$ be a given integer. Then for particular choices of u and v there exist non-commuting elements $a, b, c \in G$ such that*

$$a^2 b^2 c^r = 1.$$

Note that the group G in the theorem does act freely on some Λ-tree.

10. The Classification of One Relator CSA Groups

Fully residually free groups must be commutative transitive (CT) and CSA. CSA stands for conjugately separated abelian and signifies that maximal abelian subgroups are malnormal. CSA implies commutative transitive but not conversely in general. However they are equivalent in the presence of full residual freeness. Fine, Myasnikov, grosse Rebel and Rosenberger [**FMgRR**] recently gave a classification of the one-relator CT and one-relator CSA groups building upon previous results of Gildenhuys, Kharlampovich and Myasnikov [**GKM**]. This classification requires an additional class of groups introduced earlier by Fine and Rosenberger [**FR 1**] and independently by Cohen and Lustig [**CL**] in connection with very small group actions on $\mathbb{R}$-trees.

DEFINITION 10.1. *A group G is a **restricted Gromov group** or **RG-group** if for any $g, h \in G$ then either the subgroup $<g, h>$ is cyclic or there exists a positive integer k with $g^k \neq 1 \neq h^k$ such that $<g^k, h^k> = <g^k> \star <h^k>$, the free product of $<g^k>$ and $<h^k>$.*

It is known that torsion-free hyperbolic groups are RG. The classification is then given in the following three theorems that tie together for one-relator groups the concepts of CT, CSA, RG and the existence of Baumslag-Solitar subgroups.

THEOREM 10.1. *(**[FMgRR]**, **[FRR 2]**) Let G be a one-relator group with torsion. Then the following are equivalent:*
  (1) G is CSA
  (2) G is RG
  (3) G does not contain a copy of the infinite dihedral group

$$<x, y; x^2 = y^2 = 1>$$

  (3) G satisfies the Lyndon property LZ.

For torsion-free one-relator groups the situation is different. A torsion-free one-relator group fails to be a CSA group if and only if its contains a copy of some nonabelian Baumslag-Solitar group $B_{1,n}$ with $n \neq 1$ or a copy of the direct product of a free group of rank 2 and an infinite cyclic group, $F_2 \times \mathbb{Z}$, (see **[FMgRR]**). In **[FMgRR]** it was proved that a one-relator group fails to be an RG-group if and only if it contains a copy of one of the Baumslag-Solitar groups $B_{1,n}$ with $n \neq 0$. Recall that $B_{1,1}$ is a free abelian group of rank 2. It follows that if G is a torsion-free one-relator group which does not contain a free abelian subgroup of rank 2 then the following are equivalent.
  (1) G is a CSA group
  (2) G is an RG-group
  (3) G does not contain a copy of some $B_{1,n}$ with $n \neq -1, 0, 1$.

Further if G is a torsion-free one-relator group then G is CT if and only if it does not contain a copy of $F_2 \times \mathbb{Z}$ or a copy of the Klein bottle group $B_{1,-1}$. Combining all these we get the following results.

THEOREM 10.2. *(**[FMgRR]**) Let G be a torsion-free one-relator group which does not contain a copy of $\mathbb{Z} \times \mathbb{Z} = B_{1,1}$. Then the following are equivalent:*
  (1) G is CSA
  (2) G is RG
  (3) G does not contain a copy of one of the Baumslag-Solitar group

$$B_{1,m} = <x, y : yxy^{-1} = x^m>$$

with $m \in \mathbb{Z} \setminus \{-1, 0, 1\}$.

THEOREM 10.3. *(**[FMgRR]**) Let G be a torsion-free one-relator group. Then G is CT if and only if G does not contain a copy of $F_2 \times \mathbb{Z}$ or a copy of the Baumslag-Solitar group $B_{1,-1} = <x, y; yxy^{-1} = x^{-1}>$ (the Klein-bottle group).*

Notice that since one-relator groups with torsion are commutative transitive, this fact together with Theorem 10.2 provides the total classification of one-relator commutative transitive groups.

The proofs of these results combine an analysis of the Magnus breakdown (see section 6) of the one-relator group combinaed then with Nielsen cancellation methods considered within HNN groups. Recall that the Magnus breakdown embeds the one-relator group in an HNN extension.

References

[AFR] P.Ackermann,B.Fine and G.Rosenberger, On Surface Groups: Motivating Examples in Combinatorial Group Theory, **Groups St. Andrews 2005**. Cambridge University Press, 2007, 96–129 MR2327318 (2008j:20122)

[GB 1] G. Baumslag, On generalized free products, **Math. Z.**, 78, 1962, 423-438 MR0140562 (25:3980)

[GB 2] G.Baumslag, Groups with the same lower central sequence as a relatively free group. I. The groups. **Trans. Amer. Math. Soc.** , 129, 1967, 308–321 MR0217157 (36:248)

[GB 3] G.Baumslag, Residually finite one-relator groups **Bull.Amer. Math. Soc.**, 73, 1967, 618–620 MR0212078 (35:2953)

[BT] G.Baumslag and D. Troeger, Virtually Free-by-Cyclic Groups I, in **Aspects of Infinite Groups** World Scientific Press, 2009

[BFMT] G.Baumslag, B.Fine, C. Miller and D. Troeger, Virtual Properties of Cyclically Pinched One-Relator Groups **Int, J. of Alg. and Comp.**, 19, 2009 1–15 MR2512551 (2010c:20034)

[BFR] G.Baumslag, B.Fine, and G.Rosenberger, One-Relator Groups: A Systematic Approach, to appear

[BeF 1] M. Bestvina and M. Feighn, A combination theorem for negatively curved groups, **J. Diff. Geom.** , 35, 1992, 85-101 MR1152226 (93d:53053)

[BeF 2] M. Bestvina and M. Feighn, Notes on Sela's Work: Limit Groups and Makanin-Razborov Diagrams, preprint

[BCMR] N.Brady, L.Ciobanu, A.Martino and S.O'Rourke, The Equation $x^p y^q = z^r$ and Groups that act freely on Λ-trees, **Trans. Amer. Math. Soc.** 361,2009, 223-236 MR2439405 (2009i:20055)

[BGSS] E. Breulillard, T.Gelander, J. Souto and T.Storm, Dense Embeddings of Surface Groups, **Geometry and Topology** , 10, 2006, 1373-1389 MR2255501 (2008b:22007)

[BBS] A.M.Brunner,R.G. Burns and D.Solitar, The Subgroup Separability of Free Products of Two Free Groups with Cyclic Amalgamation, **Cont. Math.** , 33, 1984, 90-115 MR767102 (86e:20033)

[BKM] I. Bumagin, O. Kharlampovich ,A.Myasnikov, Isomorphism Problem for Finitely Generated Fully Residually Free groups, submitted

[Ca] D. Calegari, Surface subgroups from homology. **Geometry and Topology**, 12(4), 2008, 1995-20 MR2431013 (2009d:20104)

[CFR] L. Ciabanu, B.Fine and G.Rosenberger, On Lyndon's equation in some Λ-fre groups and HNN extensions, sublittted J. Grp. Theory

[CL] M.Cohen and M.Lustig, Very Small Actions on $\mathbb{R}$-trees and Dehn Twist Automorphisms, **Topology**, 34, 1985, 575–617 MR1341810 (96g:20053)

[CoLe] D.Collins and F.Levin, Automorphisms and Hopficity of Certain Baumslag-Solitar Groups, **Archiv. Math.**, 40, 1983, 385–400 MR707725 (85b:20043)

[CER] L.Comerford, C.Edmunds,G.Rosenberger, Commutators as Powers in Free Products **Proc. of the AMS** ,122 ,1994, 47–52 MR1221722 (94k:20048)

[C] P.M. Curran, Subgroups of Finite Index in Certain Classes of Finitely Presented Groups, **J. of Alg.** , 122, 1989, 118–129 MR994940 (90d:20059)

[DGr] F. Dhamiani and D. Groves, The isomorphism problem for toral relatively hyperbolic groups, **Publ. Math, de L'IHES** , 107, 2008, 211-290 MR2434694 (2009i:20081)

[DG] F. Dhamiani and V. Guirardel, Isomorphism Problem for the Class of Hyperbolic Groups, **ArXiv:1002.2590v1**, 2010

[Dy] J.L. Dyer, Separating Conjugates in Amalgamated Free Products and HNN Extensions, **J. Austr. Math Soc. Ser. A** , 29, 1980, 35-51 MR566274 (81f:20033)

[FGS] B. Fine, A.M. Gaglione and D. Spellman, Discriminating Groups; A comprehensive Overview **Proceedings of Groups St Andrws 2010** 2011

[FR] B. Fine and G. Rosenberger, **Algebraic Generalizations of Discrete Groups**, Marcel-Dekker, 1999 MR1712997 (2000m:20049)

[FR 1] B. Fine and G. Rosenberger, On Restricted Gromov Groups, **Communications in Algebra**, 20, Number 8 , 1992,2171-2182 MR1172655 (93j:20078)

[FR 2] B. Fine and G. Rosenberger, Surface Groups within Baumslag Doubles, Proceedings Edinburgh Math Society , Vol. 54 (2011) 91-97 MR2764408 (2012b:20105)

[FR 3] B. Fine and G. Rosenberger, Faithful Representations of Hyperbolic Limit Groups Journal Groups, Complexity and Cryptology , Vol.3 No.2 (2011) 349-355

[FMR] B. Fine, A. Myasnikov and G. Rosenberger, Faithful Representations of Limit Groups to appear J. Group Theory

[FMgRR] B. Fine, A.Myasnikov, V. gr. Rebel and G. Rosenberger, A Classification of Conjugately Separated Abelian, Commutative Transitive and Restricted Gromov One-Relator Groups, **Result. Math.**, 50, 2007, 183-193 MR2343587 (2008k:20066)

[FKMRR] B. Fine, O.Kharlampovich, A.Myasnikov,V.Remeslennikov and G. Rosenberger, On the Surface Group Conjecture, **Scienta: Math Series A** , 1, 2008, 1-15 MR2367908 (2009b:20050)

[FRR 1] B. Fine, A.Rosenberger, G. Rosenberger, Quadratic Properties in Group Amalgams, J. Group Theory (2011) , in press MR2831964

[FRR 2] B. Fine, A.Rosenberger, G. Rosenberger, A Note on Lyndon Properties in One Relator Groups, Results in Math. (2011), in press

[FRR] B. Fine, F.Roehl and G. Rosenberger, A Three-Free Theorem for Certain HNN Groups, in **Infinite Groups and Group Rings** edited by J.Corson,M.Dixon,M.Evans,F.Rohl, World Scientific Press, 1993 13-37 MR1377954 (96m:20042)

[FRS 1] B.Fine, G.Rosenberger and M. Stille, Conjugacy Pinched and Cyclically Pinched One-Relator Groups, **Revista Math. Madrid** , 10, 1997, 207–227 MR1605642 (99c:20039)

[FRS 2] B.Fine, G.Rosenberger and M.Stille, Nielsen Transformations and Applications: A Survey **Groups Korea 94**, DeGruyter, 1995 MR1476950 (98g:20039)

[FGMRS] B.Fine,A. Gaglione,A. Myasnikov, G.Rosenberger and D. Spellman, A Classification of Fully Residually Free Groups of Rank Three or Less, **J. of Algebra** , 200, 1998, 571–605 MR1610668 (99b:20053)

[GS] A. Gaglione and D. Spellman, Some Model Theory of Free Groups and Free Algebras, **Houston J. Math** , 19, 1993, 327-356 MR1242423 (95m:03079)

[GW] C.Gordon and H.Wilton, Surface Subgroups of Doubles of Free Groups, to appear MR2669638 (2011k:20085)

[GKM] D.Gildenhuys, O. Kharlampovich and A. Myasnikov, CSA Groups and Separated Free Constructions, **Bull. Austral. Math. Soc.**, 52, 1995, 63-84 MR1344261 (96h:20053)

[HKS 1] A. Hoare, A. Karrass and D. Solitar, Subgroups of finite index of Fuchsian groups, **Math. Z.** , 120, 1971, 289–298 MR0285619 (44:2837)

[HKS 2] A. Hoare, A. Karrass and D. Solitar, Subgroups of infinite index in Fuchsian groups, **Math. Z.** , 125, 1972, 59–69 MR0292948 (45:2029)

[H] J.Howie, Some Results on One-Relator Surface Groups, **Boletin de la Sociedad Matematica Mexicana**, to appear MR2199352 (2006k:20072a)

[JR] A. Juhasz and G. Rosenberger, On the Combinatorial Curvature of Groups of F-type and Other One-Relator Products of Cyclics, **Cont. Math.** , 169, 1994, 373-384 MR1292912 (95i:20050)

[KM] O.Kharlamapovich and A.Myasnikov, Hyperbolic Groups and Free Constructions, **Trans. Amer. Math. Soc.**,350, 2, 1998 ,571-613 MR1390041 (98d:20041)

[Kh M] O. Kharlamapovich and A.Myasnikov, **Algebraic Geometry over Free Groups**, to appear

[KhM 1] O. Kharlampovich and A.Myasnikov, Irreducible affine varieties over a free group: I. Irreducibility of quadratic equations and Nullstellensatz, **J. of Algebra** , 200, 1998, 472-516 MR1610660 (2000b:20032a)

[KhM 2] O. Kharlampovich and A.Myasnikov, Irreducible affine varieties over a free group: II. Systems in triangular quasi-quadratic form and a description of residually free groups, **J. of Algebra** , 200, 1998, 517-569 MR1610664 (2000b:20032b)

[KhM 3] O. Kharlamapovich and A.Myasnikov, Description of fully residually free groups and Irreducible affine varieties over free groups, **Summer school in Group Theory in Banff, 1996, CRM Proceedings and Lecture notes** , 17, 1999, 71-81

[KhM 4] O.Kharlamapovich and A.Myasnikov, Hyperbolic Groups and Free Constructions, **Trans. Amer. Math. Soc.** 350, 2, 1998, 571-613 MR1390041 (98d:20041)

[KhM 5] O.Kharlamapovich and A.Myasnikov, Solution of the Tarski Problem, to appear

[KMRS] O. Kharlamapovich,A.Myasnikov, V.Remeslennikov and D.Serbin, Subgroups of fully residually free groups: algorithmic problems, **Cont. Math**, 360, 2004 MR2105437 (2006b:20046)

[KW] S.Kim and H.Wilton, Surface Subgroups of Doubles of Free Groups, to appear

[KRW] J.Konieczny, G. Rosenberger and J.Wolny, Tame Almost Primitive Elements, **Result. Math.** , 38, 2000, 116-129 MR1774284 (2001d:20021)

[Ko] Y.I. Merzlyakov, **Kourovka Notebook - Unsolved Problems in Group Theory**

[Li] S.Lipschutz, The conjugacy problem and cyclic amalgamation, **Bull. Amer. Math. Soc.** , 81, 1975, 114-116 MR0379675 (52:580)

[L] R.C. Lyndon, The equation $a^2b^2 = c^2$ in free groups, **Michigan Math J.** 6, 1959, 155-164 MR0103218 (21:1999)

[LS] R.C. Lyndon and P.E. Schupp, **Combinatorial Group Theory**, Springer-Verlag 1977 MR0577064 (58:28182)

[Mo] D.I. Moldavanskii, Certain Subgroups of Groups with One Defining Relation, **Sibirsk. Mat. Z.**, 8, 1967,1370-1384 MR0220810 (36:3862)

[Re] V.N. Remeslennikov, ∃-free groups, **Siberian Mat. J.** , 30, 1989, 998–1001 MR1043446 (91f:03077)

[RiS] E.Rips and Z.Sela, Cyclic Splittings of Finitely Presented Groups and the Canonical JSJ Decomposition, **Ann. of Math. (2)** , 146. 1997, 53-109 MR1469317 (98m:20044)

[Ro 1] G. Rosenberger, On One-Relaor Groups that are Free Products of Two Free Groups with Cyclic Amalgamation **Groups St Andrews 1981**, Cambridge University Press , 1982, 328–344 MR679174 (84i:20030)

[Ro 2] G. Rosenberger, The isomorphism problem for cyclically pinched one-relator groups, **J. Pure and Applied Algebra** , 95, 1994, 75-86 MR1289120 (95g:20040)

[Sco] G.P.Scott, Subgroups of Surface Groups are Almost Geometric, **London Math. Soc. J.** , 17, 1978, 555-565 MR0494062 (58:12996)

[Se] Z. Sela, The isomorphism problem for hyperbolic groups, **Annals of Math,** 141, 1995, 217-283 MR1324134 (96b:20049)

[Se 1] Z. Sela, Diophantine Geometry over Groups I: Makanin-Razborov Diagrams, **Publ. Math. de IHES** , 93, 2001, 31-105 MR1863735 (2002h:20061)

[Se 2] Z. Sela, Diophantine Geometry over Groups II: Completions, Closures and Formal Solutions, **Israel Jour. of Math.** , 104, 2003, 173-254 MR1972179 (2004g:20061)

[Se 3] Z. Sela, Diophantine Geometry over Groups III: Rigid and Solid Solutions, **Israel Jour. of Math.** , 147, 2005, 1-73 MR2166355 (2006j:20060)

[Se 4] Z. Sela, Diophantine Geometry over Groups IV: An Iterative Procedure for Validation of a Sentence, **Israel Jour. of Math.** , 143, 2004, 17-130 MR2106978 (2006j:20059)

[Se 5] Z. Sela, Diophantine Geometry over Groups V: Quantifier Elimination, **Israel Jour. of Math.** , 150, 2005, 1-97 MR2249582 (2007k:20088)

[Se 6] Z. Sela, Diophantine Geometry over Groups VI: The Elementary Theory of a Free Group, to appear MR2238945 (2007j:20063)

[Sh] P. Shalen, Linear representations of certain amalgamated products, **J. Pure and Applied Algebra** , 15, 1979, 187–197 MR535185 (80e:20011)

[W 1] D.Wise, Residual finiteness of positive one-relator groups, **Comment. Math. Helv.**, 76, 2001, 314-338 MR1839349 (2002d:20043)

[W 2] D.Wise, The residual finiteness of quasi positive one-relator groups with torsion, **J. of Algebra**, 66, 2002, 334-350 MR1920406 (2003f:20043)

[Z] H. Zieschang, Finite Groups of Mapping Classes of Surfaces, Springer Lecture Notes in Math., 875, 1981 MR643627 (86g:57001)

[ZVC] H. Zieschang, H.Vogt and E.Coldeway, Surfaces and Planar Discontinuous Groups, Springer Lecture Notes in Math., 876, 1981

DEPARTMENT OF MATHEMATICS, FAIRFIELD UNIVERSITY, FAIRFIELD, CONNECTICUT 06430
E-mail address: `fine@fairfield.edu`

FACHBEREICH MATHEMATIK, UNIVERSITY OF HAMBURG, HAMBURG, GERMANY
E-mail address: `gerhard.rosenberger@math,uni-hamburg.de`

Contemporary Mathematics
Volume 582, 2012
http://dx.doi.org/10.1090/conm/582/11555

Discrimination and Separation in the Metabelian Variety

Anthony M. Gaglione, Seymour Lipschutz, and Dennis Spellman

ABSTRACT. In this paper, we will treat relativization to the variety $\mathcal{M}$ of all metabelian groups. If F is an absolutely free group, i.e., a group free in the variety of all groups, and if F'' is the second term in its derived series, then F/F'' is a free metabelian group, i.e., a group free in the variety $\mathcal{M}$. We shall show that certain theorems about discrimination, codiscrimination, and free products of fully residually free groups are false when relativized to $\mathcal{M}$. In the course of doing this, we also show the failure of certain free constructions in $\mathcal{M}$. The question of whether or not Benjamin Baumslag's main theorem about fully residually free groups remains true when relativized to $\mathcal{M}$ still remains open.

1. Introduction

Fully residually free groups were introduced by Benjamin Baumslag [**B**] in 1967 and characterized by him as precisely those residually free groups in which the centralizer of every nontrivial element is abelian. For the sake of completeness formal definitions follow.

Convention: *If G is a group, then $[G]$ is the isomorphism class of G.*

DEFINITION 1.1. [**BMR2**]: Let $\mathcal{X}$ be a nonempty class of groups closed under isomorphism. Let Γ be a group. $\mathcal{X}$ **separates** Γ provided to every $\gamma \in \Gamma \backslash \{1\}$ there is a group $G_\gamma \in \mathcal{X}$ and a homomorphism $\varphi_\gamma : \Gamma \to G_\gamma$ such that $\varphi_\gamma(\gamma) \neq 1$. $\mathcal{X}$ **discriminates** Γ provided to every finite nonempty subset $S \subseteq \Gamma \backslash \{1\}$ there is a group $G_S \in \mathcal{X}$ and a homomorphism $\varphi_S : \Gamma \to G_S$ such that $\varphi_S(\gamma) \neq 1\ \forall \gamma \in S$. In the event that $\mathcal{X} = [G]$ is the isomorphism class of a group G we say that G **separates** (**discriminates**) Γ provided $\mathcal{X}$ does. A group G is **discriminating** provided it discriminates every group it separates.

It was proven in [**BMR2**] that the infinite cyclic group is discriminating.

DEFINITION 1.2. Let $\mathcal{X}$ be a nonempty class of groups closed under isomorphism. Let Γ be a group. Γ is **residually-** $\mathcal{X}$ provided to every $\gamma \in \Gamma \backslash \{1\}$ there is a group $G_\gamma \in \mathcal{X}$ and a surjective homomorphism $\varphi_\gamma : \Gamma \to G_\gamma$ such that $\varphi_\gamma(\gamma) \neq 1$. Γ is **fully residually-** $\mathcal{X}$ provided to every finite nonempty subset $S \subseteq \Gamma \backslash \{1\}$

2010 *Mathematics Subject Classification.* Primary 20E26, 03C07; Secondary 20F19, 20F05.

Key words and phrases. Discrimination, metabelian groups, commutative transitive (CT), residually free, fully residually free, conjugately separated abelian (CSA), separation, free rank 1 extensions of centralizers.

©2012 American Mathematical Society

129

there is a group $G_S \in \mathcal{X}$ and a surjective homomorphism $\varphi_S : \Gamma \to G_S$ such that $\varphi_S(\gamma) \neq 1 \; \forall \gamma \in S$.

Remark: Γ is residually-$\mathcal{X}$ iff it is isomorphic to a subdirect product of a family $(G_i)_{i \in I}$ of groups in $\mathcal{X}$.

DEFINITION 1.3. Let $\mathcal{V}$ be a nontrivial variety of groups and let Γ be a group in $\mathcal{V}$. Γ is $\mathcal{V}$-**residually free** if it is residually free in $\mathcal{V}$. Γ is $\mathcal{V}$-**freely discriminated** provided it is discriminated by the family of groups free in $\mathcal{V}$. Γ is $\mathcal{V}$-**fully residually free** provided it is fully residually free in $\mathcal{V}$.

Remark: Postulating $\Gamma \in \mathcal{V}$ is redundant since varieties are closed under subgroups and direct products.

Convention: When $\mathcal{V} = \mathcal{O}$ is the variety of all groups, $\mathcal{O}$ may be omitted in the notation and terminology.

In this paper, we are interested in the variety $\mathcal{M}$ of all metabelian goups. If G is any group, $x, y \in G$, then the commuatator of x and y is defined as $[x,y] = x^{-1}y^{-1}xy$. The variety $\mathcal{M}$ is then defined by the law $[[x,y],[z,t]] = 1$. See [**N**] for more details. If F is an absolutely free group, i.e., a group free in the variety of all groups $\mathcal{O}$ and F'' is the second term in its derived series, then F/F'' is a free metabelian group.

Example: When $\mathcal{V} = \mathcal{M}$ is the variety of all metabelian groups, a group $H \in \mathcal{M}$ that is $\mathcal{M}$-freely discriminated need not be $\mathcal{M}$-fully residually free. Let G be freely generated by $\{\alpha_1, \alpha_2\}$ relative to $\mathcal{M}$. Let H be the subgroup $\langle \beta_1, \beta_2 \rangle$ where $\beta_1 = \alpha_1$ and $\beta_2 = [\alpha_2, \alpha_1]$. The inclusion map $H \to G$ will not annihilate any element of any $S \subseteq H \backslash \{1\}$ but H is not $\mathcal{M}$-residually free. This is so since under any epimorphism φ onto a group free in $\mathcal{M}$ not annihilating $[\beta_2, \beta_1]$, we would have $\varphi(\beta_2)$ commuting with $[\varphi(\beta_2), \varphi(\beta_1)]$ - a contradiction.

Observe that the residually free group $\langle a_1, a_2; \; \rangle \times \langle b; \; \rangle$ is not fully residually free since any epimorphism φ onto a free group F annihilating neither $[a_2, a_1]$ nor b would have $\varphi(a_1), \varphi(a_2) \in C_F(\varphi(b))$ forcing $\varphi(a_1)\varphi(a_2) = \varphi(a_2)\varphi(a_1)$ - a contradiction. (Here $C_F(\varphi(b))$ is the centralizer in F of $\varphi(b)$.) Benjamin Baumslag showed in [**B**] that the obstruction to a residually free group Γ being fully residually free is containing a copy of $\langle a_1, a_2; \; \rangle \times \langle b; \; \rangle$ as a subgroup. For this not to occur it is necessary and sufficient for Γ to satisfy the following universal sentence

$$\forall x, y, z (((y \neq 1) \wedge (xy = yx) \wedge (yz = zy)) \to (xz = zx)).$$

That in turn is equivalent to the condition that the centralizer of every nontrivial element be abelian - necessarily maximal abelian. Note moreover that if M_1 and M_2 are distinct maximal abelian subgroups of such a group, then $M_1 \cap M_2 = \{1\}$.

DEFINITION 1.4. A group in which the centralizer of every nontrivial element is abelian is **commutative transitive** or **CT**.

DEFINITION 1.5. A group in which every maximal abelian subgroup is malnormal is **conjugately separated abelian** or **CSA**.

Recall that a subgroup M of a group G is malnormal in G if $g \notin M$ implies that $g^{-1}Mg \cap M = \{1\}$.

Remark: Free groups are CSA.

It is well-known and easy to prove that every CSA-group is CT. Here is a facile counterexample to the converse. Let $A = \langle a; a^2 = 1\rangle$ and $B = \langle b; b^2 = 1\rangle$ be cyclic of order 2. Then their free product $A * B$ is CT but not CSA. That $A * B$ is CT follows from the fact that free products preserve commutative transitivity (See Problem 28, p. 196 of [**MKS**]). However, it is easy to show that no nonabelian CSA-group can contain any nontrivial normal abelian subgroup whereas, in $A * B$, the commutator subgroup is infinite cyclic with generator $[a, b]$.

Benjamin Baumslag showed in [**B**] that every CT residually free group is already CSA so that the concepts coincide for residually free groups. Here is a proof. Assume that Γ is residually free and CT. If Γ is abelian then Γ is certainly CSA. Assume that Γ is nonabelian. Could there be a maximal abelian subgroup M and an element $g \in \Gamma$ such that $g^{-1}Mg \cap M \neq \{1\}$ but $g \notin M$? Suppose $1 \neq m \in g^{-1}Mg \cap M$. If $g \notin M$ then $[g, m] \neq 1$. Thus, there is a free group F and an epimorphism $\varphi : \Gamma \to F$ such that $[\varphi(g), \varphi(m)] \neq 1$. But $1 \neq \varphi(m) \in \varphi(g)^{-1}C_F(\varphi(m))\varphi(g) \cap C_F(\varphi(m))$ and free groups are CSA so $\varphi(g) \in C_F(\varphi(m))$ -a contradiction.

We can now give a proof of Benjamin Baumslag's Theorem.

THEOREM 1.6. *(The Discrimination Theorem [**B**]): Let Γ be a residually free group which is CT. Then Γ is fully residually free.*

Proof: If Γ is abelian the result follows from the fact that the infinite cyclic group is discriminating. So assume that Γ is nonabelian. For a positive integer n call Γ **n-residually free** provided for every nonempty subset $S \subseteq \Gamma \backslash \{1\}$ of cardinal at most n there is a free group F and an epimorphism $\varphi_S : \Gamma \to F$ such that $\varphi_S(\gamma) \neq 1 \ \forall \gamma \in S$. The proof is by induction on n. Clearly Γ is 1-residually free. So suppose the result is true for n. Let $S = \{\gamma_1, ..., \gamma_n, \gamma_{n+1}\} \subseteq \Gamma \backslash \{1\}$. Suppose first that $\langle \gamma_1, ..., \gamma_n, \gamma_{n+1}\rangle$ is nonabelian. We may assume that $[\gamma_n, \gamma_{n+1}] \neq 1$. Then we may replace S with $T = \{\gamma_1, ..., \gamma_{n-1}, [\gamma_n, \gamma_{n+1}]\}$ and the result follows by the inductive hypothesis. Now suppose the γ_i commute in pairs. Since Γ is CSA there is $\gamma \in \Gamma$ such that $[\gamma^{-1}\gamma_n\gamma, \gamma_{n+1}] \neq 1$. Replacing S with $T = \{\gamma_1, ..., \gamma_{n-1}, [\gamma^{-1}\gamma_n\gamma, \gamma_{n+1}]\}$ the result again follows from the inductive hypothesis. ∎

At this point it will be convenient to avail ourselves of the notion of a G-group first introduced in [**BMR1**].

DEFINITION 1.7. Let G be a group. A group Γ containing a distinguished copy of G as a subgroup is a G-**group**. If Γ is a G-group it will be convenient to commit the abuse of viewing the injective homomorphism $G \to \Gamma$ as an inclusion map. With that in mind a **G-subgroup** $\Gamma_0 \leq \Gamma$ is a subgroup such that $G \leq \Gamma_0 \leq \Gamma$ and, if Γ_1 and Γ_2 are G-groups, a G-**homomorphism** $\varphi : \Gamma_1 \to \Gamma_2$ is a homomorphism $\varphi : \Gamma_1 \to \Gamma_2$ which restricts to the identity map on G. If Γ is a G-group then the kernel $Ker(\varphi) \leq \Gamma$ of a G-homomorphism with domain Γ is a G-**ideal** in Γ. Equivalently, a subgroup K normal in Γ which intersects G trivially is a G-ideal in Γ.

The proof of Theorem 1.6 can be modified to show the following.

THEOREM 1.8. *(The Discrimination Theorem with Constants): Let F be a nonabelian free group. Let Γ be an F-group which is separated by a family of* retractions $\Gamma \to F$. If Γ is CT then Γ is discriminated by a family of retractions $\Gamma \to F$.

A little thought will convince one that the proof of Theorem 1.6 can also be modified to prove the following.

THEOREM 1.9. *(The Codiscrimination Theorem)*: Let F be a nonabelian free group. Let Γ be an F-group which is CT. If Γ is separated by some family of retractions $\Gamma \to F$ then Γ is discriminated by every separating family of retractions $\Gamma \to F$.

Using the fact that CT is a necessary condition for full residual freeness we have the following immediate corollary.

THEOREM 1.10. *(The Codiscrimination Corollary)*: Let F be a nonabelian free group. Let Γ be an F-group. If Γ is discriminated by some family of retractions $\Gamma \to F$ then Γ is discriminated by every separating family of retractions $\Gamma \to F$.

In the decades since Benjamin Baumslag's paper [**B**] we have learned that nonabelian fully residually free groups are universally free and the nonabelian finitely generated examples coincide with the finitely generated universally free groups. Moreover, the finitely generated fully residually free groups are precisely the coordinate groups of systems of equations in finitely many unknowns over free groups and have played a pivotal role in the positive solution to the celebrated conjecture of Tarski that the nonabelian free groups are elementarily equivalent. None of this was known in 1967. However, below is a selling point for the concept which was presented in [**B**]. Consider the residually free group $\Gamma = \langle a_1, a_2; \rangle \times \langle b; \rangle$ which is the obstruction to full residual freeness, and let $C = \langle c; \rangle$ be infinite cyclic. We claim that the free product $\Gamma * C$ is not residually free. To see that consider the element

$$\begin{aligned} g &= [[a_2, a_1, c], [b, c]] \\ &= [c, [a_2, a_1]] [c, b][a_2, a_1, c][b, c] \\ &= c^{-1}[a_1, a_2]c[a_2, a_1]c^{-1}b^{-1}cb[a_1, a_2]c^{-1}[a_2, a_1]cb^{-1}c^{-1}bc \neq 1. \end{aligned}$$

If $\Gamma * C$ were residually free there would be a free group F and an epimorphism $\varphi : \Gamma * C \to F$ such that $\varphi(g) \neq 1$. In that event $[\varphi(a_2), \varphi(a_1)] \neq 1$ and $\varphi(b) \neq 1$. But $\varphi(a_1), \varphi(a_2) \in C_F(\varphi(b))$ - contradicting the commutative transitivity of F. Benjamin Baumslag proved in [**B**] that whenever $\Gamma_1 \neq \{1\}$ and $\Gamma_2 \neq \{1\}$ are residually free one has $\Gamma_1 * \Gamma_2$ residually free iff both Γ_1 and Γ_2 are fully residually free. Coupling that with the fact that free products preserve commutative transitivity we deduce the following preservation theorem.

THEOREM 1.11. *(The Free Product Theorem)*: If Γ_1 and Γ_2 are fully residually free then so is their free product $\Gamma_1 * \Gamma_2$.

In this paper we shall treat relativization to the variety $\mathcal{M}$ of all metabelian groups. We shall show by exhibiting explicit counterexamples that the analogues of The Codiscrimination Theorem (Theorem 1.9), The Codiscrimination Corollary (Theorem 1.10) and The Free Product Theorem (Theorem 1.11) are all false when relativized to $\mathcal{M}$. We do not know whether or not the analogue of Benjamin

Baumslag's main theorem (The Discrimination Theorem - a.k.a. Theorem 1.6) remains true when relativized to $\mathcal{M}$.

2. Relativization to the Metabelian Variety

Mal'cev proved in [**Mal**] that free metabelian groups are CT. From this it follows that, exactly as in the absolute case, commutative transitivity is a necessary condition for full residual freeness.

LEMMA 2.1. *If $\Gamma \in \mathcal{M}$ is $\mathcal{M}$-fully residually free then Γ is CT.*

Proof: Suppose not to deduce a contradiction. Let Γ be $\mathcal{M}$-fully residually free and assume that Γ is not CT. Then there are $a, b, c \in \Gamma$ with $b \neq 1$ such that $ab = ba$ and $bc = cb$ but $ac \neq ca$. It follows that there is a group G free in $\mathcal{M}$ and an epimorphism $\varphi : \Gamma \to G$ such that $\varphi(b) \neq 1$ and $\varphi([a, c]) = [\varphi(a), \varphi(c)] \neq 1$. But that contradicts Mal'cev's Theorem since $\varphi(a), \varphi(c) \in C_G(\varphi(b))$. ∎

Remark: The proof actually shows that if $\mathcal{X}$ is any nonempty class of commutative transitive groups closed under isomorphism then any group discriminated by $\mathcal{X}$ must itself be commutative transitive.

In that same paper Mal'cev showed that if G is nonabelian and free in $\mathcal{M}$ then its derived group G' is maximal abelian in G and that, moreover, every other maximal abelian subgroup of G is infinite cyclic.

LEMMA 2.2. *Any infinite cyclic maximal abelian subgroup of a commutative transitive group without elements of order 2 is malnormal in the group.*

Proof: Suppose $M = \langle m \rangle$ is maximal abelian in a CT group without elements of order 2 where m has infinite order. Suppose $g^{-1}Mg \cap M \neq \{1\}$. Since $g^{-1}Mg$ is also maximal abelian we must have $g^{-1}Mg = M$ and the inner automorphism $x \mapsto g^{-1}xg$ must restrict to an automorphism of M. It follows that $g^{-1}mg = m$ or $g^{-1}mg = m^{-1}$. If $g^{-1}mg = m$ then g commutes with m so $g \in M$ and we are finished. Assume to deduce a contradiction that $g^{-1}mg = m^{-1}$. (In particular $g \neq 1$.) Then $g^{-2}mg^2 = g^{-1}m^{-1}g = m$ and so g^2 commutes with m and $g^2 \in M$. Now $g^2 \neq 1$ since we postulated no elements of order 2. Moreover, g commutes with $g^2 \neq 1$ which commutes with m; so, by commutative transitivity, g commutes with m. But that contradicts our assumption that $g^{-1}mg = m^{-1}$. ∎

Since free metabelian groups are torsion free it follows that, if G is nonabelian and free in $\mathcal{M}$, then every maximal abelian subgroup $M \neq G'$ in G is malnormal in G. Recalling that any subgroup containing the derived group is necessarily normal, we have the following result.

LEMMA 2.3. *Let $\Gamma \in \mathcal{M}$ be a nonabelian CT group which is $\mathcal{M}$-residually free. Then*

(1) *$C_\Gamma(\Gamma')$ is maximal abelian and normal in Γ.*
(2) *If $M \neq C_\Gamma(\Gamma')$ is maximal abelian in Γ, then M is malnormal in Γ.*

Proof: (1) Since $\Gamma' \leq C_\Gamma(\Gamma')$, we have that $C_\Gamma(\Gamma')$ is normal in Γ. By commutative transitivity it is maximal abelian in Γ since $\Gamma' \neq \{1\}$ is a nontrivial abelian subgroup of Γ.

(2) Assume the contrary to deduce a contradiction. Suppose $g^{-1}Mg \cap M \neq \{1\}$ but $g \notin M$. By commutative transitivity g cannot commute with any nontrivial element of M. In particular, if $1 \neq m \in g^{-1}Mg \cap M$, then g cannot commute with m. So $[g,m] \neq 1$. Now M and $\Gamma' \leq C_\Gamma(\Gamma')$ intersect trivially since Γ is CT. Hence, m cannot commute with $[g,m]$ and thus $[g,m,m] \neq 1$. It follows that there is a group G free in $\mathcal{M}$ and an epimorphism $\varphi : \Gamma \to G$ such that $\varphi([g,m,m]) = [[\varphi(g),\varphi(m)],\varphi(m)] \neq 1$. Thus, $[\varphi(g),\varphi(m)]$ and $\varphi(m)$ are nontrivial elements of G. Moreover, since $\varphi(m)$ does not commute with $[\varphi(g),\varphi(m)]$, it cannot lie in the maximal abelian subgroup G' of G. Thus, $C_G(\varphi(m))$ is malnormal in G. But from $m \in g^{-1}Mg \cap M$ we get $1 \neq \varphi(m) \in \varphi(g)^{-1}C_G(\varphi(m))\varphi(g) \cap C_G(\varphi(m))$ and so $\varphi(g) \in C_G(\varphi(m))$ and $\varphi(g)$ commutes with $\varphi(m)$ - contrary to the assumption that $[\varphi(g),\varphi(m)] \neq 1$. ∎

From (1) we see that if $\Gamma \in \mathcal{M}$ is a nonabelian group which is CT and $\mathcal{M}$-residually free, then the maximal abelian subgroup $C_\Gamma(\Gamma')$ cannot be malnormal in Γ. Thus such Γ are not CSA and Benjamin Baumslag's proof breaks down when one tries to relativize it to $\mathcal{M}$.

3. A Counterexample to the Codiscrimination Theorem and Corollary

If A is a set of symbols and R is a subset of the set $(A \cup A^{-1})^*$ of all words on the alphabet $A \cup A^{-1}$ then the **relative presentation** $\langle A; R \rangle_\mathcal{M}$ defines the quotient of the metabelian group free on A in $\mathcal{M}$ modulo the normal closure of R. Returning to the variety $\mathcal{O}$ of all groups, if Γ is a CT group and M is maximal abelian in Γ, then the group presented as

$$\langle \Gamma, z; rel(\Gamma), z^{-1}mz = m \forall m \in M \rangle$$

is the **free rank 1 extension of the centralizer M**. Free rank 1 centralizer extensions preserve full residual freeness. We shall see in the subsequent section that the analogue of that result is false when relativized to the variety $\mathcal{M}$ of all metabelian groups.

Let G be free on $\{\alpha_1, \alpha_2\}$ in $\mathcal{M}$. Equivalently, G has relative presentation $\langle \alpha_1, \alpha_2; \rangle_\mathcal{M}$. Suppose $g \in G \backslash \{1\}$ is not a proper power in G. Let

$$\widehat{G} = \langle \alpha_1, \alpha_2, z; \ z^{-1}cz = c \forall c \in C_G(g) \rangle_\mathcal{M}$$

be the free rank 1 extension of $C_G(g)$ relative to $\mathcal{M}$. In [**GLS**] the authors introduce a matrix representation ρ of $\widehat{G}$ which is based on the classical Magnus representation of G. We here describe ρ in the case $g \in G' \backslash \{1\}$ is not a proper power. Suppose A is a multiplicatively written free abelian group with basis $\{\overline{\alpha_1}, \overline{\alpha_2}\}$. Then, from [**Mag**], we get a faithful representation of G into the group of matrices of the form

$$\begin{bmatrix} a_{11} & a_{12} \\ 0 & 1 \end{bmatrix}$$

where $a_{11} \in A$ and a_{12} lies in the free rank 2 module over the integral group-ring $\mathbb{Z}A$ with $\mathbb{Z}A$-basis $\{t_1, t_2\}$ via

$$\alpha_i \mapsto \begin{bmatrix} \overline{\alpha_i} & t_i \\ 0 & 1 \end{bmatrix}, i = 1, 2.$$

In this representation g must have the form $\begin{bmatrix} 1 & L \\ 0 & 1 \end{bmatrix}$ for some $L \neq 0$. Now let θ be an indeterminate over the ring $\mathbb{Z}$ of integers. We get a matrix representation ρ of $\widehat{G}$ into the group of all matrices of the form

$$\begin{bmatrix} a_{11} & a_{12} \\ 0 & 1 \end{bmatrix}$$

where $a_{11} \in A$ and a_{12} lies in the free rank 2 module over the group-ring $\mathbb{Z}[\theta]A$ with $\mathbb{Z}[\theta]A$-basis $\{t_1, t_2\}$ via

$$\alpha_i \mapsto \begin{bmatrix} \overline{\alpha_i} & t_i \\ 0 & 1 \end{bmatrix}, i = 1, 2,$$

$$z \mapsto \begin{bmatrix} 1 & \theta L \\ 0 & 1 \end{bmatrix}.$$

It is shown in [**GLS**] that the family of evaluation maps $\mathbb{Z}[\theta] \to \mathbb{Z}$, $f(\theta) \mapsto f(n)$ ($n \in \mathbb{Z}$) induces a discriminating family of retractions $\rho(\widehat{G}) \to G$ when we identify G with its Magnus matrix representation. Hence, $\rho(\widehat{G})$ is $\mathcal{M}$-fully residually free. Now we view $\widehat{G}$ as a G-group since G embeds in it. Note that if $\varphi: \widehat{G} \to H$ is any surjective G-homomorphism then $H = \langle G, \varphi(z) \rangle$ so any G-homomorphism $H \to G$ is determined by its effect on $\varphi(z)$. If, for $n \in \mathbb{Z}$, the assignment $\varphi(z) \mapsto g^n$ extends to a G-homomorphism $\psi: H \to G$, then we say that ψ is a $g^{\mathbb{Z}}$-**map**. Let κ be the family of all G-ideals K in $\widehat{G}$ such that $\widehat{G}/K$ is separated by a family of $g^{\mathbb{Z}}$-maps. Let

$$Rad(\widehat{G}) = \bigcap_{K \in \kappa} K.$$

It is easy to see that any $g^{\mathbb{Z}}$-map $\widehat{G}/Rad((\widehat{G}) \to G$ factors through $\rho(\widehat{G})$ and that $\widehat{G}/Rad((\widehat{G})$ is itself separated by $g^{\mathbb{Z}}$-maps. $\widehat{G}/Rad((\widehat{G})$ is the unique (up to G-isomorphism) largest G-homomorphic image of $\widehat{G}$ separated by a family of $g^{\mathbb{Z}}$-maps. It is called in [**GLS**] the $g^{\mathbb{Z}}$-**residualization of** $\widehat{G}$. If $x \notin Rad((\widehat{G})$ then there is a $g^{\mathbb{Z}}$-map $\widehat{G}/Rad((\widehat{G}) \to G$ which doesn't annihilate $xRad((\widehat{G})$. The map factors through $\rho(\widehat{G})$ so the image $xKer(\rho)$ of $xRad((\widehat{G})$ in $\rho(\widehat{G})$ is nontrivial. It follows that $Ker(\widehat{G}/Rad((\widehat{G}) \to \rho(\widehat{G}))$ is trivial and $\widehat{G}/Rad((\widehat{G})$ is G-isomorphic to $\rho(\widehat{G})$.

Now, for each fixed prime p, G is residually a finite p-group. Moreover, $\rho(\widehat{G})$ is separated (even discriminated) by $g^{\mathbb{Z}}$-maps $\rho(\widehat{G}) \to G$. It follows that $\rho(\widehat{G})$ is also, for each fixed prime p, residually a finite p-group and is, in particular, Hopfian. It then easily follows from the Hopficity of $\rho(\widehat{G})$ that

$$Ker(\rho) = Rad(\widehat{G}).$$

Now let $B = \langle \beta_1, \beta_2, \beta_3; \ \rangle_{\mathcal{M}}$ and let H be the subgroup of the direct square G^2 generated by $(\alpha_1, \alpha_1), (\alpha_2, \alpha_2)$, and $(g, 1)$ where $g = [\alpha_2, \alpha_1]$. Let $\rho_i : G^2 \to G$ be projection onto the i-th coordinate, $i = 1, 2$. Given $w(\alpha_1, \alpha_2) \in G$ surely $w((\alpha_1, \alpha_1), (\alpha_2, \alpha_2)) = (w(\alpha_1, \alpha_2), w(\alpha_1, \alpha_2))$ projects to $w(\alpha_1, \alpha_2)$ under each of ρ_1 and ρ_2. Thus H is a subdirect power of G and so is $\mathcal{M}$-residually free. Now let $w(\beta_1, \beta_2, \beta_3) \in B$. Suppose

$$w(\beta_1, \beta_2, \beta_3) \equiv \beta_1^{e_1} \beta_2^{e_2} \beta_3^{e_3} \pmod{B'}.$$

Consider the element
$$(h_1, h_2) = w((\alpha_1, \alpha_1), (\alpha_2, \alpha_2), (g, 1)) = (w(\alpha_1, \alpha_2, g), w(\alpha_1, \alpha_2, 1))$$
of H. We claim that $h_1 \equiv h_2 (\mod G')$. This is so since
$$w(\alpha_1, \alpha_2, g) \equiv \alpha_1^{e_1} \alpha_2^{e_2} g^{e_3} \equiv \alpha_1^{e_1} \alpha_2^{e_2} \equiv w(\alpha_1, \alpha_2, 1) (\mod G').$$

We next claim that H is CT. Suppose $a = (a_1, a_2)$ commutes with $b = (b_1, b_2) \neq 1$ and b commutes with $c = (c_1, c_2)$ in H. Suppose first that $b_1 \neq 1$ and $b_2 \neq 1$. Then from commutative transitivity in each coordinate we have that a commutes with c; so, we have to treat the case where exactly one of the coordinates of b is nontrivial. We may assume that $b_2 = 1$ so that $b_1 \neq 1$. But $b_1 \equiv b_2 \equiv 1 (\mod G')$. Since a_1 and c_1 commute with $b_1 \in G' \setminus \{1\}$ they must each lie in G'. Then $a_2 \equiv a_1 \equiv 1 (\mod G')$ and $c_2 \equiv c_1 \equiv 1 (\mod G')$ so a_2 and c_2 must also commute and we are finished. A similar argument allows us to pull through if $b_1 = 1$ so that $b_2 \neq 1$.

Now H is a G-group since it contains the diagonal $\{(x, x) \in G^2 \mid x \in G\}$. Thus $(g, g) \in H$ and $(1, g) = (g, 1)^{-1} \cdot (g, g) \in H$. Letting $\Phi = \{\varphi_1, \varphi_2\}$ where $\varphi_i = \rho_i |_H, i = 1, 2$, we see that Φ is a separating family of retractions $H \to G$ when we identify G with its image in H. Now $\varphi_1((1, g)) = 1$ and $\varphi_2((g, 1)) = 1$ so that Φ cannot discriminate H. Thus H is a counterexample to the analogue of the Codiscrimination Theorem (Theorem 1.9).

We shall next show that H is G-isomorphic to $\rho(\widehat{G})$. Assuming that for the moment, H would be discriminated by a family of retractions $H \to G$ so that H is also a counterexample to the analogue of the Codiscrimination Corollary (Theorem 1.10). The strategy is as follows. Suppose we could show that H is separated by a family of $g^{\mathbb{Z}}$-maps $H \to G$ determined by $(g, 1) \mapsto g^n$ as n varies over $\mathbb{Z}$. Then $H \in \kappa$ and we could mimic the proof that $\widehat{G}/Rad((\widehat{G})$ is G-isomorphic to $\rho(\widehat{G})$ to show that $\widehat{G}/Rad((\widehat{G})$ is G-isomorphic to H.

To that end, for each fixed $n \in \mathbb{Z}$, let K_n be the normal closure in H of the element $(g, 1) \cdot (g, g)^{-n} = (g^{1-n}, g^{-n})$. If we could show that K_n intersects the diagonal $\{(x, x) \in G^2 \mid x \in G\}$ trivially, then the family of canonic epimorphisms $H \to H/K_n$ would be a family of $g^{\mathbb{Z}}$-maps $H \to G, (g, 1) \to g^n$ after making the obvious identifications. Furthermore, the subfamily $\{(g, 1) \to g^1 = g, (g, 1) \to g^0 = 1\}$ coincides with $\{\varphi_1 = \rho_1 \mid H, \varphi_2 = \rho_2 \mid_H\}$ so separates H. We could represent G^2 as a group of ordered pairs

$$\left(\begin{bmatrix} a_{11} & a_{12} \\ 0 & 1 \end{bmatrix}, \begin{bmatrix} b_{11} & b_{12} \\ 0 & 1 \end{bmatrix} \right)$$

of Magnus matrices; however, it will be more convenient to replace the ordered pair of matrices with their "tensor product"

$$\begin{bmatrix} (a_{11}, b_{11}) & (a_{12}, b_{12}) \\ 0 & 1 \end{bmatrix}.$$

Clearly we get an isomorphic group in this fashion. Now if $h = (h_1, h_2) \in H$, then $h_1 \equiv h_2 (\mod G')$. Since the upper left hand entry of a Magnus matrix depends on the coset modulo G' only, an element of H has the form

$$\begin{bmatrix} (a, a) & (L_1, L_2) \\ 0 & 1 \end{bmatrix}.$$

This will lie in the diagonal iff $L_1 = L_2$. Suppose $\begin{bmatrix} 1 & L \\ 0 & 1 \end{bmatrix}$ represents g (so $L \neq 0$). If $\varepsilon = \pm 1$, then $(g^{1-n}, g^{-n})^\varepsilon$ is represented by $\begin{bmatrix} 1 & (\varepsilon(1-n)L, -\varepsilon L) \\ 0 & 1 \end{bmatrix}$. Conjugating by an element $h = \begin{bmatrix} (a,a) & (L_1, L_2) \\ 0 & 1 \end{bmatrix}$ of H, we get

$$h^{-1}(g^{1-n}, g^{-m})^\varepsilon h = \begin{bmatrix} 1 & (\varepsilon(1-n)a^{-1}L, -\varepsilon n a^{-1}L) \\ 0 & 1 \end{bmatrix}.$$

A product $\prod_{i=1}^{k} h_i^{-1}(g^{1-n}, g^{-n})^\varepsilon h_i$, of such conjugates is then

$$\begin{bmatrix} 1 & \left(((1-n)\sum_{i=1}^{k} \varepsilon_i a_i^{-1})L, (-n \sum_{i=1}^{k} \varepsilon_i a_i^{-1})L \right) \\ 0 & 1 \end{bmatrix}.$$

If this lies on the diagonal, then $((1-n)\sum_{i=1}^{k} \varepsilon_i a_i^{-1})L = (-n\sum_{i=1}^{k} \varepsilon_i a_i^{-1})L$, from which we get $\left(\sum_{i=1}^{k} \varepsilon_i a_i^{-1} \right) L = 0$ - forcing the above matrix to be $\begin{bmatrix} 1 & 0 \\ 0 & 1 \end{bmatrix} = 1$ and we are finished!

4. Failure of Free Constructions

Let $\coprod$ be the coproduct in the category $\mathcal{M}$. That is $\coprod$ is the relatively free product as described in Section 1.8 of Hanna Neumann's book [N]. Let $G = \langle \alpha_1, \alpha_2; \ \rangle_\mathcal{M}$ and let $G^+ = G \coprod \langle \alpha_3; \ \rangle = \langle \alpha_1, \alpha_2, \alpha_3; \ \rangle_\mathcal{M}$. We may view $G^+ = G \coprod \langle \alpha_3; \ \rangle$ as a G-group. Let $g = [\alpha_2, \alpha_1]$ and let

$$\widehat{G} = \langle \alpha_1, \alpha_2, z; \ z^{-1}cz = c \forall c \in G' \rangle_\mathcal{M}$$

be the free rank 1 extension of $C_G(g) = G'$ relative to $\mathcal{M}$. We get a G-epimorphism $\varphi : G^+ \to \widehat{G}$ determined by $\alpha_3 \mapsto z$. $K = Ker(\varphi)$ is a G-ideal and may be characterized as the normal closure in G^+ of

$$\{[c, \alpha_3] \mid c \in G'\}.$$

We claim that $\widehat{G}$ is not $\mathcal{M}$-fully residually free nor even CT. To see that let Ω be free on the set $\{\omega_1, \omega_2, \omega_3\}$ in the variety $\mathcal{N}_3$ of all groups nilpotent of class at most 3. Then Ω is metabelian so we get an epimorphism $\psi : G^+ \to \Omega$ determined by $\alpha_i \mapsto \omega_i$, $i = 1, 2, 3$. Let $Z = \langle [\omega_2, \omega_1, \omega_3] \rangle$. Z is central in Ω. Let $\pi : \Omega \to \Omega/Z$ be the canonical epimorphism. Composing (where we write our maps to the left of their arguments) we get an epimorphism $\pi\psi : G^+ \to \Omega/Z$. We claim that $K \leq Ker(\pi\psi)$ so that we get an induced epimorphism $\lambda : \widehat{G} \to \Omega/Z$. To see that let $\Omega_0 = \langle \omega_1, \omega_2 \rangle$. Then ψ maps G' onto Ω_0' and maps α_3 to ω_3. Every element of Ω_0' is congruent modulo the center of Ω to an integral power of $[\omega_2, \omega_1]$. Thus, if $c \in G'$, then for some $n \in \mathbb{Z}$, $\psi([c, \alpha_3]) = [\omega_2, \omega_1; \omega_3]^n \in Z$ and the claim is established.

We next claim that z does not commute with $[z, \alpha_1]$. If it did then $[z, \alpha_1, z]$ would be trivial and so $[\lambda(z), \lambda(\alpha_1), \lambda(z)]$ would be trivial in Ω/Z. But
$$\psi([\alpha_3, \alpha_1, \alpha_3]) = [\omega_3, \omega_1.\omega_3] \notin Z$$
by uniqueness of the basic commutator representation. Hence $[\lambda(z), \lambda(\alpha_1), \lambda(z)] = \pi\psi([\alpha_3, \alpha_1, \alpha_3]) \neq 1$ in Ω/Z. Therefore, $[z, \alpha_1, z] \neq 1$ in $\widehat{G}$ and z does not commute with $[z, \alpha_1]$. But that contradicts commutative transitivity since the commutators $g = [\alpha_2, \alpha_1]$ and $[z, \alpha_1]$ commute in the metabelian group $\widehat{G}$ and z commutes with g. Thus, a relativized free rank 1 centralizer extension of a group $\mathcal{M}$-fully residually free need not be $\mathcal{M}$-fully residually free nor even CT.

We next claim that the coproduct of two $\mathcal{M}$-fully residually free groups need not be $\mathcal{M}$-fully residually free nor even CT. Let G and G^+ be as above and let $g = [\alpha_2, \alpha_1]$ and $\widehat{G}$ be as above. Let H be the subgroup of G^2 generated by $(\alpha_1, \alpha_1), (\alpha_2, \alpha_2)$ and $(g, 1)$ as in the previous section. We shall show that $G \coprod H$ is not CT. G is $\mathcal{M}$-free so certainly $\mathcal{M}$-fully residually free. Recall that H is G-isomorphic to $\rho(\widehat{G})$ so is $\mathcal{M}$-fully residually free. Let $\gamma_3(G) = [G', G]$ be the third term of the lower central series of G. Recall that if $(h_1, h_2) \in H$ then $h_1 \equiv h_2 \pmod{G'}$. Let $N = \{(h_1, h_2) \in H \mid h_1 \equiv h_2 \pmod{\gamma_3(G)}\}$. It is straightforward to show that N is a subgroup of H. We claim that N is normal in H. It will suffice to show that N is invariant under conjugation by the generators $(\alpha_1, \alpha_1), (\alpha_2, \alpha_2)$ and $(g, 1)$ and their inverses. Let $\varepsilon = \pm 1$. If $h_1 \equiv h_2 \pmod{\gamma_3(G)}$ then certainly $\alpha_i^{-\varepsilon} h_1 \alpha_i^{\varepsilon} \equiv \alpha_i^{-\varepsilon} h_2 \alpha_i^{\varepsilon} \pmod{\gamma_3(G)}$, $i = 1, 2$. Since $g = [\alpha_2, \alpha_1]$ is central modulo $\gamma_3(G)$ we get $g^{-\varepsilon} h_1 g^{\varepsilon} \equiv h_1 \equiv h_2 \equiv 1^{-\varepsilon} h_2 1^{\varepsilon} \pmod{\gamma_3(G)}$ whenever $h_1 \equiv h_2 \pmod{\gamma_3(G)}$. The claim is therefore verified.

We next claim that H/N is infinite cyclic. Clearly $(\alpha_1, \alpha_1) \in N$ and $(\alpha_2, \alpha_2) \in N$ so that H/N is cyclic generated by $(g, 1)N$. Now $G/\gamma_3(G)$ is torsion free so, for every integer $n > 0$, $g^n \neq 1 \pmod{\gamma_3(G)}$ and therefore $(g, 1)N$ has infinite order. By the universal property characterizing the coproduct, any pair of homomorphisms $\sigma_1 : G \to G^+$ and $\sigma_2 : H \to G^+$ will extend uniquely to a homomorphism $\sigma : G \coprod H \to G^+$. Let $\sigma_1 : G \to G^+$ be the inclusion map and let $\sigma_2 : H \to G^+$ be the composition of the maps
$$H \to H/N \to \langle \alpha_3 \rangle \to G^+.$$
Observe that $(g, g) \in H \backslash \{1\}$; furthermore, $(g, 1) \in H$ commutes with (g, g) in H so certainly also in $G \coprod H$.
$$(g, g) = ([\alpha_2, \alpha_1], [\alpha_2, \alpha_1]) = [(\alpha_2, \alpha_2), (\alpha_1, \alpha_1)]$$
lies in $H' \leq (G \coprod H)'$. Now $g \in G' \leq (G \coprod H)'$. Since $G \coprod H$ is metabelian g commutes with (g, g) in $G \coprod H$. Now $\sigma(g) = \sigma_1(g) = g = [\alpha_2, \alpha_1]$ in G^+. Since $\sigma(g)$ and $\sigma((g, 1))$ do not commute in G^+, g and $(g, 1)$ do not commute in $G \coprod H$ and $G \coprod H$ is not CT. Thus the analogue of the Free Product Theorem (Theorem 1.11) is false when relativized to $\mathcal{M}$.

5. Epilogue

Let B be the multiplicatively written free $\mathbb{Z}[\theta]$-module with basis $\{\overline{\alpha_1}, \overline{\alpha_2}\}$ where θ is an indeterminate over the ring $\mathbb{Z}$ of integers. The authors showed in

[**GLS**] that the group-ring $\mathbb{Z}[\theta]B$ is an integral domain. Let Φ be its field of quotients. Consider the group of all matrices of the form $\begin{bmatrix} a_{11} & a_{12} \\ 0 & 1 \end{bmatrix}$ where $a_{11} \in B$ and a_{12} lies in the vector space over Φ with basis $\{t_1, t_2\}$. We introduce exponents from $Z[\theta]$ into this group by defining

$$\begin{bmatrix} 1 & a_{12} \\ 0 & 1 \end{bmatrix}^{f(\theta)} = \begin{bmatrix} 1 & f(\theta)a_{12} \\ 0 & 1 \end{bmatrix}$$

and, if $a_{11} \neq 1$,

$$\begin{bmatrix} a_{11} & a_{12} \\ 0 & 1 \end{bmatrix}^{f(\theta)} = \begin{bmatrix} a_{11}^{f(\theta)} & \left(\frac{a_{11}^{f(\theta)}-1}{a_{11}-1}\right) a_{12} \\ 0 & 1 \end{bmatrix}.$$

In view of Magnus' faithful matrix representation of $G = \langle \alpha_1, \alpha_2; \rangle_\mathcal{M}$, we may identify the subgroup generated by

$$\begin{bmatrix} \overline{\alpha_i} & t_i \\ 0 & 1 \end{bmatrix},$$

$i = 1, 2$ with G. Let $G^{\mathbb{Z}[\theta]}$ be the $\mathbb{Z}[\theta]$-subgroup, generated as a $\mathbb{Z}[\theta]$-group by G. We have that

$$G^{\mathbb{Z}[\theta]} = \bigcup_{n=0}^{\infty} G_n$$

is the union of an increasing chain $G_0 \leq G_1 \leq ... \leq G_n \leq ...$ of ordinary subgroups. Here the **level** n subgroup G_n is defined inductively by $G_0 = G$ and G_{k+1} is the subgroup generated as a group by all elements of the form $x^{f(\theta)}$ as x varies over G_k and $f(\theta)$ varies over $\mathbb{Z}[\theta]$. The family of evaluation maps $\mathbb{Z}[\theta] \to \mathbb{Z}$, $\theta \mapsto n$, will induce a discriminating family of group retractions $G_1 \to G$. (For level higher than the first, one could conceivably have a factor of the form $x^{f(\theta)}-1$ in the denominator of the upper right hand entry. If n is a root of $f(\theta)$, then the evaluation map $\theta \mapsto n$ will not induce a group homomorphism in any obvious natural way.) Thus, the G-subgroups of $G_1 \leq G^{\mathbb{Z}[\theta]}$ provide a rich trove of examples of $\mathcal{M}$-fully residually free groups. $G^{\mathbb{Z}[\theta]}$ looks suspiciously like the analogue in $\mathcal{M}$ of Lyndon's free exponential group $F^{\mathbb{Z}[\theta]}$.

For the remainder of this section G will denote an arbitrary group. We outline below a possible approach to attempt a proof of the analogue of The Discrimination Theorem with Constants (Theorem 1.8) relativized to $\mathcal{M}$- at least in the case of finite generation.

Let $r \geq 2$ be an integer and let $F_r(\mathcal{M})$ be a group free of rank r relative to $\mathcal{M}$. Let Γ be a finitely generated metabelian $F_r(\mathcal{M})$-group which is commutative transitive and assume that Γ is separated by a family of retractions $\Gamma \to F_r(\mathcal{M})$. Following [**BMR1**] a G-group H is G-**equationally Noetherian** provided every system of equations over H in any fixed finite number of variables admitting constants from G is equivalent to some finite subsystem. A group H is **equationally Noetherian** provided it is $\{1\}$-equationally Noetherian. It was observed in [**BMR1**] that linear groups are equationally Noetherian. But free metabelian groups of finite rank are linear as per the following.

THEOREM 5.1. $F_r(\mathcal{M}) = \langle \alpha_1, ..., \alpha_r; \rangle$ *is linear.*

Proof. Let A_r be the rank r multiplicatively written free abelian group with basis $\{\alpha_1, ..., \alpha_r\}$. Since every nontrivial subgroup of A_r has the infinite cyclic group as a homomorphic image, i.e., it is locally indicable, it follows from a classical result of Graham Higman [**H**] that the integral group-ring $\mathbb{Z}A_r$ is an integral domain. Let $\mathcal{Q}(\mathbb{Z}A_r)$ be its field of fractions. Let t be an indeterminate over $\mathcal{Q}(\mathbb{Z}A_r)$ and let k be the transcendental field extension $\mathcal{Q}(\mathbb{Z}A_r)(t)$. Then
$$\{1, t, ..., t^{r-1}\}$$
is linearly independent over $\mathcal{Q}(\mathbb{Z}A_r)$. It follows from the Magnus' matrix representation that the assignment
$$\alpha_j \mapsto \begin{bmatrix} \overline{\alpha_j} & t^{j-1} \\ 0 & 1 \end{bmatrix}, \; j = 1, ..., r$$
extends to a faithful representation of $F_r(\mathcal{M})$ in $GL_2(k)$. ∎

REMARK 5.2. We note that the above is not a new result. See, for example, p. 26 of WehrFritz [**W**] where it is states that Levič and V.N. Remeslennikov have shown that any finitely generated, torion-free, metabelian group has a faithful representation over the complex numbers.

Hence $F_r(\mathcal{M})$ is equationally Noetherian. We presume the stronger condition that $F_r(\mathcal{M})$ is $F_r(\mathcal{M})$-equationally Noetherian is also correct. Modulo that presumption, consider Corollary 5, p. 251 of [**MR2**] which we quote verbatim:

> Let H and K be G-groups and suppose that at least one of them is G-equationally Noetherian. Then H is G-universally equivalent to K if and only if H is locally G-discriminated by K and K is locally G-discriminated by H.

Here G-universal equivalence means satisfying precisely the same universal sentences in the first order language with equality appropriate for groups and admitting constants from G. Thus, modulo the presumption, it will suffice to show that Γ is $F_r(\mathcal{M})$-universally equivalent to $F_r(\mathcal{M})$. Since $F_r(\mathcal{M})$ is a subgroup of Γ, we are halfway there because universal sentences are preserved in subgroups. So we are left with showing that every universal sentence true in $F_r(\mathcal{M})$ must also be true in Γ. Recall that a **quasi-identity** is a universal sentence of the form
$$\forall \vec{x} \left(\bigwedge_i (u_i(\vec{x}) = 1) \to (v(\vec{x}) = 1) \right).$$

It is well-known and easy to prove that quasi-identities are preserved in subdirect products. Since Γ is separated by a family of retractions $\Gamma \to F_r(\mathcal{M})$ it is isomorphic to a subdirect power of $F_r(\mathcal{M})$. Thus every quasi-identity true in $F_r(\mathcal{M})$ must also be true in Γ. We are getting closer. We need to know a set of universal sentences that, taken together with the quasi-identities true in $F_r(\mathcal{M})$, axiomatizes the universal theory of $F_r(\mathcal{M})$. Here again the material in [**MR2**] comes to the rescue. Section 7.2 (specifically the material in pp. 263 - 265) gives such a set (when the models are restricted to $F_r(\mathcal{M})$-groups) there denoted $\Delta_{F_r(\mathcal{M})}$ in our situation. The bottom line is that to prove that Γ is discriminated by a family of retractions $\Gamma \to F_r(\mathcal{M})$ it will suffice to establish the veracity of the following two assertions:

(1) $F_r(\mathcal{M})$ is $F_r(\mathcal{M})$-equationally Noetherian.

(2) The set of universal sentences $\Delta_{F_r(\mathcal{M})}$ is implied by commutative transitivity.

6. Questions

In the Questions below a commutative transitive metabelian group Γ will be called $\mathcal{M}$-**regular** provided it is nonabelian and $C_\Gamma(\Gamma') = \Gamma'$.

(1) Must the analogue of Benjamin Baumslag's main theorem (The Discrimination Theorem, a.k.a. Theorem 1.6) be true when relativized to $\mathcal{M}$?

(2) We know that the relative centralizer extension of an $\mathcal{M}$-fully residually free group need not be CT. Must it be $\mathcal{M}$-residually free?

(3) If Γ is $\mathcal{M}$-regular and $g \in \Gamma \backslash \Gamma'$ must the relative centralizer extension
$$\widehat{\Gamma} = \langle \Gamma, z;\ rel(\Gamma),\ z^{-1}mz = m \forall m \in C_\Gamma(g) \rangle_\mathcal{M}$$
also be $\mathcal{M}$-regular?

(4) If Γ is $\mathcal{M}$-regular and $\mathcal{M}$-fully residually free and $g \in \Gamma \backslash \Gamma'$ must the relative centralizer extension
$$\widehat{\Gamma} = \langle \Gamma, z;\ rel(\Gamma),\ z^{-1}mz = m \forall m \in C_\Gamma(g) \rangle_\mathcal{M}$$
also be $\mathcal{M}$-regular and $\mathcal{M}$-fully residually free?

(5) We know the coproduct of two $\mathcal{M}$-fully residually free groups need not be CT. Must it be $\mathcal{M}$-residually free?

(6) Must the coproduct of two $\mathcal{M}$-regular groups also be $\mathcal{M}$-regular?

(7) Must the coproduct of two $\mathcal{M}$-regular, $\mathcal{M}$-fully residually free groups also be $\mathcal{M}$-regular and $\mathcal{M}$-fully residually free?

(8) Must every finitely generated $\mathcal{M}$-fully residually free group embed in $G^{\mathbb{Z}[\theta]}$?

(9) Is $G^{\mathbb{Z}[\theta]}$ a free object in the category $\mathcal{M}^{\mathbb{Z}[\theta]}$ of all metabelian groups admitting exponents from the ring $\mathbb{Z}[\theta]$?

(10) Is $G^{\mathbb{Z}[\theta]}$ the tensor $\mathbb{Z}[\theta]$-completion of G in the category $\mathcal{O}$ of all groups?

(11) Is $G^{\mathbb{Z}[\theta]}$ the tensor $\mathbb{Z}[\theta]$-completion of G in the category $\mathcal{M}$ of all metabelian groups?

(12) Is $F_r(\mathcal{M})$ an $F_r(\mathcal{M})$-equationally Noetherian group?

(13) Does commutative transitivity imply the sentences $\Delta_{F_r(\mathcal{M})}$?

References

[B]   B. Baumslag, "Residually free groups," Proc. London Math. Soc. (3), 17 (1967), 402-418. MR0215903 (35:6738)

[BMR1] G. Baumslag, A.G. Myasnikov and V.N. Remeslennikov, "Algeraic geometry over groups. I. Algebraic sets and ideal theory," J. Algebra 219 (1999), 16 - 79. MR1707663 (2000j:14003)

[BMR2] G. Baumslag, A.G. Myasnikov and V.N. Remeslennikov, "Discriminating and co-discriminating groups," J. GroupTheory 3 (2000), 467 - 479. MR1790342 (2001i:20062)

[GLS]   A.M. Gaglione, S. Lipschutz and D. Spellman, "Big powers and metabelian groups," Preprint.

[H]   G. Higman, "The units of group-rings," Proc. London Math. Soc (2) 46 (1940), 231-248. MR0002137 (2:5b)

[Mag]   W. Magnus, "On a theorem of Marshall Hall," Ann. of Math. 40 (1939), 764 - 768. MR0000262 (1:44b)

[Mal]   A.I. Mal'cev, "On free soluble groups," Dokl. Nauk. SSSR 130 (1960), 495 - 498 (Russian); translated as Soviet Math. Dokl. 1, 65 - 68. MR0117274 (22:8056)

[MKS] W. Magnus, A. Karrass and D. Solitar, Combinatorial Group Theory, John - Wiley, New York, 1966.
[MR1] A.G. Myasnikov and V.N. Remeslennikov, "Exponential groups. II. Extensions of centralizers and tensor completion of CSA - groups," Internat. J. Comput. 6, No. 6 (1996), 687 - 711. MR1421886 (97j:20039)
[MR2] A.G. Myasnikov and V.N. Remeslennikov, "Algebraic geometry over groups. II. Logical foundations," J. Algebra 234 (2000), 225 - 276.G. Higman MR1799485 (2001i:14001)
[N] H. Neumann, Varieties of Groups, Springer - Verlag, New York, 1967. MR0215899 (35:6734)
[W] B.A.F. Wehrfritz, Infinite Linear Groups, Springer - Verlag, New York, 1973. MR0335656 (49:436)

DEPARTMENT OF MATHEMATICS, U.S. NAVAL ACADEMY, ANNAPOLIS, MARYLAND 21402
E-mail address: amg@usna.edu
URL: http://www.usna.edu

DEPARTMENT OF MATHEMATICS, TEMPLE UNIVERSITY, PHILADELPHIA, PENNSYLVANIA 19132
E-mail address: seymour@temple.edu

5147 WHITAKER AVENUE, PHILADELPHIA, PENNSYLVANIA 19124

A Secret Sharing Scheme Based on Group Presentations and the Word Problem

Maggie Habeeb, Delaram Kahrobaei, and Vladimir Shpilrain

ABSTRACT. A (t, n)-threshold secret sharing scheme is a method to distribute a secret among n participants in such a way that any t participants can recover the secret, but no $t - 1$ participants can. In this paper, we propose two secret sharing schemes using non-abelian groups. One scheme is the special case where all the participants must get together to recover the secret. The other one is a (t, n)-threshold scheme that is a combination of Shamir's scheme and the group-theoretic scheme proposed in this paper.

1. Introduction

Suppose one would like to distribute a secret among n participants in such a way that any t of the participants can recover the secret, but no group of $t - 1$ participants can. A method that allows one to do this is called a (t, n)-threshold scheme. Shamir [7] back in 1979 offered a (t, n)-threshold scheme that utilizes polynomial interpolation. In Shamir's scheme the secret is an element $x \in \mathbb{Z}_p$. In order to distribute the secret, the dealer begins by choosing a polynomial f of degree $t - 1$ such that $f(0) = x$. Then he sends the value $x_i = f(i)$ secretly to participant P_i. In order for the participants to recover the secret, they use polynomial interpolation to recover f and hence the secret $f(0)$. In this setting, no $t - 1$ participants can gain any information about the secret while any t of them can (see [7] or [8]).

The field of non-commutative group-based cryptography has produced many new cryptographic protocols over the last decade or so. We refer an interested reader to [4] or [1] for a survey of developments in this area. Recently, Panagopoulos [6] suggested a (t, n)-threshold scheme using group presentations and the word problem. His scheme is two-stage: at the first stage, long-term private information (defining relations of groups) is distributed to all participants over secure channels. At the second stage, shares of the actual secret are distributed to participants over open channels. Thus, an advantage of Panagopoulos' scheme over Shamir's scheme is that the actual secret need not be distributed over secure channels. This is obviously useful in various real-life scenarios. On the other hand, his scheme is not quite practical because in his scheme, it takes time exponential in n to distribute the shares of a secret to n participants.

2010 *Mathematics Subject Classification.* Primary 68P25, 94A60; Secondary 20F05, 20F10.

©2012 American Mathematical Society

In this paper, we present two practical secret sharing schemes based on group presentations and the word problem. The first one is a two-stage scheme, like the one due to Panagopoulos mentioned above, but it is just an (n, n)-threshold scheme. Our second scheme is a (t, n)-threshold scheme, $0 < t \le n$. Both our schemes are designed for the scenario where the dealer and participants initially are able to communicate over secure channels, but afterwards they communicate over open channels.

We first consider the special (trivial) case where $t = n$; that is, we propose a scheme in which all participants are needed to recover the secret. Then, we propose a hybrid scheme that combines Shamir's scheme and the idea of the (n, n)-threshold scheme we proposed. This combined scheme has the same distributed secret as Shamir's scheme does, but rather than sending $f(i) = x_i$ over secure channels we send the integers in disguise, and then participants use group-theoretic methods to recover the integers. This scheme has the following useful advantages over Shamir's scheme:

- The actual secret need not be distributed over secure channels and, furthermore, once the long-term private information is distributed to all participants, several different secrets can be distributed without updating the long-term private information.
- While recovering the secret, participants do not have to reveal their shares to each other if they do not want to.

2. Very Brief Background on Group Theory

In this section, we give a minimum of information and notation from group theory necessary to understand our main Sections 3 and 4. Further facts from group theory that are used, explicitly or implicitly, in this paper are collected in Sections 6 and 7.

A free group $F = F_m$ of rank m, generated by $X = \{x_1, \ldots, x_m\}$, is the set of all reduced words in the alphabet $\{x_1^{\pm 1}, \ldots, x_m^{\pm 1}\}$, where a word is called reduced if it does not have subwords of the form $x_i x_i^{-1}$ or $x_i^{-1} x_i$.

Any m-generated group G is a factor group of F_m by some normal subgroup N. If there is a recursive (better yet, finite) set of elements $\{r_1, \ldots, r_k, \ldots, \}$ of N that generate N as a normal subgroup of F_m, then we use a compact description of G by generators and defining relators:

$$G = \langle x_1, \ldots, x_m \mid r_1, \ldots, r_k, \ldots, \rangle$$

and call $\{x_1, \ldots, x_m\}$ generators of G, and $\{r_1, \ldots, r_k, \ldots, \}$ defining relators of G. Here "generate N as a normal subgroup of F_m" means that every element u of N can be written as a (finite) product of conjugates of relators r_i: $u = \prod_k h_k^{-1} r_{i_k} h_k$.

Given a presentation of a group G as above, the *word problem* for this presentation is: given a word $w = w(x_1^{\pm 1}, \ldots, x_m^{\pm 1})$ in the generators of G, find out whether or not $w \in N$. (If $w \in N$, we say that $w = 1$ in G.)

3. An (n, n)-threshold Scheme

Suppose we would like the dealer to distribute a k-column $C = \begin{pmatrix} c_1 \\ c_2 \\ \vdots \\ c_k \end{pmatrix}$ consisting of 0's and 1's among n participants in such a way that the vector can be retrieved only when all n participants cooperate. We begin by making a set of group generators $X = \{x_1, \ldots, x_m\}$ public. The scheme is as follows:

(1) The dealer distributes over a secure channel to each participant P_j a set of words R_j in the alphabet $X^{\pm 1} = \{x_1^{\pm 1}, \ldots, x_m^{\pm 1}\}$ such that each group $G_j = \langle x_1, \ldots, x_m \mid R_j \rangle$ has efficiently solvable word problem.
(2) The dealer splits the secret bit column C (the actual secret to be shared) into a sum $C = \sum_{j=1}^{n} C_j$ modulo 2 of n bit columns; these are secret shares to be distributed.
(3) The dealer then distributes words $w_{1j}, \ldots, w_{kj}$ in the generators $x_1, \ldots, x_m$ over an open channel to each participant P_j, $1 \leq j \leq n$. The words are chosen so that $w_{ij} \neq 1$ in G_j if $c_i = 0$ and $w_{ij} = 1$ in G_j if $c_i = 1$.
(4) Participant P_j then checks, for each i, whether the word $w_{ij} = 1$ in G_j or not. After that, each participant P_j can make a column of 0's and 1's,
$$C_j = \begin{pmatrix} c_{1j} \\ c_{2j} \\ \vdots \\ c_{kj} \end{pmatrix},$$
by setting $c_{ij} = 1$ if $w_{ij} = 1$ in G_j and 0 otherwise.
(5) The participants then construct the secret by forming the column vector $C = \sum_{j=1}^{n} C_j$, where the sum of the entries is taken modulo 2.

In Step (5) of the above protocol, the participants can use secure computation of a sum as proposed in [2] if they do not want to reveal their individual column vectors, and therefore their individual secret shares, to each other. In order to implement the protocol to compute a secure sum, the participants should be able to communicate over secure channels with one another. These secure channels should be arranged in a circuit, say, $P_1 \to P_2 \to \ldots \to P_n \to P_1$. Then the protocol to compute a secure sum is as follows:

(1) P_1 begins the process by choosing a random column vector N_1. He then sends to P_2 the sum $N_1 + C_1$.
(2) Each P_i, for $2 \leq i \leq n-1$, does the following. Upon receiving a column vector C from participant P_{i-1}, each participant P_i chooses a random column vector N_i and adds $N_i + C_i$ to C and sends the result to P_{i+1}.
(3) Participant P_n chooses a random column vector N_n and adds $N_n + C_n$ to the column he has received from P_{n-1} and sends the result to P_1. Now P_1 has the column vector $\sum_{i=1}^{n}(N_i + C_i)$.
(4) Participant P_1 subtracts N_1 from what he got from P_n; the result now is the sum $S = \sum_{1 \leq i \leq k} C_i + \sum_{2 \leq i \leq k} N_i$. (This step is needed for P_1 to preserve privacy of his N_1, and therefore of his C_1, since P_2 knows $N_1 + C_1$.) Then P_1 broadcasts S to other participants.
(5) The participants then pool together to recover the secret. They do this by each subtracting his random column vector N_i, $2 \leq i \leq n$, from S.

Thus, by using n secure channels between the participants, the participants are able to compute a secure sum in this secret sharing scheme. For more on the computation of a secure sum see [2].

3.1. Efficiency.

We note that the dealer can efficiently build a word w in the normal closure of R_i as a product of arbitrary conjugates of elements of R_i, so that $w = 1$ in G_i. Furthermore, if G_i is a *small cancellation group* (see our Section 6), then it is also easy to build a word w such that $w \neq 1$ in G_i: it is sufficient to take care that w does not have more than half of any cyclic permutation of any element of R_i as a subword. See our Section 6 for more details. Finally, we note that in small cancellation groups (these are the platform groups that we propose in this paper), the word problem has a very efficient solution, namely, given a word w in the generators of a small cancellation group G, one can determine, in linear time in the length of w, whether or not $w = 1$ in G.

4. A (t,n)-threshold Scheme

Here we propose a scheme that combines Shamir's idea with our scheme in Section 3. As in Shamir's scheme, the secret is an element $x \in \mathbb{Z}_p$, and the dealer chooses a polynomial f of degree $t - 1$ such that $f(0) = x$. In addition the dealer determines integers $y_i = f(i) \pmod{p}$ that are to be distributed to participants P_i, $1 \leq i \leq n$. A set of group generators $\{x_1, \ldots, x_m\}$ is made public. We assume here that all integers x and y_i can be written as k-bit columns. Then the scheme is as follows.

(1) The dealer distributes over a secure channel to each participant P_j a set of relators R_j such that each group $G_j = \langle x_1, \ldots, x_m | R_j \rangle$ has efficiently solvable word problem.

(2) The dealer then distributes over open channels k-columns
$$b_j = \begin{pmatrix} b_{1j} \\ b_{2j} \\ \vdots \\ b_{kj} \end{pmatrix}, 1 \leq j \leq n,$$
of words in $x_1, \ldots, x_m$ to each participant. The words b_{ij} are chosen so that, after replacing them by bits (as usual, "1" if $b_{ij} = 1$ in the group G_j and "0" otherwise), the resulting bit column represents the integer y_j.

(3) Participant P_j then checks, for each word b_{ij}, whether or not $b_{ij} = 1$ in his/her group G_j, thus obtaining a binary representation of the number y_j, and therefore recovering y_j.

(4) Each participant now has a point $f(i) = y_i$ of the polynomial. Using polynomial interpolation, any t participants can now recover the polynomial f, and hence the secret $x = f(0)$.

If $t \geq 3$, then the last step of this protocol can be arranged in such a way that participants do not have to reveal their individual shares y_i to each other if they do not want to. Indeed, from the Lagrange interpolation formula we see that

$$f(0) = \sum_{i=1}^{t} y_i \prod_{1 \leq j \leq t,\ j \neq i} \frac{-j}{i - j}.$$

Thus, $f(0)$ is a linear combination of private y_i with publicly known coefficients $c_i = \prod_{1 \leq j \leq t, j \neq i} \frac{-j}{i-j}$. If $t \geq 3$, then this linear combination can be computed without revealing y_i, the same way the sum of private numbers was computed in our Section 3.

In the special case $t = 2$, this yields an interesting problem. Note that in the original Shamir's scheme, pairs $(i, f(i))$ of coordinates are sent to participants over secure channels, so that the second coordinates are private, whereas the first coordinates are essentially public because they just correspond to participants' numbers in an ordering that could be publicly known. This, however, does not have to be the case, i.e., the first coordinates can be made private, too, so that the dealer sends private points $(x_i, f(x_i))$ to participants. Then, for $t = 2$, we have the following problem of independent interest:

PROBLEM 1. Given that two participants, P_1 and P_2, each has a point (x_i, y_i) in the plane, is it possible for them to exchange information in such a way that at the end, they both can recover an equation of the line connecting their two points, but neither of them can recover precise coordinates of the other participant's point?

5. Why Use Groups?

One might ask a natural question at this point: "What is the advantage of using groups in this scheme? Why not use just sets of elements R_j as long-term secrets, and then distribute elements w_{ij} that either match some elements of R_j or not?" The disadvantage of this procedure is that it will eventually compromise the secrecy of R_j because matching elements will have to be repeated sooner or later. On the other hand, there are infinitely many different words that are equal to 1 in a given group G. For example, if $w = 1$ in G, then also $\prod_i h_i^{-1} w h_i = 1$ for any words h_i. Thus, the dealer can send as many words $w_{ij} = 1$ in G to the participants as he likes, without having to repeat any word or update the relators R_j.

The question that still remains is whether some information about relators R_j may be leaked, even though the words distributed over open channels will never match any words in R_j. This is an interesting question of group theory; we address it, to some extent, in our Section 7.

6. Platform Group

In order for our scheme to be practical, we need each participant to have a finite presentation of a group with efficiently solvable word problem. Here we suggest small cancellation groups as a platform for the protocol. For more information on small cancellation groups see e.g. [**3**].

Let $F(X)$ be the free group on generators $X = \{x_1, \ldots, x_n\}$.
A word $w(x_1, \ldots, x_n) = x_{i_1}^{\epsilon_1} \cdots x_{i_n}^{\epsilon_n}$, where $\epsilon_i = \pm 1$ for $1 \leq i \leq n$, is called *cyclically reduced* if it is a reduced word and $x_{i_1}^{\epsilon_1} \neq x_{i_n}^{-\epsilon_n}$.

A set R containing cyclically reduced words is called *symmetrized* if it is closed under taking cyclic permutations and inverses. Given a set R of relators, a non-empty word $w \in F(X)$ is called a *piece* if there are two distinct relators $r_1, r_2 \in R$ such that w is an initial segment of both r_1 and r_2; that is, $r_1 = wv_1$ and $r_2 = wv_2$ for some $v_1, v_2 \in F(X)$ and there is no cancellation between w and v_1 or w and v_2.

In the definition below, $|w|$ denotes the lexicographic length of a word w.

DEFINITION 6.1. Let R be a symmetrized set of relators, and let $0 < \lambda < 1$. A group $G = \langle X; R \rangle$ with the set X of generators and the set R of relators is said to satisfy the small cancellation condition $C'(\lambda)$ if for every $r \in R$ such that $r = wv$ and w is a piece, one has $|w| < \lambda |r|$. In this case, we say that G belongs to the class $C'(\lambda)$.

We propose groups that satisfy the small cancellation property because groups in the class $C'(\frac{1}{6})$ have the word problem efficiently solvable by Dehn's algorithm. The algorithm is straightforward: given a word w, look for a subword of w which is a piece of a relator from R of length more than a half of the length of the whole relator. If no such piece exists, then $w \neq 1$ in G. If there is such a piece, say u, then $r = uv$ for some $r \in R$, where the length of v is smaller than the length of u. Replace the subword u by v^{-1} in w, and the length of the resulting word will become smaller than that of w. Thus, the algorithm must terminate in at most $|w|$ steps. This (original) Dehn's algorithm is therefore easily seen to have at most quadratic time complexity with respect to the length of w. We note that there is a slightly more elaborate version of Dehn's algorithm that has linear time complexity.

We also note that a generic finitely presented group is a small cancellation group (see e.g. [5]); this means, a randomly selected set of relators will define a small cancellation group with overwhelming probability. Therefore, to randomly select a small cancellation group, the dealer in our scheme can just take a few random words of length > 6 and check whether the corresponding symmetrized set satisfies the condition for $C'(\frac{1}{6})$. If not, then repeat.

7. Tietze transformations: elementary isomorphisms

This section is somewhat more technical than the previous ones. Our goal here is to show how to break long defining relators in a given group presentation into short pieces by using simple isomorphism-preserving transformations. This is useful because in a small cancellation presentation (see our Section 6) defining relators tend to be long and, moreover, a word that is equal to 1 in a presentation like that should contain a subword which is a piece of a defining relator of length more than a half of the length of the whole relator. Therefore, exposing sufficiently many words that are equal to 1 in a given presentation may leak information about defining relators. On the other hand, if all defining relators are short (of length 3, say), a word that is equal to 1 in such a presentation is indistinguishable from random.

Long time ago, Tietze introduced isomorphism-preserving elementary transformations that can be applied to groups presented by generators and defining relators (see e.g. [3]). They are of the following types.

(T1): *Introducing a new generator*: Replace $\langle x_1, x_2, \ldots \mid r_1, r_2, \ldots \rangle$ by $\langle y, x_1, x_2, \ldots \mid ys^{-1}, r_1, r_2, \ldots \rangle$, where $s = s(x_1, x_2, \ldots)$ is an arbitrary word in the generators $x_1, x_2, \ldots$.

(T2): *Canceling a generator* (this is the converse of (T1)): If we have a presentation of the form $\langle y, x_1, x_2, \ldots \mid q, r_1, r_2, \ldots \rangle$, where q is of the form ys^{-1}, and $s, r_1, r_2, \ldots$ are in the group generated by $x_1, x_2, \ldots$, replace this presentation by $\langle x_1, x_2, \ldots \mid r_1, r_2, \ldots \rangle$.

(T3): *Applying an automorphism*: Apply an automorphism of the free group generated by $x_1, x_2, \ldots$ to all the relators $r_1, r_2, \ldots$.

(T4): *Changing defining relators*: Replace the set $r_1, r_2, \ldots$ of defining relators by another set $r_1', r_2', \ldots$ with the same normal closure. That means, each of $r_1', r_2', \ldots$ should belong to the normal subgroup generated by $r_1, r_2, \ldots$, and vice versa.

Tietze has proved (see e.g. [3]) that two groups $\langle x_1, x_2, \ldots \mid r_1, r_2, \ldots \rangle$ and $\langle x_1, x_2, \ldots \mid s_1, s_2, \ldots \rangle$ are isomorphic if and only if one can get from one of the presentations to the other by a sequence of transformations (T1)–(T4).

For each Tietze transformation of the types (T1)–(T3), it is easy to obtain an explicit isomorphism (as a mapping on generators) and its inverse. For a Tietze transformation of the type (T4), the isomorphism is just the identity map. We would like here to make Tietze transformations of the type (T4) recursive, because *a priori* it is not clear how one can actually apply these transformations. Thus, we are going to use the following recursive version of (T4):

(T4′) In the set $r_1, r_2, \ldots$, replace some r_i by one of the: r_i^{-1}, $r_i r_j$, $r_i r_j^{-1}$, $r_j r_i$, $r_j r_i^{-1}$, $x_k^{-1} r_i x_k$, $x_k r_i x_k^{-1}$, where $j \neq i$, and k is arbitrary.

Now we explain how the dealer can break down given defining relators into short pieces. More specifically, he can replace a given presentation by an isomorphic presentation where all defining relators have length at most 3. This is easily achieved by applying transformations (T1) and (T4′), as follows. Let Γ be a presentation $\langle x_1, \ldots, x_k; r_1, \ldots, r_m \rangle$. We are going to obtain a different, isomorphic, presentation by using Tietze transformations of types (T1). Specifically, let, say, $r_1 = x_i x_j u$, $1 \leq i, j \leq k$. We introduce a new generator x_{k+1} and a new relator $r_{m+1} = x_{k+1}^{-1} x_i x_j$. The presentation $\langle x_1, \ldots, x_k, x_{k+1}; r_1, \ldots, r_m, r_{m+1} \rangle$ is obviously isomorphic to Γ. Now if we replace r_1 with $r_1' = x_{k+1} u$, then the presentation $\langle x_1, \ldots, x_k, x_{k+1}; r_1', \ldots, r_m, r_{m+1} \rangle$ will again be isomorphic to Γ, but now the length of one of the defining relators (r_1) has decreased by 1. Continuing in this manner, one can eventually obtain a presentation where all relators have length at most 3, at the expense of introducing more generators.

We conclude this section with a simple example, just to illustrate how Tietze transformations can be used to cut relators into pieces. In this example, we start with a presentation having two relators of length 5 in 3 generators, and end up with a presentation having 5 relators of length 3 in 6 generators. The symbol $\cong$ below means "is isomorphic to".

Example.
$\langle x_1, x_2, x_3 \mid x_1^2 x_2^3, \ x_1 x_2^2 x_1^{-1} x_3 \rangle \cong \langle x_1, x_2, x_3, x_4 \mid x_4 = x_1^2, \ x_4 x_2^3, \ x_1 x_2^2 x_1^{-1} x_3 \rangle$
$\cong \langle x_1, x_2, x_3, x_4, x_5 \mid x_5 = x_1 x_2^2, \ x_4 = x_1^2, \ x_4 x_2^3, \ x_5 x_1^{-1} x_3 \rangle$
$\cong \langle x_1, x_2, x_3, x_4, x_5, x_6 \mid x_5 = x_1 x_2^2, \ x_4 = x_1^2, \ x_6 = x_4 x_2, \ x_6 x_2^2, \ x_5 x_1^{-1} x_3 \rangle.$

We note that this procedure of breaking relators into pieces of length 3 increases the total length of relators by at most the factor of 2.

8. Conclusions

We have proposed a two-stage (t, n)-threshold secret sharing scheme where long-term secrets are distributed to participants over secure channels, and then shares of the actual secret can be distributed over open channels.

Our scheme has the same distributed secret as Shamir's scheme does, but rather than sending shares of a secret over secure channels, we send the integers in disguise (as tuples of words in a public alphabet) over open channels, and then participants use group-theoretic methods to recover the integers. This scheme has the following

useful advantages over Shamir's original scheme:
- The actual secret need not be distributed over secure channels and, furthermore, once the long-term private information is distributed to all participants (over secure channels), several different secrets can be distributed without updating the long-term private information.
- While recovering the secret, participants do not have to reveal their shares to each other if they do not want to.

References

[1] B. Fine, M. Habeeb, D. Kahrobaei, and G. Rosenberger. Survey and open problems in noncommutative cryptography. *JP Journal of Algebra, Number Theory and Applications*, 21:1–40, 2011. MR2840945 (2012f:94107)
[2] D. Grigoriev and V. Shpilrain. Unconditionally secure multiparty computation and secret sharing. *preprint*.
[3] R. Lyndon and P. Schupp. *Combinatorial Group Theory*. Classics in Mathematics. Springer, 2001. MR1812024 (2001i:20064)
[4] A. Myasnikov, V. Shpilrain, and A. Ushakov. *Group-based cryptography*. Advanced Courses in Mathematics. CRM Barcelona. Birkhäuser Verlag, Basel, 2008. MR2437984 (2009d:94098)
[5] A. Yu. Ol'shanskii. Almost every group is hyperbolic. *Internat. J. Algebra Comput.*, 2:1–17, 1992. MR1167524 (93j:20068)
[6] D. Panagopoulos. A secret sharing scheme using groups. Preprint, 2010. Available at http://arxiv.org/PScache/arxiv/pdf/1009/1009.0026v1.pdf.
[7] A. Shamir. How to share a secret. *Comm. ACM*, 22:612–613, 1979. MR549252 (80g:94070)
[8] D. R. Stinson. *Cryptography: Theory and Practice*. Chapman and Hall, 2006. MR2182472 (2007f:94060)

CUNY GRADUATE CENTER, CITY UNIVERSITY OF NEW YORK
E-mail address: MHabeeb@GC.Cuny.edu

CUNY GRADUATE CENTER, CITY UNIVERSITY OF NEW YORK
E-mail address: DKahrobaei@GC.Cuny.edu

THE CITY COLLEGE OF NEW YORK AND CUNY GRADUATE CENTER
E-mail address: shpil@groups.sci.ccny.cuny.edu

Authenticated Key Agreement with Key Re-Use in the Short Authenticated Strings Model

Stanislaw Jarecki and Nitesh Saxena

ABSTRACT. Serge Vaudenay [19] introduced a notion of Message Authentication (MA) protocols in the Short Authenticated String (SAS) model. A SAS-MA protocol authenticates arbitrarily long messages sent over insecure channels as long as the sender and the receiver can additionally send a very short, e.g. 20 bit, *authenticated* message to each other. The main practical application of a SAS-MA protocol is Authenticated Key Agreement (AKA) in this communication model, i.e. SAS-AKA, which can be used for so-called "pairing" of wireless devices. Subsequent work [8, 11, 9] showed three-round SAS-AKA protocols. However, the Diffie-Hellman (DH) based SAS-AKA protocol of [9] requires choosing fresh DH exponents in each protocol instance, while the generic SAS-AKA construction given by [11] applies only to AKA protocols which have no *shared state* between protocol sessions. Therefore, both prior works exclude the most efficient, although not perfect-forward-secret, AKA protocols that re-use private keys (for encryption-based AKAs) or DH exponents (for DH-based AKAs) across multiple protocol sessions.

In this paper, we propose a novel three-round *encryption-based* SAS-AKA protocol, using non-malleable commitments and CCA-secure encryption as tools, which we show secure (but without perfect-forward secrecy) if each player re-uses its private/public key across protocol sessions. The cost of this protocol is dominated by a single public key encryption for one party and a decryption for the other, assuming the Random Oracle Model (ROM). When implemented with RSA encryption the new SAS-AKA protocol is especially attractive if the two devices being paired have asymmetric computational power (e.g., a desktop and a keyboard).

1. Introduction

Serge Vaudenay [19] introduced a notion of a message authentication protocol (MA) based on so-called short authenticated strings (SAS). Such a protocol allows authenticating messages of arbitrary sizes (sent over insecure channel) making use of an auxiliary channel which can authenticate short, e.g. 20-bit, messages. It is assumed that an adversary has complete control over the insecure channel, i.e., it can eavesdrop, delay, drop, replay, inject and/or modify messages, while the only restriction on the auxiliary channel is that the adversary cannot modify or inject messages on it, but it can eavesdrop, delay, drop, or replay them. Crucially, no

---

2010 *Mathematics Subject Classification*. Primary 94A60.

Key words and phrases. Short authenticated strings, authentication, authenticated key agreement.

other infrastructure assumptions are made, i.e. the players do not share any keys or passwords, and there is no Public Key Infrastructure they can use. The only leverage for establishing security is this bandwidth-restricted, public but authenticated "SAS channel" connecting every pair of players.

The primary application of SAS-MA protocols is to enable SAS-based authenticated key agreement (SAS-AKA) between devices with no reliance on key pre-distribution or a public-key infrastructure. A perfectly fitting and urgently needed application of SAS-AKA protocols is establishing secure communication channels between two devices communicating over a publicly-accessible medium (such as Bluetooth, WiFi), which in addition can also send short authenticated messages to each other (and are hence equipped with a SAS channel), given some amount of manual supervision or involvement from the users. (This problem is referred to as "device pairing" in the systems literature.) Implementations of such SAS channels have been proposed for a variety of device types, assuming various user interfaces and different type of manual supervision. In the simplest example of two cell-phones, phone owners can be asked to type a 20 bit string (6 digits) displayed by one phone into the keypad of the other. The systems proposed in [18, 1, 13, 7, 15, 17, 12] show that the same effect can be accomplished with more primitive devices (e.g., with no keypads) or with less user involvement (e.g. relying on sound, blinking LED lights, cameras on the phones, etc). In all of these schemes, it is desirable to have SAS-AKA protocols which are inexpensive both in computation and communication, since the underlying devices might have limited computation and battery power, and which provably achieve an optimal $2^{-k}+\epsilon$ bound on the probability of adversary's attack given a k-bit SAS channel, where ϵ is a negligible factor in the security parameter independent of k. The SAS-AKA protocol we propose in this paper significantly improves upon the first goal compared to the previous work, at the expense of achieving a slightly weaker bound on adversary's attack, namely $2^{-k+1}+\epsilon$.

1.1. Prior Work on SAS-MA Protocols. Following [19, 11], we refer to a bi-directional message authentication protocol in the SAS model as SAS-MCA, which stands for "message *cross*-authentication". Note that two instances of a SAS-MA protocol run in each direction always yield such SAS-MCA scheme, but at twice the cost of the underlying SAS-MA scheme. A straightforward solution for a SAS-MCA was suggested by Balfanz, et al. [1]: Devices A and B exchange the messages m_A, m_B over the insecure channel, and the corresponding hashes $H(m_A)$ and $H(m_B)$ over the SAS channel. Although non-interactive, the protocol requires H to be a collision-resistant hash function and therefore it needs at least 160 bits of the SAS bandwidth in each direction. Pasini and Vaudenay [10] showed a non-interactive protocol which weakens the requirement on the hash function to weak (i.e. second-preimage) collision resistance, and reduces the SAS bandwidth to 80-bits. The 'MANA' protocols in Gehrmann et al. [6] reduce the SAS bandwidth to any k bits while assuring the 2^{-k} bound on attack probability,[1] but these protocols require a stronger assumption on the SAS channel, namely the adversary is assumed to be incapable of delaying or replaying the SAS messages, which in practice requires

---

[1]Formally, by "2^{-k} bound on attack probability" we mean that the probability that any adversary that runs in time polynomial in a security parameter n, which is independent of the SAS-bandwidth k, succeeds against a single instance of the protocol is upper-bounded by $2^{-k}+\epsilon(n)$, where $\epsilon(n)$ is negligible in n.

synchronization between the two devices, e.g. one device never abandons one session and restarts another session without the other device also doing the same.

In [19], Vaudenay presented the first SAS-MA scheme, called V-MA and depicted in Figure 1, with the analysis that bounds the attack probability by 2^{-k} for a k-bit SAS channel. In [19] this protocol is shown secure under the assumption that the commitment scheme satisfies what Vaudenay refers to as "extractable commitment", and subsequently [8] pointed out that this proof goes through under the more standard and possibly weaker assumption of a non-malleable commitment. The bi-directional SAS-MCA protocol presented in [19] results from running two instances of the V-MA protocol, one for each direction, but with each player $P_{i/j}$ using the same challenge $R_{i/j}$ in both protocol instances. This SAS-MCA scheme requires 4 communication rounds over the insecure channel and was shown to give a 2^{-k} security bound.

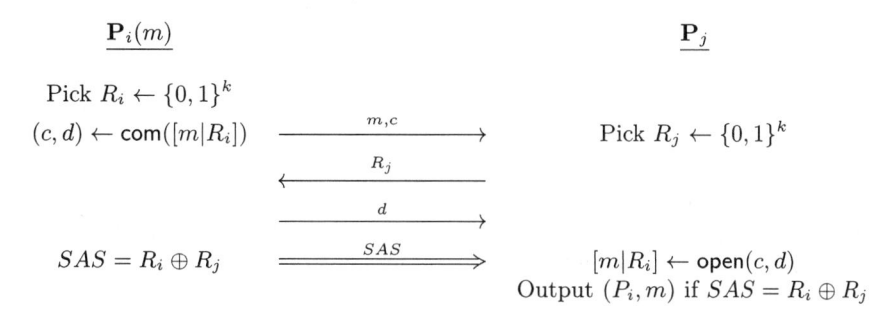

$\mathbf{P}_i(m)$ $\qquad\qquad\qquad\qquad\qquad\qquad\qquad\qquad$ $\mathbf{P}_j$

Pick $R_i \leftarrow \{0,1\}^k$
$(c,d) \leftarrow \mathsf{com}([m|R_i])$ $\qquad \xrightarrow{m,c} \qquad$ Pick $R_j \leftarrow \{0,1\}^k$
$\qquad\qquad\qquad\qquad \xleftarrow{R_j}$
$\qquad\qquad\qquad\qquad \xrightarrow{d}$
$SAS = R_i \oplus R_j \quad \xrightarrow{SAS} \quad [m|R_i] \leftarrow \mathsf{open}(c,d)$
$\qquad\qquad\qquad\qquad\qquad\qquad$ Output (P_i, m) if $SAS = R_i \oplus R_j$

FIGURE 1. V-MA : unidirectional SAS-MA authentication (P_i to P_j) based on non-malleable commitments

In subsequent work, Laur, Asokan, and Nyberg [8, 9] and Pasini and Vaudenay [11] independently gave three-round SAS-MCA protocols. Both schemes are modifications of the V-MA protocol of Figure 1, and both employ (although differently) a universal hash function in computation of the SAS message. Both of these protocols make just a few symmetric key operations if the commitment scheme is implemented using a cryptographic hash function modeled as a Random Oracle. Both protocols claim the 2^{-k} security bound at least in the ROM model, although the scheme of [8, 9] was analyzed only in a "synchronized" setting where the same pair of players never execute multiple parallel protocol instances with each other [2] (see Theorem 3, Note 5 of [9]).

1.2. Prior Work on SAS-AKA Protocols. Pasini and Vaudenay [11] argue that one can construct a 3-round SAS-based key agreement protocol (SAS-AKA), from any 3-round SAS-based message cross-authentication protocol (SAS-MCA) like the SAS-MCA protocol presented in [11], and any 2-round key agreement scheme (KA) which is secure over authenticated links, e.g. a Diffie-Hellman or an encryption-based KA scheme. The idea is to run the 2-round KA protocol

---

[2]While in practice it might often be the case that a pair of players is not *supposed* to execute several protocol instances concurrently, a man-in-the-middle adversary can cause that several instances of the protocol between the same pair of players are effectively alive, if he manages to force one device to time-out and start a new session while the other device is still waiting for an answer.

over an insecure channel, and authenticate the two messages m_1, m_2 produced by the KA protocol using the SAS-MCA protocol. (To achieve a 3-round SAS-AKA protocol, the KA messages m_1, m_2 are piggybacked with the SAS-MCA protocol messages.) This compilation is significantly different from the standard compilation from a protocol secure over authenticated links to a protocol secure over insecure channels, which works by running a separate unidirectional message-authentication sub-protocol (MA) for each message of the underlying protocol, e.g. as in Canetti and Krawczyk's MA + KA $\rightarrow$ AKA compilation in [4]. If the SAS-MA authentication protocol has k rounds then this compilation would result in a $2k$-round SAS-AKA scheme, because the responder cannot, in general, send the second KA message m_2 before successful completion of the SAS-MA sub-protocol that authenticates the first KA message m_1. In contrast, to achieve a $(k+1)$-round SAS-AKA protocol, the compilation given in [11] prescribes that the second message of the KA protocol, m_2, is sent by the responder straight away, i.e. on the basis of the first KA message m_1, which at this moment has not been authenticated yet.

The compilation of Pasini and Vaudenay does result in secure 3-round SAS-AKA schemes, but only when it utilizes a KA scheme which does not keep shared state between different instances of the KA protocol run by the same player. (This was indeed the implicit assumption taken by the proof of security for this compilation given in [11].) Moreover, such SAS-MCA + KA $\rightarrow$ SAS-AKA compilation cannot be applied to KA schemes which *do* share state between instances. For a simple counter-example, consider a 2-round KA protocol secure in the authenticated links model, which is amended so that (1) the computed session key is sent in the last message encrypted using responder P_j's long term public key pk_{ij} chosen for a particular initiator P_i, and (2) the responder P_j reveals the corresponding private key sk_{ij} if the initiator P_i's first message is a special symbol which is never used by an honest sender. Such protocol remains secure in the authenticated links model (in the static corruption case), because only a dishonest sender P_i can trigger P_j to reveal sk_{ij}. However, this protocol is insecure when compiled using the method above, because when P_j computes its response it does not know if the message sent by P_i is authentic, and thus a man-in-the-middle adversary can trigger P_j to reveal sk_{ij} by replacing P_i's initial message in the KA protocol with that special symbol. This way the adversary's interference in a single protocol session leads to revealing the keys on *all* sessions shared between the same pair of players, and thus the compiled protocol is not a secure SAS-AKA. (We elaborate on this counter-example in more detail in Appendix B.)

Independently, Laur and Nyberg also proposed a SAS-AKA protocol [9], based on their own SAS-MCA protocol [8]. In this (Diffie-Hellman based) SAS-AKA protocol, the Diffie-Hellman exponents are picked afresh in each protocol instance, and so this protocol also does not support key re-use across multiple sessions.

1.3. Limitations of SAS-AKA Protocols without Key Re-Use.
The key agreement protocols that do not share state between sessions, and thus in particular do not allow for re-use of private keys, are by definition Perfect-Forward Secret (PFS) but they are also significantly more expensive than non-PFS key agreement protocols. Specifically, the standard Diffie-Hellman PFS KA requires two exponentiations per player, while the encryption-based PFS KA requires generation of a (public,private) key pair and a decryption operation by one player, and a public key encryption by the other player. These are also the dominant costs

of the corresponding SAS-AKA schemes implied by the above results of [8, 11]. In contrast, the non-PFS Diffie-Hellman with fixed exponents costs only one exponentiation per player, and the encryption-based KA costs one decryption for one player and one encryption for the other. Note that in practice the efficiency of the non-PFS KA schemes often takes precedence over the stronger security property offered by perfect forward secret KA schemes. For example, even though SSL supports PFS version of Diffie-Hellman KA, almost all commercial SSL sessions run the non-PFS encryption-based KA using RSA encryption, since this mode offers dramatically faster client's time (and twice faster server's time). Also, just as the asymmetric division of work in the RSA-encryption based key agreement was attractive for the SSL applications, the same asymmetric costs in the RSA-encryption based SAS-AKA could be attractive for "pairing" of devices with unequal computational power, e.g. a PC and a keyboard, a PC and a cell-phone, or a cell-phone and an earset speaker.

Other applications could also benefit from SAS-AKA protocols which allow for re-use of public keys across multiple protocol sessions. One compelling application is in secure initialization of a sensor network [16]. Sensor initialization can be achieved by the base station simultaneously executing an instance of the SAS-AKA protocol with each sensor. However, since the number of sensors can be large, generating fresh (RSA or DH) encryption keys per protocol instance would impose a large overhead on the base station. An encryption-based SAS-AKA protocol with re-usable public key would be especially handy because it would minimize sensors' computation to a single RSA encryption, and the base station would pick one RSA key pair and then perform one RSA decryption per each sensor. Another application where key re-use in SAS-AKA offers immediate benefits is protection against so-called "Evil Twin" attacks in a cyber-cafe, where multiple users run SAS-AKA protocols to associate their devices with one central access point [14].

1.4. Our Contributions. In this work, we present a provably secure and minimal cost SAS-AKA scheme which re-uses public key pairs across protocol sessions and thus presents a lower-cost but non-PFS alternative to the perfect-forward secret SAS-AKA protocols of [9, 11]. Our SAS-AKA relies on a non-malleable commitments just like the SAS-AKA schemes of [19, 8, 11], but unlike the previous schemes it is built directly on CCA-secure encryption, and it relies on encryption not just for key-establishment but also for authentication security. As a consequence, the new SAS-AKA is somewhat simpler than the previous SAS-AKA's which were built on top of the three-round SAS-MCA's of [8, 11], and in particular it does not need to use universal hash functions.[3] However, the most important contribution of the new SAS-AKA scheme is that it remains secure if each player uses a permanent public key, and hence shares a state across all protocol sessions it executes. This leads to two minimal-cost 3-round non-PFS SAS-AKA protocols where the same public/private key pair or the same Diffie-Hellman random contribution is re-used across protocol instances. Specifically, when instantiated with the hash-based commitment and the CCA-secure OAEP-RSA, this implies a 3-round SAS-AKA

---

[3]On the other hand, it might help to clarify that even though our SAS-AKA protocol implies also a new SAS-MCA scheme, we do not claim that our scheme is interesting *as SAS-MCA*, because it relies on a public-key encryption and is therefore much more expensive than the SAS-MCA's of [8, 11] which can be implemented using only symmetric-key cryptography, at least in ROM.

protocol secure under the RSA assumption in ROM, with the cost of a single RSA encryption for the responder and a single RSA decryption for the initiator. When instantiated with the randomness-reusing CCA-secure version of ElGamal [3] this implies a 3-round SAS-AKA protocol secure under the DH assumption in ROM, with the cost of one exponentiation per player. In other words, the costs of the SAS-AKA protocols implied by our result are (for the first time) essentially the same as the costs of the corresponding basic unauthenticated key agreement protocols. By contrast, previously known *PFS* SAS-AKA protocols require two exponentiations per player if they are based on DH [11, 9] or a generation of fresh public/private RSA key pair for each protocol instance if the general result of [11] is instantiated with an RSA-based key agreement.

We note that the SAS-MCA/AKA protocol we show secure is very similar to the SAS-AKA protocols of [19, 8, 11], and it is indeed only a new variant of the same three-round commitment-based SAS-MA protocol analyzed in [19], which also forms a starting point for protocols of [8, 11].However, prior to our work there was no argument that such SAS-AKA scheme remains secure when players re-use their public/private key pairs across multiple sessions. Moreover, as we explain above, it is unlikely that such result can be proven using a modular argument similar to the one used by [11] for KA protocols that do *not* keep state between protocol instances, which is also why our analysis of the proposed protocol proceeds "from scratch" rather than proceeding in a modular fashion based on already known properties of Vaudenay's SAS-MA scheme. Secondly, our analysis shows that the SAS-AKA protocol can be simpler than even a standard encryption-based (and ke-reusing) KA protocol executed over the 3-round SAS-MCA protocol of [8] or [11]. In fact, our protocol consists of a single instance of the basic *unidirectional* SAS-MA scheme of [19], shown in Figure 1, which authenticates only the initiator's message, but this message includes the initiator's (long-term) public key, which the responder uses to encrypt its message. It turns out that this encryption not only transforms this protocol to a SAS-AKA scheme but also authenticates responder's message, thus yielding not just a cheaper but also a simpler three-round SAS-AKA protocol.

Paper Organization. Section 2 contains our cryptographic tools. Section 3 contains the communication and adversarial models for SAS-MCA and SAS-AKA protocols. We propose our SAS-MCA / SAS-AKA protocol in Section 4. In the same section we argue that this protocol is a secure SAS-MCA scheme, and then we extend this argument to an argument that (essentially the same protocol) is a secure SAS-AKA scheme in Section 5.

2. Preliminaries

Public-key Encryption. A public-key encryption scheme is a tuple of algorithms (KeyGen, Enc, Dec), where KeyGen on input of a security parameter produces a pair of public and secret keys (pk, sk), $\text{Enc}_{pk}(m)$ outputs ciphertext c for message m, and $\text{Dec}_{sk}(c)$ decrypts m from $c = \text{Enc}_{pk}(m)$. In the SAS-MCA/AKA protocol construction, the encrypted messages come from a special space $\mathcal{M}_{\overline{m}} = \{[\overline{m}|R] \text{ s.t. } R \in \{0,1\}^k\}$ where $\overline{m}$ is some (adversarially chosen) string. Since this message space contains 2^k elements, a chosen-ciphertext secure encryption ensures that an adversary who is given an encryption of a random message in this space can predict this message with probability at most negligibly higher than 2^{-k}. Namely, the following is a simple fact about CCA-secure encryption. For completeness we give the standard definition of CCA security and a proof of this fact in appendix A.

FACT 1. *If an encryption scheme is (T, ϵ)-SS-CCA then for every T-bounded algorithm $\mathcal{A}$ and every m,*

$$\Pr[\mathcal{A}^{\mathsf{Dec}^C_{sk}(\cdot)}(pk, C) = \hat{m} \mid (pk, sk) \leftarrow \mathsf{KeyGen}, \ m \leftarrow \mathcal{M}_{\overline{m}},$$
$$C \leftarrow \mathsf{Enc}_{pk}(m)] \leq \ 2^{-k} + \epsilon$$

where $\mathsf{Dec}^C_{sk}(\cdot)$ is a decryption oracle except it outputs $\bot$ on C.

Commitment Schemes. Similarly to the SAS-channel message authentication protocols given before by [19, 8, 11], the protocols here are also based on commitment schemes with some form of non-malleability. In fact, the assumption on commitment schemes we make is essentially the same as in the SAS-MCA protocols of [19, 11], but we slightly relax (and re-name) this property of commitment schemes here, so that, in particular, it is satisfied by a very efficient hash-based commitment scheme in the ROM model for a hash function.

The commitment scheme consists of following three functions: **gen** generates a public parameter K_p on input a security parameter, $\mathsf{com}_{K_p}(m)$, on input of message m, outputs a pair of a "commitment" c and "decommitment" d, and $\mathsf{open}_{K_p}(c, d)$, on input (c, d), either outputs some value m' or rejects. This triple of algorithms must meet a completeness property, namely for any K_p generated by **gen** and for any m, if (c, d) is output by $\mathsf{com}_{K_p}(m)$ then $\mathsf{open}_{K_p}(c, d)$ outputs m. We assume a *common reference string* (CRS) model, where a trusted third party generates the commitment key K_p and this key is then embedded in every instance of the protocol. Therefore, we will use a simplified notation, and write $\mathsf{com}(m)$ and $\mathsf{open}(c, d)$ without mentioning the public parameter K_p explicitly. For simplicity of notation in the SAS-MCA/AKA protocols, we sometimes use $m_2 \leftarrow \mathsf{open}(m_1, c, d)$ do denote a procedure which first does $m \leftarrow \mathsf{open}(c, d)$ and then compares if m is of the form $m = [m_1|m_2]$ for the given m_1. If it is, the modifed **open** procedure outputs m_2, and otherwise it rejects.

Non-Malleable Commitment Scheme. In our protocols, we use the same notion of non-malleable commitments as in [8], adopted from [5]. An adversary is a quadruple $\mathcal{A} = (\mathcal{A}_1, \mathcal{A}_2, \mathcal{A}_3, \mathcal{A}_4)$ of efficient algorithms interacting with Challenger. $(\mathcal{A}_1, \mathcal{A}_2, \mathcal{A}_3)$ represents an active part of the adversary that creates and afterwards tries to open related commitments and $\mathcal{A}_4$ represents a distinguisher. Challenger is initialized to be in either of two environments, called "World$_0$" and "World$_1$". $\mathcal{A}$ succeeds if $\mathcal{A}_4$ can distinguish between these two environments World$_0$ and World$_1$.

Challenger first runs **gen** to produce K_p and sends it to $\mathcal{A}_1$. $\mathcal{A}_1$ outputs a message space $\mathcal{M}$ along with state σ and sends it back to Challenger. Challenger picks two messages m_0 and m_1 at random from $\mathcal{M}$ and computes a challenge commitment $(c, d) = \mathsf{com}_{K_p}(m_1)$ and sends c to $\mathcal{A}_2$. $\mathcal{A}_2$ in turn responds with a commitment c^*. Challenger aborts if $c^* = c$, and otherwise sends d to $\mathcal{A}_3$. Now, $\mathcal{A}_3$ must output a valid decommitment d^*. Challenger computes $y^* = \mathsf{open}_{K_p}(c^*, d^*)$. If $y^* = \bot$, then $\mathcal{A}$ is halted. Finally, in the environment World$_0$, Challenger invokes $\mathcal{A}_4$ with inputs (m_0, y^*), whereas in World$_1$, it invokes $\mathcal{A}_4$ with inputs (m_1, y^*). A commitment scheme is (T, ϵ)-NM (non-malleable) iff for any t time adversary $\mathcal{A}$,

$Adv^{\mathsf{NM}}_{com}(\mathcal{A}) = |\Pr[\mathcal{A}_4 = 1|\mathsf{World}_1] - \Pr[\mathcal{A}_4 = 1|\mathsf{World}_0]| \leq \epsilon$.

For notational convenience, we give a specialization of this non-malleability notion to message space $\mathcal{M}_{\overline{m}} = \{[\overline{m}|R] \text{ s.t. } R \in \{0,1\}^k\}$, which our SAS-MCA/AKA protocol deals with, and to a particular simple type of tests which our reductions use to distinguish between the two distributions above. Namely, we say that the commitment scheme is (T, ϵ)-NM if for every T-limited adversary $\mathcal{A} = (\mathcal{A}_1, \mathcal{A}_2, \mathcal{A}_3)$, the following holds:

$Pr[m^* \oplus m = \sigma \mid K_P \leftarrow \mathsf{gen}, (\overline{m}, s) \leftarrow \mathcal{A}_1(K_P), m \leftarrow \mathcal{M}_{\overline{m}}, (c, d) \leftarrow \mathsf{com}_{K_P}(m),$
$(c^*, \sigma) \leftarrow \mathcal{A}_2(c, s), d^* \leftarrow \mathcal{A}_3(c, d, s),$
$m^* = \mathsf{open}_{K_P}(c^*, d^*)] \leq 2^{-k} + \epsilon$

Non-Malleable Commitment from SS-CCA Encryption. (T, ϵ)-NM commitment scheme can be created from any (T, ϵ)-SS-CCA encryption scheme (KeyGen, Enc, Dec) [5]. The (K_s, K_p) is a private/public key pair (sk, pk) of the encryption scheme. $\mathsf{com}_{pk}(m)$ picks a random string r and outputs $c = \mathsf{Enc}_{pk}(m; r)$ and $d = (m, r)$, where $\mathsf{Enc}_{pk}(\cdot; r)$ denotes the (randomized) encryption procedure with randomness r. Procedure $\mathsf{open}_{pk}(c, (m, r))$ outputs m if $c = \mathsf{Enc}_{pk}(m; r)$ and $\bot$ otherwise.

Non-Malleable Commitment in the Random Oracle Model (ROM). One can make a fast and simple commitment scheme using a hash function $H : \{0,1\}^* \rightarrow \{0,1\}^{l'}$ modeled as a random oracle, where the adversary's advantage in the NM-Security game can be set arbitrarily low at very little cost. Generator gen in this scheme is a null procedure, $\mathsf{com}(m)$ picks $r \in \{0,1\}^l$ and returns $c = H(m, r)$ and $d = (m, r)$, $\mathsf{open}(c, (m, r))$ returns m if $c = H(m, r)$ and $\bot$ otherwise. This scheme is (T, ϵ)-NM for $\epsilon = q_H 2^{-l} + q_H^2 2^{-l'}$, where q_H is the number of H-function queries that can be made by a T-bounded adversary $\mathcal{A}$. This is because the probability that $\mathcal{A}_2$ learns anything about the value committed by the challenger is $q_H 2^{-l}$ because the only information $\mathcal{A}_2$ can get on m chosen by the challenger is by querying hash function H for some $m \in \mathcal{M}$ and r used by the challenger, but the probability that $\mathcal{A}$ hits the same r as the challenger is bounded by $q_H 2^{-l}$. Moreover, the probability that $\mathcal{A}_3$ is able to decommit to more than one value is bounded by $q_H^2 2^{-l'}$, because this is the probability that within q_H queries to H, the adversary gets a pair of values which collide.

3. Communication and Adversarial Model

3.1. Network and Communication Setting. We consider the same model as in [19, 8, 11], but we explicitly cast it in the multi-player/multi-session world. In other words, we consider a network consisting of n players $P_1, \cdots, P_n$. Each ordered pair of players (P_i, P_j) is connected by two unidirectional point-to-point communication channels: (1) an insecure channel, e.g. internet or a Bluetooth or a WiFi channel, over which an adversary has complete control by eavesdropping, delaying, dropping, replaying, and/or modifying messages, and (2) a low-bandwidth out-of-band authenticated (but not secret) channel, referred to as a *SAS channel* from here on, which preserves the integrity of messages and also provides source and target authentication. In other words, on the insecure channel, an adversary can behave arbitrarily, but it is *not* allowed to modify (or inject) messages sent on the SAS channel (which we'll call *SAS messages* for short), although it can still read them, as well as delay, drop, or re-order them.

3.2. SAS-MCA and its Security.

Our security model follows the Canetti-Krawczyk model for authenticated key exchange protocols [4], and the earlier work of [2], which allows modeling concurrent executions of multiple protocol instances. While in practice it will very often be the case (e.g. in the device pairing application) that a single player is not *supposed* to execute several protocol instances concurrently, a man-in-the-middle adversary can cause that several instances of the protocol between the same pair of players are effectively alive, if he manages to force device A to time-out and start a new SAS-AKA protocol session, while device B is still waiting for an answer. In this case the adversary can choose which messages to forward to device B among the messages sent on the different sessions started by device A.

A SAS-MCA protocol is a "cross-party" message authentication protocol, executed between two players P_i and P_j, whose goal is for P_i and P_j to send authenticated messages to one another. We denote the τ-th protocol instance run by a player P_i as Π_i^τ, where τ is a locally unique index. The inputs of Π_i^τ are a tuple $(\text{role}_i^\tau, P_j, m_i^\tau)$ where role_i^τ designates P_i as either the initiator (*"init"*) or a responder (*"resp"*) in this instance of the SAS-MCA protocol, P_j identifies the communication partner for this protocol instance, i.e. it identifies a pair of SAS channels $(P_i \to P_j)$ and $(P_i \leftarrow P_j)$ with an entity (P_j) with whom P_i's application wants to communicate, and m_i^τ is the message to be sent to P_j in this session. With each session Π_i^τ there is associated a unique string sid_i^τ, which is a concatenation of all messages sent and received on this session, including the messages on the SAS channel. We denote input P_j on session Π_i^τ as $\text{Peer}(\Pi_i^\tau)$. We say that sessions Π_i^τ and Π_j^η executed by two different players are **matching** if $\text{Peer}(\Pi_i^\tau) = P_j$, $\text{Peer}(\Pi_j^\eta) = P_i$, and $\text{role}_j^\eta \neq \text{role}_i^\tau$. We say that the sessions are **partnered** if they are matching and their messages are properly exchanged between them, i.e. $sid_i^\tau = sid_j^\eta$. By the last requirement, and by inclusion of random nonces in the protocol, we ensure that except of negligible probability each session can be partnered with at most one other session. The output of Π_i^τ can be either a tuple $(\text{Peer}(\Pi_i^\tau), m_i^\tau, \hat{m}_i^\tau, sid_i^\tau)$, for some $\hat{m}_i^\tau$, or a rejection. Similarly, Π_j^η can either output $(\text{Peer}(\Pi_j^\eta), m_j^\eta, \hat{m}_j^\eta, sid_j^\eta)$ or reject. The SAS-MCA protocol should satisfy the following **correctness** condition: If sessions Π_i^τ and Π_j^η are partnered then both sessions accept and output the messages sent by the other player, i.e. $\hat{m}_i^\tau = m_j^\eta$ and $\hat{m}_j^\eta = m_i^\tau$.

We model the **security** of a SAS-MCA protocol via a following game between the challenger performing the part of the honest players $P_1, ..., P_n$, and the adversary $\mathcal{A}$. We consider only the *static* corruption model, where the adversary does not adaptively corrupt initially honest players. The challenger and the adversary communicate by exchanging messages as follows: At the beginning of the interaction, the challenger initializes the long-term private state of every player P_i, e.g. by generating a public/private key pair for each player. In the rest of the interaction, the challenger keeps the state of every initialized protocol instance and follows the SAS-MCA protocol on its behalf. $\mathcal{A}$ can trigger a new protocol instance Π_i^τ on inputs (role, P_j, m) by issuing a query $\text{launch}(\Pi_i^\tau, \text{role}, P_j, m)$. The challenger responds by initializing the state of session Π_i^τ and sending back to $\mathcal{A}$ the message this session generates. If $\mathcal{A}$ issues a query $\text{send}(\Pi_i^\tau, M)$ for any previously initialized Π_i^τ and any M, the challenger delivers message M to session Π_i^τ and responds by following the SAS-MCA protocol on its behalf, handing the response of Π_i^τ on M to $\mathcal{A}$. However, if Π_i^τ's next message is a SAS message, the challenger hands this message to $\mathcal{A}$ and adds it to a multiset $\mathsf{SAS}(i,j)$, for $P_j = \mathsf{Peer}(\Pi_i^\tau)$, which models the unidirectional SAS channel from P_i to P_j, denoted $\mathsf{SAS}(P_i \to P_j)$. $\mathcal{A}$ can issue a $\mathsf{SAS\text{-}send}(\Pi_j^\tau, M)$ query for any message M in set $\mathsf{SAS}(i,j)$, where $P_i = \mathsf{Peer}(\Pi_j^\tau)$. The challenger then removes element M from $\mathsf{SAS}(i,j)$ and delivers M on the $\mathsf{SAS}(P_j \to P_i)$ channel to Π_i^τ. This models the fact that the adversary can see, stall, delete, and re-order messages on each $\mathsf{SAS}(P_i \to P_j)$ channel, but $\mathcal{A}$ cannot modify, duplicate, or add to any of the messages on such channel.

We say that $\mathcal{A}$ *wins* in attack against SAS-MCA if there exists session Π_i^τ which outputs (P_j, m_i, m_j, sid) but there is no session Π_j^η which ran on inputs $(*, P_i, m_j)$. In other words, if Π_i^τ outputs a message m_j as sent by P_j but P_j did not send m_j to P_i on any session. We call a SAS-MCA protocol (T, ϵ)-**secure** if for every adversary $\mathcal{A}$ running in time T, $\mathcal{A}$ wins with probability at most ϵ. Note that in the SAS-MCA game the adversary can launch multiple concurrent sessions among every pair of players. To make our security results concrete in the multi-player setting, we will consider an (n, τ_t, τ_c)-attacker $\mathcal{A}$ against the SAS-MCA protocol, where the above game is restricted to n players P_i, at mosts τ_t total number of sessions per player, and at mosts τ_c sessions that can be concurrently held by any *pair* of players, i.e. $\mathsf{SAS}(i,j) \leq \tau_c$ for all i,j. We note that the τ_c bound is determined by how long the adversary can lag the SAS messages, how many sessions he can cause to re-start at one side, and how long he can keep alive a session waiting for its SAS message on the other side. In many applications it will be rather small, but it is important to realize that in many applications it is greater than 1.

3.3. SAS-AKA and its Security. SAS-AKA is an Authenticated Key Agreement (AKA) protocol in the SAS model. The inputs to the protocol are as in the SAS-MCA but with no messages. Each instance Π_i^τ outputs either a rejection or a tuple $(\mathsf{Peer}(\Pi_i^\tau), K, sid)$, where K is a fresh, authenticated, and secret key which P_i hopes to have shared with $P_j = \mathsf{Peer}(\Pi_i^s)$, and sid is a locally unique session id. An SAS-AKA scheme protects the secrecy of keys output by honest players on sessions involving other uncorrupted player. The correctness property for a SAS-AKA protocol is that if two sessions Π_i^τ and Π_j^η are partnereed then both sessions accept and output the same key $K_i^\tau = K_j^\eta$.

We model **security** of the SAS-AKA protocol similarly as in the SAS-MCA case, by an interaction between the (n, τ_t, τ_c)-attacker $\mathcal{A}$ and the challenger that operates the network of n players $P_1, ..., P_n$. In this game, however, the challenger has a *private input* of bit b. The rules of communication model between the challenger and $\mathcal{A}$ and the set-up of all honest players are the same as in the SAS-MCA game above, and the challenger services $\mathcal{A}$'s requests launch, send, and SAS-send in the same way as in the SAS-MCA game, except that there's no message in inputs to the launch request. In addition, $\mathcal{A}$ can issue a query of the form $\mathsf{reveal}(\Pi_i^\tau)$ for any Π_i^τ, which gives him the key K_i^τ output by Π_i^τ if this session computed a key, and a null value otherwise. Finally, on one of the sessions Π_i^τ subject to the constraints specified below, the adversary can issue a $\mathsf{Test}(\Pi_i^\tau)$ query. If Π_i^τ has not completed, the adversary gets a null value. Otherwise, if $b = 1$ then $\mathcal{A}$ gets the key K_i^τ, and if $b = 0$ then $\mathcal{A}$ gets a random bitstring of the same length. The constraint on the tested session Π_i^τ is that the adversary issues no $\mathsf{reveal}(\Pi_i^\tau)$ query *and* no $\mathsf{reveal}(\Pi_j^\eta)$ query for any Π_j^η which is partnered with Π_i^τ. After testing a session, the adversary can then keep issuing the launch, send, SASsend and reveal commands, except it cannot reveal the tested session or a session that is partnered with it. Eventually $\mathcal{A}$

outputs a bit $\hat{b}$. We say that an adversary has *advantage* ϵ in the SAS-AKA attack if the probability that $\hat{b} = b$ is at most $1/2 + \epsilon$. We say that the SAS-AKA protocol is (T, ϵ)-**secure** if for all $\mathcal{A}$'s bounded by time T this advantage is at most ϵ.

We note that the above model includes only *static* corruption patterns. Indeed, the protocols we present here do *not* have perfect forward secrecy, since we are interested in provable security of minimal-cost AKA protocols in which players re-use their private key material across all protocol sessions.

4. Encryption-based SAS Message Authentication Protocol

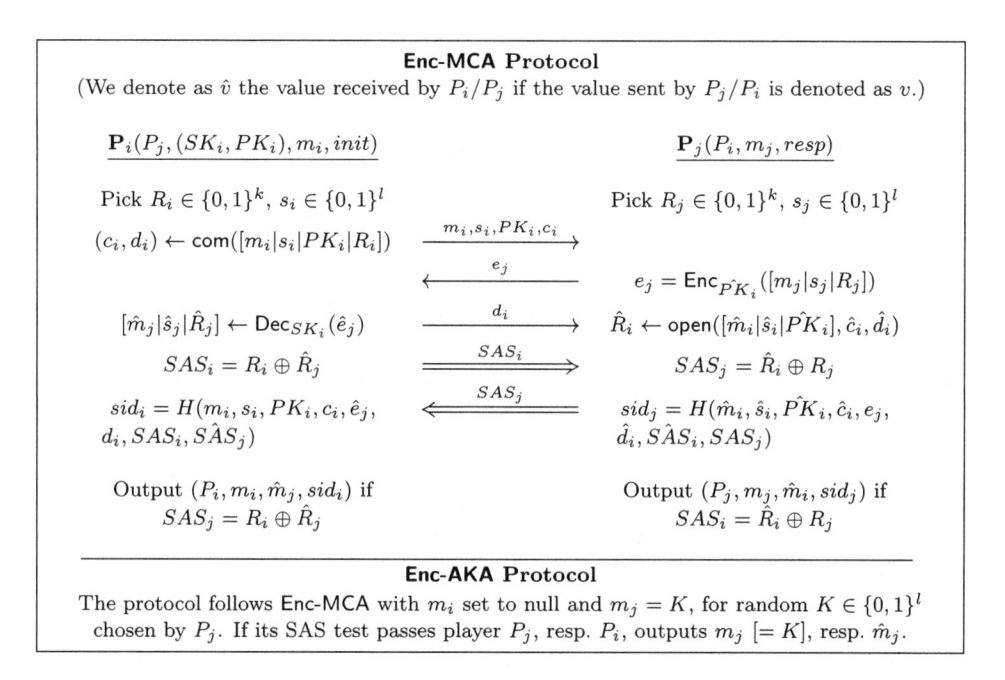

Enc-MCA Protocol
(We denote as $\hat{v}$ the value received by P_i/P_j if the value sent by P_j/P_i is denoted as v.)

$\mathbf{P}_i(P_j, (SK_i, PK_i), m_i, init)$ $\qquad\qquad\qquad$ $\mathbf{P}_j(P_i, m_j, resp)$

Pick $R_i \in \{0,1\}^k$, $s_i \in \{0,1\}^l$ $\qquad\qquad$ Pick $R_j \in \{0,1\}^k$, $s_j \in \{0,1\}^l$

$(c_i, d_i) \leftarrow \text{com}([m_i|s_i|PK_i|R_i])$ $\xrightarrow{m_i, s_i, PK_i, c_i}$

$\xleftarrow{e_j}$ $\qquad\qquad$ $e_j = \text{Enc}_{\hat{PK}_i}([m_j|s_j|R_j])$

$[\hat{m}_j|\hat{s}_j|\hat{R}_j] \leftarrow \text{Dec}_{SK_i}(\hat{e}_j)$ $\xrightarrow{d_i}$ $\hat{R}_i \leftarrow \text{open}([\hat{m}_i|\hat{s}_i|\hat{PK}_i], \hat{c}_i, \hat{d}_i)$

$SAS_i = R_i \oplus \hat{R}_j$ $\xRightarrow{SAS_i}$ $SAS_j = \hat{R}_i \oplus R_j$

$sid_i = H(m_i, s_i, PK_i, c_i, \hat{e}_j,$ $\xLeftarrow{SAS_j}$ $sid_j = H(\hat{m}_i, \hat{s}_i, \hat{PK}_i, \hat{c}_i, e_j,$
$\quad d_i, SAS_i, \hat{SAS}_j)$ $\qquad\qquad\qquad$ $\quad \hat{d}_i, \hat{SAS}_i, SAS_j)$

Output $(P_i, m_i, \hat{m}_j, sid_i)$ if $\qquad\qquad$ Output $(P_j, m_j, \hat{m}_i, sid_j)$ if
$\quad SAS_j = R_i \oplus \hat{R}_j$ $\qquad\qquad\qquad\qquad$ $\quad SAS_i = \hat{R}_i \oplus R_j$

Enc-AKA Protocol
The protocol follows Enc-MCA with m_i set to null and $m_j = K$, for random $K \in \{0,1\}^l$ chosen by P_j. If its SAS test passes player P_j, resp. P_i, outputs m_j [$= K$], resp. $\hat{m}_j$.

FIGURE 2. Encryption-based SAS-MCA protocol (Enc-MCA) and SAS-AKA protocol (Enc-AKA)

In this section, we present a novel 3-round encryption-based bidirectional SAS-MCA protocol denoted **Enc-MCA**. The protocol is depicted in Figure 2. It runs between the initiator P_i, who intends to authenticate a message m_i, and the responder P_j, who intends to authenticate a message m_j. (SK_i, PK_i) denotes P_i's private/public key pair of an IND-CCA encryption scheme, which w.l.o.g. we assume to be permanent. The protocol assumes the CRS model where the instance K_P of the CCA-Secure commitment scheme is globally chosen. The protocol is based on the *unidirectional* message-authentication V-MA protocol of Vaudenay [19], Figure 1. The only difference is that P_i adds to its message m_i its public key PK_i and a random nonce $s_i \in \{0,1\}^l$, and the responder P_j sends its randomness R_j *encrypted* under PK_i, together with its message m_j and a random nonce $s_j \in \{0,1\}^l$. In other words, P_i sends (m_i, s_i, PK_i) along with a commitment c_i to (m_i, s_i, PK_i, R_i) where R_i is a random k-bit bitstring. P_j replies with an encryption of m_j, s_j, and a random value $R_j \in \{0,1\}^k$. Finally P_i sends to P_j its decommitment d_i to c_i, and P_i and P_j exchange over the SAS channel values

$SAS_i = R_i \oplus R_j$, where P_i obtains R_j by decrypting e_j, and $SAS_j = R_i \oplus R_j$, where P_j obtains R_i by opening the commitment c_i. The players accept if the SAS values match, and reject otherwise. P_i and P_j also output session identifiers sid_i and sid_j, respectively, which are outputs of a collision-resistant hash function H on the concatenation of all messages sent (received resp.) and received (sent resp.) on this session, including the messages on the SAS channel. (This is done only for simplicity of security analysis: In fact the same security argument goes through if $sid_i = sid_j = [s_i|s_j]$.) The following theorem states the security of this protocol against an (n, τ_t, τ_c)-adversary:

THEOREM 1 (**Security of Enc-MCA**). *If the commitment scheme is (T_C, ϵ_C)-NM and the encryption scheme is (T_E, ϵ_E)-SS-CCA, then the Enc-MCA protocol is (T, p)-secure against (n, τ_t, τ_c)-attacker for $p \geq 2n\tau_t\tau_c(2^{-k} + max(\epsilon_C, \epsilon_E))$ and $T \leq min(T_C, T_E) - \mu$, for a small constant μ.*

Note on the Security Claim and the Proof Strategy. The $n\tau_t\tau_c 2^{-k}$ security bound would be optimally achievable in the context of (n, τ_t, τ_c)-adversary because this is the probability, for $n\tau_t\tau_c \ll 2^{-k}$, that the k-bit SAS messages are equal on some two matching sessions, even though the adversary substitutes sender's messages on every session, since there are $n\tau_t$ sessions, each of which can succeed if the SAS message it requires to complete is present among τ_c SAS messages produced by the sessions concurrently executed by its peer player. We note that if adversary's goal is to attack any *particular* player and session, the same theorem applies with values $n = \tau_t = 1$.

However, the security bound $n\tau_t\tau_c 2^{-k+1}$ we show is factor of 2 away from the optimal. This factor is due to the fact that the reduction has to guess whether the adversary essentially attacks the encryption or the commitment tool used in our protocol. This also accounts for the essential difference between our proof and those of [8, 11]. Even assuming the simplest $n = \tau_t = \tau_c = 1$ case, there are several patterns of attack, corresponding to three possibilities for interleaving messages and other decisions the adversary can make (in our case the crucial switch is whether or not the adversary modifies the initiator's payload m, s, PK). For each pattern of attack, we provide a reduction, which given an attack that breaks the SAS-MCA/AKA scheme with probability $2^{-k} + \epsilon$, *conditioned on this attack type being chosen*, attacks either the commitment or the encryption scheme with probability ϵ.[4] However, it is not clear how to use such reductions to show any better security bound than $q * 2^{-k}$ where q is the number of such attack cases. Fortunately, we manage to group these attack patterns into just two groups, with two reductions, the first translating *any* attack in the first group into an encryption attack, the second translating *any* attack in the second group into a commitment attack. Crucially, both reductions are non-rewinding, and hence they are security-preserving. However, faced with an adversary which adaptively decides which group his attack will fall in we still need to guess which reduction to follow, hence the bound on attacker's probability we show for our SAS-MCA/AKE scheme is a factor of 2 away from the optimal.

Proof: We prove the above by showing that if there exists (n, τ_t, τ_c)-adversary $\mathcal{A}$ which can attack the proposed protocol in time $T < min(T_C, T_E) - \mu$ and

---

[4] While some of these component reductions are identical to those shown for the same underlying SAS-MA protocol by Vaudenay in [19], others are different because they attack encryption and/or commitment because we need to structure the attack cases differently.

probability $p > 2n\tau_t\tau_c(2^{-k} + max(\epsilon_C, \epsilon_E))$, then there exists *either* a $T + \mu < T_C$ adversary $\mathcal{B}_C$ which breaks NM security of the commitment scheme with probability better than $2^{-k} + \epsilon_C$, *or* there exists a $T + \mu < T_E$ adversary $\mathcal{B}_E$ which wins the SS-CCA game for the encryption scheme with probability better than $2^{-k} + \epsilon_E$.

$\mathcal{A}$ succeeds if it can find a player P_i and a session Π_i^s with a peer party P_j, such that Π_i^s accepts message $\hat{m}_j^{(s)}$ but the adversary never launches an instance of P_j on message $\hat{m}_j^{(s)}$. To achieve this $\mathcal{A}$ in particular has to route to Π_i^s a SAS message $SAS_j^{(s')}$ originated by *some* session $\Pi_j^{s'}$ s.t. $\mathsf{Peer}(\Pi_j^{s'}) = P_i$. By inspection of the protocol, Π_i^s accepts only if $R_i^{(s)} \oplus \hat{R}_j^{(s)} = \hat{R}_i^{(s')} \oplus R_j^{(s')}$, or equivalently, $SAS_i^{(s)} = SAS_j^{(s')}$.

Note that this condition must hold regardless whether the attacked session Π_i^s is an initiator or a responder. This allows us to simplify the notation and in the remainder of the proof we assume Π_i^s is the initiator, $\Pi_j^{s'}$ is the responder, and we assume that *either* $\hat{m}_i^{(s)} \neq m_i^{(s)}$ or $\hat{m}_j^{(s')} \neq m_j^{(s')}$.

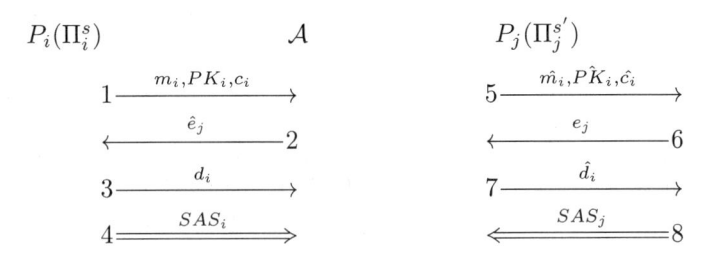

$$
\begin{array}{ccc}
P_i(\Pi_i^s) & \mathcal{A} & P_j(\Pi_j^{s'}) \\
& & \\
1 \xrightarrow{m_i, PK_i, c_i} & & 5 \xrightarrow{\hat{m}_i, \hat{PK}_i, \hat{c}_i} \\
\xleftarrow{\hat{e}_j} 2 & & \xleftarrow{e_j} 6 \\
3 \xrightarrow{d_i} & & 7 \xrightarrow{\hat{d}_i} \\
4 \xRightarrow{SAS_i} & & \xLeftarrow{SAS_j} 8
\end{array}
$$

FIGURE 3. Adversarial Behavior in the Enc-MCA protocol

In Figure 3 we show adversary's interactions as a man in the middle between Π_i^s and $\Pi_j^{s'}$. Note that $\mathcal{A}$ can control the *sequence* in which the messages received by these two players are interleaved, and $\mathcal{A}$ has a choice of the following three possible sequences:

Interleaving pattern I : $(1 \prec 5 \prec 6 \prec 2 \prec 3 \prec 4 \prec 7 \prec 8)$

Interleaving pattern II : $(1 \prec 5 \prec 6 \prec 7 \prec 8 \prec 2 \prec 3 \prec 4)$

Interleaving pattern III : $(1 \prec 2 \prec 3 \prec 4 \prec 5 \prec 6 \prec 7 \prec 8)$

In each of these three message interleaving patterns we consider two subcases, depending on whether the pair $(\hat{m}_i, \hat{PK}_i)$ that the adversary delivers to $\Pi_j^{s'}$ in message #5 (see Figure 3) is equal to (m_i, PK_i) that Π_i^s sends in message #1.

Let's denote the event that adversary succeeds in an attack as AdvSc, the event that $(\hat{m}_i, \hat{PK}_i) = (m_i, PK_i)$ *and* that the attack succeeds as SM, the event that $(\hat{m}_i, \hat{PK}_i) \neq (m_i, PK_i)$ *and* that the attack succeeds as NSM, and we'll use Int[1], Int[2], Int[3] to denote events when the adversary follows, respectively, the 1st, 2nd, or 3rd message interleaving pattern. We divide the six possible patterns which the successful attack must follow into the following two cases:

Case1 = NSM ∨ (AdvSc ∧ Int[2]) & Case2 = SM ∧ (Int[1] ∨ Int[3])

We construct two reduction algorithms, $\mathcal{B}_C$ and $\mathcal{B}_E$, which attack respectively the NM property of the commitment, and the SS-CCA property of the encryption scheme used in the Enc-MCA protocol. Both algorithms $\mathcal{B}_C$ and $\mathcal{B}_E$ use the

Enc-MCA attacker $\mathcal{A}$ as a black box, and both reductions have only constant computational overhead which we denote as μ, hence both $\mathcal{B}_C$ and $\mathcal{B}_E$ run in time at most $T - \mu < min(T_C, T_E)$. We will show that if $\Pr[\text{Case1}] \geq p/2$ then $\mathcal{B}_C$ wins the NM game with probability greater than $2^{-k} + \epsilon_C$, and if $\Pr[\text{Case2}] \geq p/2$ then $\mathcal{B}_E$ wins the SS-CCA game with probability greater than $2^{-k} + \epsilon_E$. This will complete the proof because $\text{AdvSc} = \text{Case1} \cup \text{Case2}$, and therefore if $\Pr[\text{AdvSc}] = p$ then either $\Pr[\text{Case1}] \geq p/2$ or $n\tau_t\tau_c$) or $\Pr[\text{Case2}] \geq p/2$.

Both $\mathcal{B}_C$ and $\mathcal{B}_E$ proceed by first guessing the sessions Π_i^s and $\Pi_j^{s'}$ involved in $\mathcal{A}$'s attack. The probability that the guess is correct is at least $1/n\tau_t\tau_c$ because $\mathcal{A}$ runs at most $n\tau_t$ sessions and each session can have at most τ_c concurrently running peer sessions. Since the probability of a correct guess is independent of adversary's view, for either $i = 1$ or $i = 2$, the probability that the guess is correct *and* Case$i$ happens is at least $p/2 * 1/n\tau_t\tau_c > 2^{-k} + max(\epsilon_C, \epsilon_E)$. We show that if $i = 1$ then $\mathcal{B}_C$ wins in the NM game, and hence its probability of winning is greater than $2^{-k} + \epsilon_C$, and if $i = 2$ then $\mathcal{B}_E$ wins the SS-CCA game, and hence its probability of winning is greater than $2^{-k} + \epsilon_E$.

It remains for us to construct algorithms $\mathcal{B}_C$ and $\mathcal{B}_E$ with the properties claimed above. Algorithm $\mathcal{B}_C$, depending on the behavior of $\mathcal{A}$, executes one of the following sub-algorithms:

> If $(\hat{m}_i, \hat{s}_i, \hat{PK}_i) \neq (m_i, s_i, PK_i)$ and $\mathcal{A}$ chooses interleaving pattern I or III, then $\mathcal{B}_C$ executes sub-algorithms, respectively, $\mathcal{B}_C[1]$ and $\mathcal{B}_C[3]$.
> If $\mathcal{A}$ chooses interleaving pattern II, $\mathcal{B}_C$ executes $\mathcal{B}_C[2]$.
> Otherwise, i.e. if $\mathcal{A}$ sends $(\hat{m}_i, \hat{s}_i, \hat{PK}_i) = (m_i, s_i, PK_i)$ and $\mathcal{A}$ follows patterns I or III, $\mathcal{B}_C$ fails.

Similarly, based on the behavior of $\mathcal{A}$, algorithm $\mathcal{B}_E$ proceeds in one of the following ways:

> If $(\hat{m}_i, \hat{s}_i, \hat{PK}_i) = (m_i, s_i, PK_i)$ and $\mathcal{A}$ chooses interleaving pattern I, $\mathcal{B}_E$ executes $\mathcal{B}_E[1]$.
> If $(\hat{m}_i, \hat{s}_i, \hat{PK}_i) = (m_i, s_i, PK_i)$ and $\mathcal{A}$ chooses interleaving pattern III, $\mathcal{B}_E$ executes $\mathcal{B}_E[2]$.
> Otherwise, i.e. if $\mathcal{A}$ sends $(\hat{m}_i, \hat{s}_i, \hat{PK}_i) \neq (m_i, s_i, PK_i)$ or $\mathcal{A}$ follows interleaving pattern II, $\mathcal{B}_E$ fails.

We show algorithms $\mathcal{B}_C[1], \mathcal{B}_C[2]$ and $\mathcal{B}_C[3]$ in Figures 7, 8, and 9, respectively. Note that if $(\hat{m}_i, \hat{s}_i, \hat{PK}_i) \neq (m_i, s_i, PK_i)$ then $\mathcal{A}$ essentially attacks the V-MA protocol of Vaudenay, because pair (m_i, PK_i) in the Enc-MCA protocol plays a role of the message in the V-MA protocol, so this event in the Enc-MCA protocol is equivalent to P_j accepting the wrong message in the V-MA protocol. Therefore, the three reduction (sub)algorithms $\mathcal{B}_C[1]$, $\mathcal{B}_C[2]$, and $\mathcal{B}_C[3]$, essentially perform the same attacks on the NM game of the commitment scheme as the corresponding three reductions given by Vaudenay for the V-MA protocol. The only difference is that our reductions put a layer of encryption on the messages sent by P_j, as is done in our protocol Enc-MCA. As in Vaudenay's reductions, we extend the NM game so that the challenger, at the end of the game sends to the attacker the decommitment d corresponding to the challenge commitment c. Since this happens after the attacker sends its R, the difficulty of the NM game remains the same. However, if the $\mathcal{B}_C$ reduction gets the decommitment d from the NM challenger, the reduction can complete the view of the protocol to $\mathcal{A}$, which makes it easier (esp. in case of $\mathcal{B}_C[4]$) to compare the probability of $\mathcal{A}$'s success with the probability of success of $\mathcal{B}_C$.

For completeness, we show these three subcases of the reduction to an NM attack in Appendix C. By inspection of the figures, note that each of these subcases of the $\mathcal{B}_C$ reduction at first follows the same protocol with the NM challenger, and that $\mathcal{B}_C$ can decide which path to follow, namely whether to switch to subalgorithm $\mathcal{B}_C[1,2]$ or $\mathcal{B}_C[3]$, based on the first message it receives from $\mathcal{A}$. In this case, $\mathcal{B}_C$ switches to $\mathcal{B}_C[3]$ if $\mathcal{A}$ first sends message $\hat{e}_j$, and otherwise $\mathcal{B}_C$ follows $\mathcal{B}_C[1,2]$. Similarly, in the latter case, $\mathcal{B}_C$ switches to either $\mathcal{B}_C[1]$ or $\mathcal{B}_C[2]$ based on $\mathcal{A}$'s next response. Therefore the three pictures represent not different algorithms $\mathcal{B}_C[1\text{-}3]$ but just three subcases of a single algorithm $\mathcal{B}_C$.

By inspection of Figure 7, note that $\mathcal{B}_C[1]$ wins in the NM game in the case of event $\mathsf{NSM} \wedge \mathsf{Int}[1]$. Note that an extraction of $\hat{c}_i$ is allowed because $(\hat{m}_i, \hat{s}_i, \hat{PK}_i)$ is different from tag (m_i, s_i, PK_i) used in commitment c_i. Similarly, by inspection of Figure 8, note that $\mathcal{B}_C[2]$ wins the NM game in the case of the event that interleaving pattern II is followed by $\mathcal{A}$. Consequently, $\mathcal{B}_C$ wins in any of these cases as well. The case of $\mathcal{B}_C[3]$ is slightly different: Here the probability that $\mathcal{A}$ wins is actually at most 2^{-k} unconditionally, as long the commitment scheme is perfectly binding. Note that $\mathcal{B}_C[3]$ has the same 2^{-k} probability of winning in this case because it just returns a randomly chosen R to the challenger. Since event $\mathsf{Case1}$ implies one of these three cases, and we have that $\mathcal{B}_C$ wins in cases $(\mathsf{NSM} \wedge \mathsf{Int}[1]) \vee \mathsf{Int}[2]$, while in the remaining case $(\mathsf{NSM} \wedge \mathsf{Int}[3])$ the probability that $\mathcal{B}_C$ is greater or equal to the probability that $\mathcal{A}$ wins (given that this case happens), it follows that the probability that $\mathcal{B}_C$ wins is at least the probability $\Pr[\mathsf{Case1}]$, as required.

The construction of $\mathcal{B}_E[1]$ and $\mathcal{B}_E[3]$ are depicted in Figures 4 and 5. The construction works as follows. Receive the public key PK of the challenger. Then, on receiving m_i, m_j from $\mathcal{A}$, pick $R_i \in \{0,1\}^k$, compute $(c_i, d_i) \leftarrow \mathsf{com}(m_i, PK, R_i)$ and forward (m_i, PK, c_i) to $\mathcal{A}$. Send m_j to the challenger and forward the received ciphertext $e_j = \mathsf{Enc}_{PK}(m_j, R_j)$ (where R_j is a random k-bit string picked by the challenger) to $\mathcal{A}$. When $\mathcal{A}$ sends $\hat{e}_j = \mathsf{Enc}_{PK}(\hat{m}_j, \hat{R}_j)$, query it to the decryption oracle to obtain the plaintext $(\hat{m}_j, \hat{R}_j)$. Note that since $\hat{m}_j$ differs from m_j, $\hat{e}_j$ must also differ from e_j, $(\mathsf{NSM} \wedge \mathsf{Int}[1])$ and therefore the query to the decryption oracle is allowed. If $\mathcal{A}$ wins, then R_j must equal $\hat{R}_j \oplus \hat{R}_i \oplus R_i$, which $\mathcal{B}_E$ sends to the challenger to win the challenger game. The same holds in the case of the $\mathcal{B}_E[2]$ reduction.

5. Encryption-based SAS Authenticated Key Agreement Protocol

We call our SAS-AKA protocol Enc-AKA. The protocol is very similar to the Enc-MCA protocol. In fact Enc-AKA is simply an instance of Enc-MCA where P_i's message m_i is set to null and P_j's message m_j is a fresh random key which P_j picks for each session. See Figure 2 for a description of both protocols.

THEOREM 2 (**Security of Enc-AKA**). *If the commitment scheme is (T_C, ϵ_C)-NM and the encryption scheme is (T_E, ϵ_E)-SS-CCA, then the Enc-AKA protocol is (T, p)-secure against (n, τ_t, τ_c)-attacker for $p \geq 2n\tau_t\tau_c(2^{-k} + max(\epsilon_C, \epsilon_E))$ and $T \leq min(T_C, T_E) - \mu$, for a small constant μ.*

Proof (Sketch): We show that if there exists a (n, τ_t, τ_c)-adversary $\mathcal{A}$ which can attack the proposed protocol in time $T \geq min(t_C, t_E) - \mu$ with probability p better than $2n\tau_t\tau_c(2^{-k} + max(\epsilon_C, \epsilon_E))$, then there exists an adversary $\mathcal{B}_C$ which can win the NM security game of the commitment scheme with a probability better than $2^{-k} + \epsilon_C$ or there exists an adversary $\mathcal{B}_E$ which can win the SS-CCA challenger game of the encryption scheme with a probability significantly better than $2^{-k} + \epsilon_E$.

$\mathcal{A}$ succeeds if it can find a pair of players P_i (initiator) and P_j (responder) both running a "partnered" session with session id s, and can distinguish the session key computed by either of them, from random. We will consider the case where $\mathcal{A}$ tests the initiator and the case when $\mathcal{A}$ tests the responder separately below.

Both reduction algorithms, $\mathcal{B}_C$ and $\mathcal{B}_E$ start by guessing some session initialized as $(P_i, init, P_j, s)$ (there are at most $n\tau_t/2$ of these). We'll call this a (P_i, s) session, but this choice determines P_j. Both reductions also pick one session at random among all sessions of the form $(P_j, resp, P_i, s')$, for the above P_i, P_j pair (that's additional τ_c guesses). Additionally, each reduction guesses whether it's (P_i, s) or (P_j, s') that will be tested. If $\mathcal{A}$ tests some other session than the one guessed by $\mathcal{B}_C$ or $\mathcal{B}_E$, either reduction outputs a random bit. Therefore, as in the reduction of Enc-MCA protocol security, Theorem 1, the success probability of this reduction deteriorates by a factor of $n\tau_t\tau_c$.

In either case (initiator or responder) considered below, the reduction considers two subcases, and if it guesses which subcase it is prepared to handle; this results in additional factor of 2 in the security degradation, thus leading to the $p \leq 2n\tau_t\tau_c * [2^{-k} + max(\epsilon_C, \epsilon_E)]$ bound on p.

(1) Consider the case when $\mathcal{A}$ attacks the initiator P_i. We first argue that $\mathcal{A}$ cannot make the initiator P_i accept a key $\hat{K}^{(s)}$ different from $K^{(s)}$ picked by P_j on the session s. This is because the success of $\mathcal{A}$ in doing so is clearly equivalent to an attack against P_j to P_i direction of the Enc-MCA protocol shown in Figure 2 and follows directly from the reductions $\mathcal{B}_C[1], \mathcal{B}_C[2], \mathcal{B}_C[3]$, and $\mathcal{B}_E[1]$ and $\mathcal{B}_E[2]$ shown in the proof of Theorem 1. Note that these reductions will also need to simulate the responses to the "reveal" queries issued by $\mathcal{A}$. In the first three reductions, our algorithm is able to perfectly simulate the responses to reveal queries by responding with the session keys that it simply picks itself or it obtains by following the protocol. While in the last two reductions, to answer the reveal queries corresponding to sessions of the initiator P_i, the reduction makes use of the decryption oracle; for any other session, where P_i is not an initiator, "revelation" of keys is done by following the protocol.

From the above argument, it follows that P_i must output the same key $K^{(s)}$ which was picked by P_j on session s. If $\mathcal{A}$ now succeeds in distinguishing this key from random, we reduce it to an attacker $\mathcal{C}_E$ against the SS-CCA game of the encryption scheme, as shown in Figure 6. The simulation and "revelation" of keys of the sessions other than the "tested" session, other than the ones corresponding to P_i and the ones where P_i is not an initiator, are done by following the protocol. While to simulate and answer the "reveal" queries corresponding to sessions of the initiator P_i, the reduction makes use of the CCA decryption oracle.

(2) Consider the case when $\mathcal{A}$ attacks the responder P_j by succeeding in sending a public key $\hat{PK}_i$ different from PK_i. In this case, we reduce $\mathcal{A}$

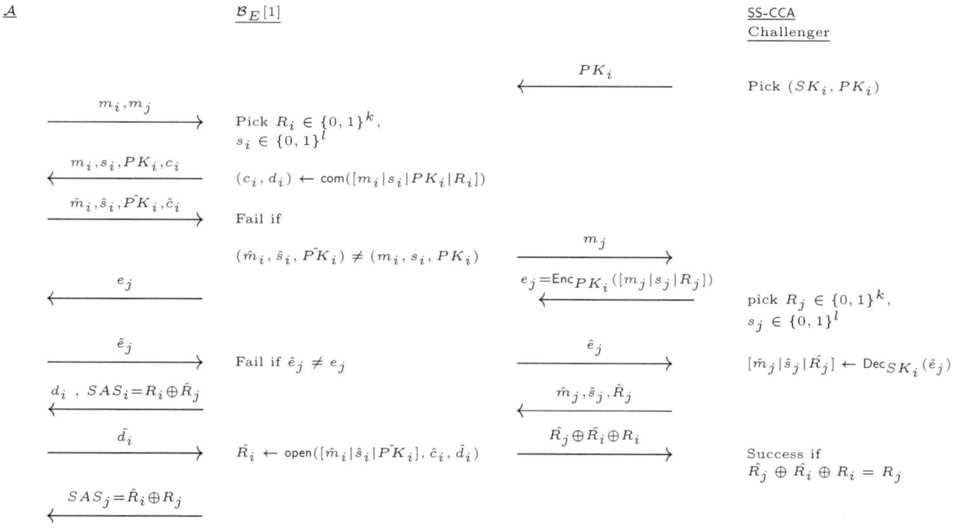

```
    A                              B_E[1]                                              SS-CCA
                                                                                       Challenger

                                                         ←——— PK_i ————                Pick (SK_i, PK_i)

    ——— m_i, m_j ———→      Pick R_i ∈ {0,1}^k,
                           s_i ∈ {0,1}^l
    ←— m_i, s_i, PK_i, c_i —  (c_i, d_i) ← com([m_i|s_i|PK_i|R_i])
    — m̂_i, ŝ_i, P̂K_i, ĉ_i →   Fail if
                           (m̂_i, ŝ_i, P̂K_i) ≠ (m_i, s_i, PK_i)
                                                         ———— m_j ————→
    ←——— e_j ————                                        ←— e_j = Enc_{PK_i}([m_j|s_j|R_j]) —
                                                                                       pick R_j ∈ {0,1}^k,
                                                                                       s_j ∈ {0,1}^l
    ——— ê_j ————→         Fail if ê_j ≠ e_j              ———— ê_j ————→                [m̂_j|ŝ_j|R̂_j] ← Dec_{SK_i}(ê_j)
    ←— d_i, SAS_i=R_i⊕R̂_j —                              ←— m̂_j, ŝ_j, R̂_j —
    ——— d̂_i ————→         R̂_i ← open([m̂_i|ŝ_i|P̂K_i], ĉ_i, d̂_i)   —— R̂_j ⊕ R̂_i ⊕ R_i →    Success if
                                                                                       R̂_j ⊕ R̂_i ⊕ R_i = R_j
    ←— SAS_j = R̂_i ⊕ R_j —
```

FIGURE 4. Construction of $\mathcal{B}_E[1]$ $((m_i, s_i, PK_i) = (\hat{m}_i, \hat{s}_i, \hat{PK}_i)$, interleaving case I)

to an attack $\mathcal{B}_C$ which executes sub-algorithms $\mathcal{B}_C[1]$, $\mathcal{B}_C[2]$ and $\mathcal{B}_C[3]$, based on the message interleaving patterns. This follows directly from the constructions $\mathcal{B}_C[1]$, $\mathcal{B}_C[2]$ and $\mathcal{B}_C[3]$, of the proof of the Theorem 1. Note that on any session except the tested session, the reduction simply follows the protocol and is therefore able to respond to the "reveal" queries by $\mathcal{A}$ with the session keys that it outputs.

Now, consider the case when $\mathcal{A}$ attacks the responder P_i, but sets $PK_i = \hat{PK}_i$. In this case, we reduce $\mathcal{A}$ to a CCA attacker similarly as shown in Figure 6 and as we argued above for the case of $\mathcal{A}$ attacking the initiator.

Figure 5

$\mathcal{A}$ $\mathcal{B}_E[3]$ SS-CCA Challenger

$\xleftarrow{PK_i}$ Pick (SK_i, PK_i)

$\xrightarrow{m_i, m_j}$ Pick $R_i \in \{0,1\}^k$, $s_i \in \{0,1\}^l$

$\xleftarrow{m_i, s_i, PK_i, c_i}$ $(c_i, d_i) \leftarrow \text{com}([m_i|s_i|PK_i|R_i])$

$\xrightarrow{\hat{e}_j}$ $\xrightarrow{\hat{e}_j}$ $[\hat{m}_j|\hat{s}_j|\hat{R}_j] \leftarrow \text{Dec}_{SK_i}(\hat{e}_j)$

$\xleftarrow{d_i,\ SAS_i = R_i \oplus \hat{R}_j}$ $\xleftarrow{\hat{m}_j, \hat{s}_j, \hat{R}_j}$

$\xrightarrow{\hat{m}_i, \hat{s}_i, \hat{PK}_i, \hat{c}_i}$ Fail if

$(\hat{m}_i, \hat{s}_i, \hat{PK}_i) \neq (m_i, s_i, PK_i)$ $\xrightarrow{m_j}$

$\xleftarrow{e_j}$ $\xleftarrow{e_j = \text{Enc}_{PK_i}([m_j|s_j|R_j])}$ pick $R_j \in \{0,1\}^k$, $s_j \in \{0,1\}^l$

$\xrightarrow{\hat{d}_i}$ $\hat{R}_i \leftarrow \text{open}([\hat{m}_i|\hat{s}_i|\hat{PK}_i], \hat{c}_i, \hat{d}_i)$ $\xrightarrow{\hat{R}_j \oplus \hat{R}_i \oplus R_i}$ Success if $\hat{R}_j \oplus \hat{R}_i \oplus R_i = R_j$

$\xleftarrow{SAS_j = \hat{R}_i \oplus R_j}$

FIGURE 5. Construction of $\mathcal{B}_E[3]$ $((m_i, s_i, PK_i) = (\hat{m}_i, \hat{s}_i, \hat{PK}_i)$, interleaving case III)

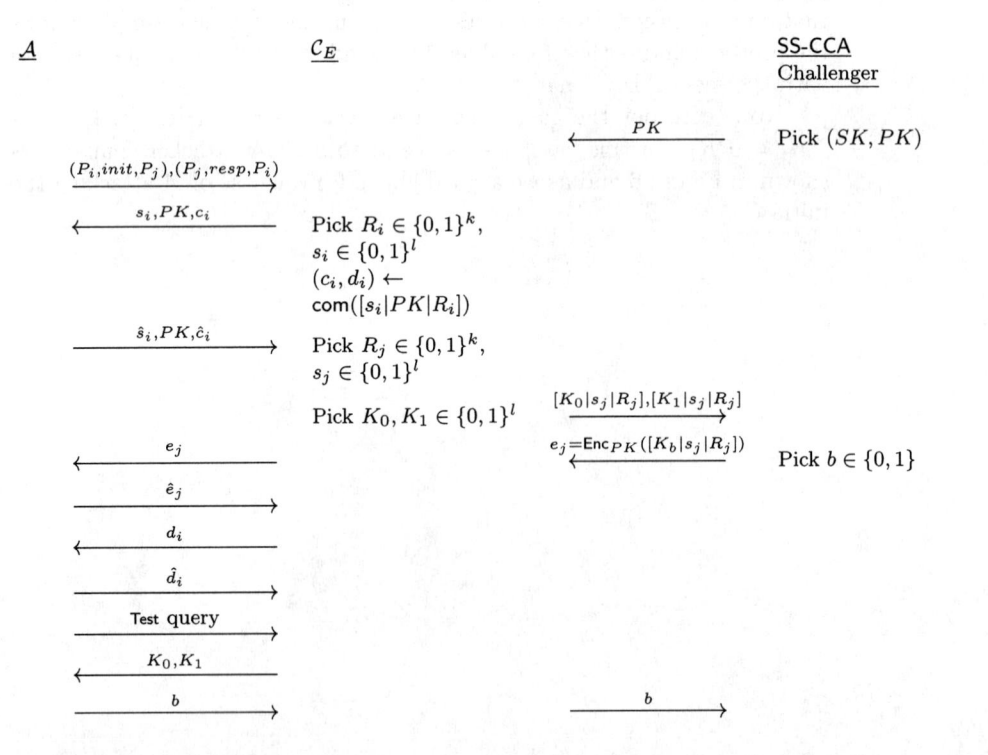

Figure 6

$\mathcal{A}$ $\mathcal{C}_E$ SS-CCA Challenger

$\xleftarrow{PK}$ Pick (SK, PK)

$\xrightarrow{(P_i, \text{init}, P_j), (P_j, \text{resp}, P_i)}$

$\xleftarrow{s_i, PK, c_i}$ Pick $R_i \in \{0,1\}^k$, $s_i \in \{0,1\}^l$ $(c_i, d_i) \leftarrow \text{com}([s_i|PK|R_i])$

$\xrightarrow{\hat{s}_i, PK, \hat{c}_i}$ Pick $R_j \in \{0,1\}^k$, $s_j \in \{0,1\}^l$

Pick $K_0, K_1 \in \{0,1\}^l$ $\xrightarrow{[K_0|s_j|R_j], [K_1|s_j|R_j]}$

$\xleftarrow{e_j}$ $\xleftarrow{e_j = \text{Enc}_{PK}([K_b|s_j|R_j])}$ Pick $b \in \{0,1\}$

$\xrightarrow{\hat{e}_j}$

$\xleftarrow{d_i}$

$\xrightarrow{\hat{d}_i}$

$\xrightarrow{\text{Test query}}$

$\xleftarrow{K_0, K_1}$

$\xrightarrow{b}$ $\xrightarrow{b}$

FIGURE 6. Construction of $\mathcal{C}_E$ from $\mathcal{A}$ for interleaving case I

Appendix A. IND-CCA Encryption

IND-CCA Encryption. We recall the standard notion of CCA-security of encryption, formalized in conrete security terms for non-uniform algorithms, and we sketch a proof of fact 1 of Section 2.

DEFINITION 1. *We call an encryption scheme (T, ϵ)-SS-CCA if for every T-bounded algorithm A and every pair of messages m_0, m_1 it holds that $p_1 - p_0 < \epsilon$ where*
$$p_b = \Pr[\mathcal{A}^{\mathsf{Dec}^C_{sk}(\cdot)}(pk, C) = 1 \mid (pk, sk) \leftarrow \mathsf{KeyGen}, \ b \leftarrow \{0,1\}$$
$$C \leftarrow \mathsf{Enc}_{pk}(m_b)]$$
where $\mathsf{Dec}^C_{sk}(\cdot)$ is a decryption oracle modified to output $\perp$ on C.

Proof of Fact 1: Let $\mathcal{M}$ be any uniform distribution over d messages which is easy to recognize and sample. (Fact 1 will be implied if $\mathcal{M} = \mathcal{M}_{\overline{m}}$ for some $\overline{m}$ and $d = 2^k$.) Let A be any T-bounded algorithm, and for any $x, y \in \mathcal{M}$, define $p_{x,y}$ as the probability that $A(\mathsf{Enc}_{pk}(x)) = y$, where the probability is taken over the randomness of the key generation, encryption, and A. Note that without loss of generality we can assume that A always outputs a message in $\mathcal{M}$, and so for every x we have $\sum_y p_{x,y} = 1$. Now, if encryption is (T, ϵ)-SS-CCA then for all x, y we have that $p_{x,x} < p_{x,y} + \epsilon$, or otherwise the SS-CCA definition would be violated for $m_0 = x$ and $m_1 = y$. Summing over all x's and y's we get $d * \sum_x p_{x,x} < \sum_{x,y} p_{x,y} + d^2 \epsilon$, and since $\sum_y p_{x,y} = 1$ for all x, we get $d * \sum x, x < d + d^2 \epsilon$, which implies the claim because
$$\Pr[A(C) = x \mid (pk, sk) \leftarrow \mathsf{KeyGen}, x \leftarrow \mathcal{M}, C \leftarrow \mathsf{Enc}_{pk}(m)]$$
$$= 1/d \sum_x p_{x,x} < 1/d + \epsilon$$

Appendix B. Difficulty in Extending the General Compilation Theorem of Pasini-Vaudenay

We give some intuition for the claim we make in the introduction, namely that the general composition theorem given by Passini and Vaudenay [11], for transforming KA protocols to SAS-AKA protocols given any SAS-MCA scheme, cannot be applied, in general, to KA schemes which share state between sessions. The theorem of [11] Consider a 2-round (non-authenticated) KA protocol. To save round complexity in the compiled SAS-AKA protocol, we would like to make the two messages generated by the KA protocol, m_i of the initiator P_i and m_j of the responder P_j, inputs to the SAS-MCA scheme, where P_i goes first, and m_j is possibly based on m_i. (The known 3-round SAS-MCA protocols allow the responder's message m_j to be picked in the second round.)

Note that at the time P_j computes his response m_j, following the algorithm of the KA protocol on the received message m_i, the message m_i is not yet authenticated by P_j. If the KA protocol does not share state between sessions, having P_j compute m_j on adversarially-chosen $\hat{m}_i$ can possibly endanger only the current session, and since the SAS-MCA subprotocol will eventually let P_j know that $\hat{m}_i$ was not sent by P_i, P_j will reject in this session anyway. (And so will P_i, because we can assume that m_j always contains the initator's own message m_i, or its hash.)

However, if P_j keeps a shared state between sessions then the information P_j reveals in m_j, computed on *unauthenticated* message $\hat{m}_i$, could potentially reveal some secret information that endangers all other sessions of player P_j, or at least all other sessions between P_j and P_i. It's easy to create a contrived example of a Key Agreement protocol which is secure in the static adverarial model when implemented over authenticated channels but yields an insecure SAS-AKA protocol when implemented with a SAS-MCA scheme in this fashion. For example, take any Key Agreement protocol, KA, secure over authenticated links, let each player P_j keep an additional long-term secret s_j and compute a per-partner secret $k_{ij} = F_{s_j}(<P_i>)$ where F is a PRF. If the initiator's message m_i contains a special symbol $\perp$, P_j sends $m_j = k_{ij}$ to P_i in the open. Otherwise, P_j follows the KA protocol to compute its response m_j, except that it attaches to it the resulting session key encrypted with a symmetric encryption scheme under k_{ij}. In the authenticated link model, and considering a static adversary, an honest player never sends the $\perp$ symbol. If the encryption is secure, encrypting the session key does not endanger its security. Also, if F is a PRF then learning values of the F function under indices corresponding to the corrupt players does not reveal any information about the values of F on indices corresponding to the honest players. On the other hand, this protocol is an insecure SAS-AKE protocol, because an adversary can inject message $\hat{m}_i = \perp$ on the insecure channel on behalf of any player P_i, and since P_j will reply with k_i, this allows the attacker to compute the keys for *all* sessions, past and future, between P_j and P_i.

This counter-example relies on an artificial KA protocol with shared session state where interference with a single session between a pair of players trivially reveals the keys on all sessions between the same players. However, this shows that the compilation technique of [11] can apply only to KA protocols with no shared state.

Of course, while this general compilation does not apply, a combination of any *particular* SAS-MCA protocol and a KA scheme with shared state can still be shown secure "from scratch", and that, with some simplifications to the SAS-MCA protocol made in the process, is exactly what we show in this paper.

Appendix C. Reductions $\mathcal{B}_C[1\text{-}3]$ in the Proof of Theorem 1

FIGURE 7. Construction of $\mathcal{B}_C[1]$ $((m_i, s_i, PK_i) \neq (\hat{m}_i, \hat{s}_i, \hat{PK}_i)$, interleaving case I)

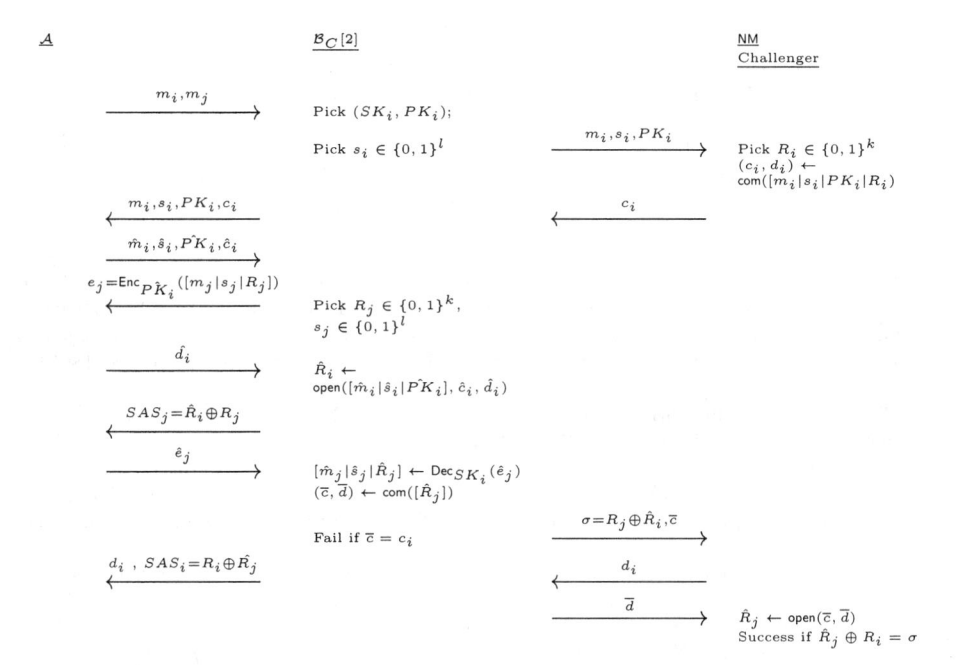

FIGURE 8. Construction of $\mathcal{B}_C[2]$ (interleaving case II)

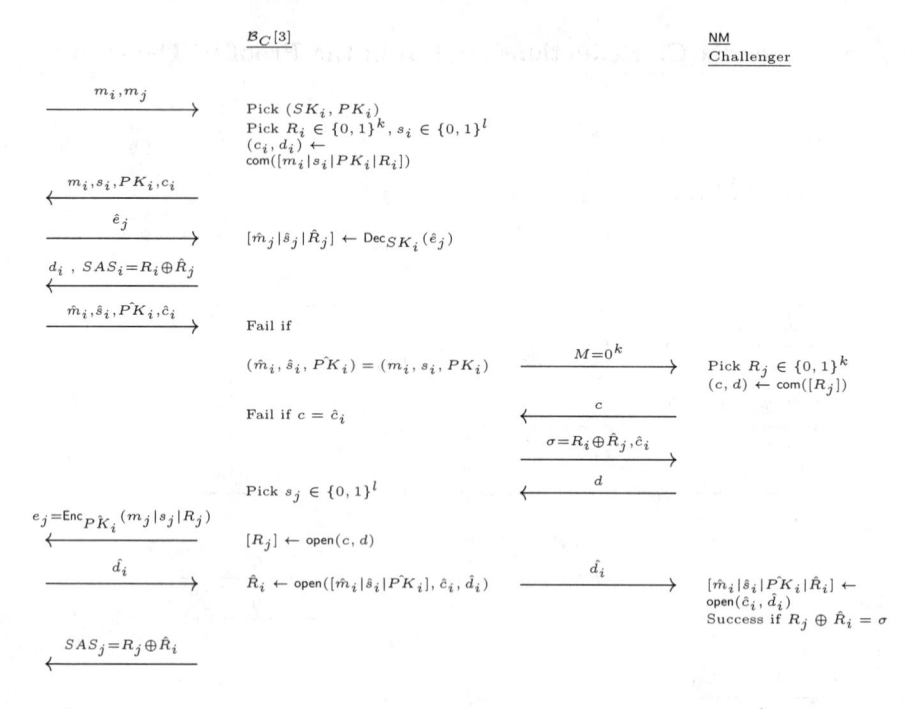

FIGURE 9. Construction of $\mathcal{B}_C[3]$ $((m_i, s_i PK_i) \neq (\hat{m}_i, \hat{s}_i, \hat{PK}_i)$, interleaving case III)

References

[1] Dirk Balfanz, Diana Smetters, Paul Stewart, and H. Chi Wong, *Talking to strangers: Authentication in ad-hoc wireless networks*, Network and Distributed System Security Symposium, 2002.
[2] M. Bellare, R. Canetti, and H. Krawczyk, *A modular approach to the design and analysis of authentication and key-exchange protocols*, Symposium on Theory of Computing, 2001.
[3] M. Bellare, T. Kohno, and V. Shoup, *Stateful public-key cryptosystems: How to encrypt with one 160-bit exponentiation*, ACM Conference on Computer and Communications Security, 2006.
[4] Ran Canetti and Hugo Krawczyk, *Analysis of key-exchange protocols and their use for building secure channels.*, EUROCRYPT, LNCS, 2001, pp. 453–474. MR1895449 (2003b:94037)
[5] Giovanni Di Crescenzo, Jonathan Katz, Rafail Ostrovsky, and Adam Smith, *Efficient and non-interactive non-malleable commitment*, EUROCRYPT, 2001, pp. 40–59. MR1895425 (2003b:94067)
[6] Christian Gehrmann, Chris J. Mitchell, and Kaisa Nyberg, *Manual authentication for wireless devices*, RSA CryptoBytes **7** (2004), no. 1, 29 – 37.
[7] Michael T. Goodrich, Michael Sirivianos, John Solis andGene Tsudik, and Ersin Uzun, *Loud and Clear: Human-Verifiable Authentication Based on Audio*, International Conference on Distributed Computing Systems (ICDCS), July 2006, Available at http://www.ics.uci.edu/ccsp/lac.
[8] Sven Laur, N. Asokan, and Kaisa Nyberg, *Efficient mutual data authentication based on short authenticated strings*, IACR Cryptology ePrint Archive: Report 2005/424 available at http://eprint.iacr.org/2005/424, November 2005.
[9] Sven Laur and Kaisa Nyberg, *Efficient mutual data authentication using manually authenticated strings*, CANS, 2006, pp. 90–107.
[10] Sylvain Pasini and Serge Vaudenay, *An optimal non-interactive message authentication protocol.*, CT-RSA, 2006. MR2243994 (2007e:94091)

[11] Sylvain Pasini and Serge Vaudenay, *SAS-Based Authenticated Key Agreement*, Workshop on Practice and Theory in Public Key Cryptograph (PKC), LNCS, April 2006. MR2423203 (2009d:94121)
[12] Ramnath Prasad and Nitesh Saxena, *Efficient device pairing using human-comparable synchronized audiovisual patterns*, Applied Cryptography and Network Security (ACNS), to appear, 2008.
[13] Michael Rohs and Beat Gfeller, *Using camera-equipped mobile phones for interacting with real-world objects*, Advances in Pervasive Computing (Vienna, Austria) (Alois Ferscha, Horst Hoertner, and Gabriele Kotsis, eds.), Austrian Computer Society (OCG), April 2004, pp. 265–271.
[14] Volker Roth, Wolfgang Polak, Eleanor Rieffel, and Thea Turner, *Simple and effective defenses against evil twin access points*, ACM Conference on Wireless Network Security (WiSec), short paper, 2008.
[15] Nitesh Saxena, Jan-Erik Ekberg, Kari Kostiainen, and N. Asokan, *Secure device pairing based on a visual channel (short paper)*, IEEE Symposium on Security and Privacy (S&P'06), May 2006.
[16] Nitesh Saxena and Borhan Uddin, *Blink 'em all: Scalable, user-friendly and secure initialization of wireless sensor nodes*, Cryptology and Network Security (CANS), December 2009.
[17] Claudio Soriente, Gene Tsudik, and Ersin Uzun, *BEDA: Button-Enabled Device Association*, International Workshop on Security for Spontaneous Interaction (IWSSI), 2007.
[18] Frank Stajano and Ross J. Anderson, *The resurrecting duckling: Security issues for ad-hoc wireless networks.*, Security Protocols Workshop, 1999, pp. 172–194.
[19] Serge Vaudenay, *Secure communications over insecure channels based on short authenticated strings*, Advances in Cryptology - CRYPTO 2005, Lecture Notes in Computer Science, no. 3621, Springer Verlag, 2005, pp. 309 – 326. MR2237314 (2007a:94245)

COMPUTER SCIENCE, UNIVERSITY OF CALIFORNIA, IRVINE
E-mail address: stasio@ics.uci.edu

COMPUTER AND INFORMATION SCIENCES, UNIVERSITY OF ALABAMA, BIRMINGHAM
E-mail address: saxena@cis.uab.edu

Publicly Verifiable Secret Sharing Using Non-Abelian Groups

Delaram Kahrobaei and Elizabeth Vidaurre

ABSTRACT. In his paper [9], Stadler develops techniques for improving the security of existing secret sharing protocols by allowing to check whether the secret shares given out by the dealer are valid. In particular, the secret sharing is executed over abelian groups. In this paper we develop similar methods over non-abelian groups.

1. Introduction to Publicly Verifiable Secret Sharing

Secret sharing is the process which involves a dealer and n participants. The dealer picks a secret and hands out to each participant an element, not equal to the secret, called a share through a secure channel. When any k of the participants come together, they can compute the secret, where k is called the threshold. Secret sharing has the property that if any $k-1$ participants come together, it is difficult for them to deduce the secret. The main example of this process is called Shamir's secret sharing scheme [10]. Stadler uses it in his first example of PVSS. The main application is the situation in which there is a bank with n managers and at least k managers have to be together to open a vault.

The method of secret sharing depends on the benevolence of the dealer because any party involved must trust that the dealer is distributing valid shares to each participant. Verifiable secret sharing adds a layer of security to the scheme by solving the problem of a cheating dealer. In other words, a verifiable secret sharing (VSS) sheme prevents the dealer from distributing a share to a participant that, together with an appropriate number of other shares, does not yield the secret.

The goal of publicly verifiable secret sharing (PVSS) is to allow anyone to verify that the participants received valid shares. In particular, P_i can check that P_j has a valid share. Applications of PVSS are software key escrow and design of electronic cash systems. An example of key escrow is Micali's fair cryptosystems [11].

In practice, the protocols proposed use a similar method for accomplishing their respective goals. In a VSS scheme, the dealer would make one or more pieces of information public as proof. Participants would then compute a value using their secret share and compare it to the public proof. In the PVSS scheme, both the

---

2010 *Mathematics Subject Classification.* Primary 68-XX, 20-XX.

The Research of the first author has been supported by PSC CUNY research foundation as well as City Tech Foundation.

The Research of the second author has been supported by NSF LS-AMP.

dealer and participants publish encrypted values of their secret information. It is preferable that the proof and/or encrypted values involve the least amount of pieces of information possible to prevent the dealer from providing proof that a fake share is valid to a particular participant (defeating the purpose of the scheme).

A VSS scheme can be *non-interactive*, meaning that the participants are not required to interact with each other in order to verify the validity of their shares. Moreover, in a PVSS scheme, the dealer distributes the shares to each participant using an assymetric key encryption algorithm. Using the public information and possible additional interaction with the dealer, any person can check that the encrypted secret share is valid. In the case that no interaction with the dealer is required, the PVSS scheme is called *non-interactive*.

This paper describes the two protocols developed by Stadler, both of which rely heavily on the well-known El Gamal encryption sheme. Next, we illustrate a new VSS sheme that uses nonabelian groups. Lastly, we attempt to mimick Stadler's schemes using the non abelian version of El Gamal's scheme, however we were unable to efficiently use all pieces of information, making the scheme insecure.

1.1. Discrete Logarithm and $\mathbb{Z}_p$ scheme. The following describes a Shamir's secret sharing scheme with an additional non-interactive VSS and PVSS protocol. In the PVSS protocol, the dealer uses El Gamal's scheme to distribute the shares and then proves to a verifier that the pair (A, B) associated to the participant P_i encrypts the discrete logarithm of a public element V. (see [9])

- Fixed: p a large prime, $q = \frac{(p-1)}{2}$ prime, $h \in \mathbb{Z}_p^*$ order q, G a group of order p, g a generator of G, $s \in \mathbb{Z}$ is the secret, k threshold.
- Public info: $S = g^s$, nonzero $x_i \in \mathbb{Z}_p$ assigned to P_i. , $F_j = g^{f_j}$ for random $f_j \in \mathbb{Z}_p$ and $j < k$.
- Private to P_i: $s_i = s + \sum_{j=1}^{k-1} f_i x_i^j \pmod{p}$.
- Secret can be recovered using Lagrange interpolation.
- VSS algorithm: P_i computes $S_i = S \prod_{j=1}^{k-1} F_j^{x_i^j}$ and if $S_i = g^{s_i}$, then P_i has a valid share.
- PVSS algorithm:
  - P_i choose a secret key $z \in \mathbb{Z}_q$ and publishes $y = h^z \pmod{p}$
  - the element $V = g^v$ of G and the pair $(A, B) = (h^\alpha, v^{-1} y^\alpha) \pmod{p}$ are made public
  - P_i can retrieve his share by calculating $m = A^z B^{-1} \pmod{p}$
  - for some fixed $l \approx 100$ and i such that $1 \leq i \leq l$, dealer/prover chooses $w_i \in \mathbb{Z}_q$ to compute $t_{hi} = h^{w_i} \pmod{p}$ and $t_{gi} = g^{y^{w_i}}$.
  - Using a cryptographically strong hash-function (for an in-depth discussion description of hash-functions see [2]), $\mathcal{H}_l : \{0,1\}^* \to \{0,1\}^l$, she publishes

  $$(c_1, \ldots, c_l) = \mathcal{H}_l(V||A||B||t_{h1}||t_{g1}||t_{h2}||t_{g2}|| \ldots ||t_{hl}||t_{gl})$$

  He/she also publishes

  $$(r_1, \ldots, r_l) = (w_1 - c_1\alpha \pmod{q}, \ldots, w_l - c_l\alpha \pmod{q})$$

– verifier would compute $t_{hi} = h^{r_i} A^{c_i}$ (mod p) and $t_{gi} = (g^{1-c_i} V^{c_i} B)^{y^{r_i}}$ and then check whether $\mathcal{H}_l(V||A||B||t_{h1}||t_{g1}||t_{h2}||t_{g2}||\ldots||t_{hl}||t_{gl})$ is $(c_1, \ldots, c_l)$.

1.2. eth root and $\mathbb{Z}_n$ scheme. For this interactive PVSS scheme, the dealer also uses El Gamal's scheme and then must prove that the pair (A, B) encrypts the e-th root of a public element M (see [9])

- secret to P_i: random $z \in \mathbb{Z}_n$
- secret to dealer: random $\alpha \in \mathbb{Z}_n$
- public: $g \in \mathbb{Z}_n^*$, $y = g^z$ (mod n), $(A, B) = (g^\alpha, my^\alpha)$, $M = m^e$
- P_i can retrieve his share by calculating $m = A^{-z}B$ (mod n)
- dealer picks $w \in \{0, \ldots, \lceil 2^l n^{l+\epsilon} \rceil\}$ and makes $t_g = g^w$ (mod n) and $t_y = y^{ew}$ public
- the verifier publishes $c \in \{0, \ldots, 2^l - 1\}$
- the dealer publishes $r = w - c\alpha$
- the verifier checks that $t_G = g^r A^c$ (mod n) and $t_y = y^{er}(B^e/M)^c$ (mod n)

2. New schemes

In recent years, non-abelian groups have been used in cryptography. One of the first cryptosystems over a non-abelian group was suggested by Anshel-Anshel-Goldfeld [1]. Conjugation in non-abelian groups is central to the cryptosystems proposed by [6]. In particular [4] and [8] proposed new secret sharing protocols using group presentations. Also [5] non-abelian El Gamal key exchange has been used. For more information on group-based cryptography see [7] and [3] In this paper we are proposing a new PVSS and VSS protocols using non-abelian groups.

2.1. Non-Commutative Key Exchange using Conjugacy. [5]

In this section, we discuss the use of conjugation in protocols over non-abelian groups as background to the new protocols proposed. Suppose G is a non-abelian group and $S, T \subset G$ such that $[S, T] = 1$. Bob takes $s \in S, b \in G$ and publishes b and $c = b^s$ as his public keys, keeping s as his private key. Here $b^s = s^{-1}bs$. If Alice wishes to send $x \in G$ as a session key to Bob, she first chooses a random $t \in T$ and sends
$$E = x^{(c^t)}$$
to Bob, along with the header
$$h = b^t.$$
Bob then calculates $(b^t)^s = (b^s)^t = c^t$ with the header. He can now compute
$$E' = (c^t)^{-1}$$
which allows him to decrypt the session key,
$$(x^{(c^t)})^{E'} = (x^{(c^t)})^{(c^t)^{-1}} = x.$$
The element $x \in G$ can now be used as a session key.

The feasibility of this protocol rests on the assumption that products and inverses of elements of G can be computed efficiently. To deduce Bob's private key from public information would require solving the equation $c = b^s$ for s, given the public values b and c. This is called the *conjugacy search problem* for G. Thus the security of this scheme rests on the assumption that there is no fast algorithm for solving the conjugacy search problem for the group G.

2.2. PVSS using non-abelian groups.
Authentication schemes described in [7] use conjugation, which of course require non-abelian groups. Although authentication serves a different purpose, the method also works for PVSS.

An algorithm analogous to one of Stadler's starts out with the non-abelian El Gamal. Each participant randomly chooses his private key $s \in S$ and publishes b and $c = b^s$. Here $b^s = s^{-1}bs$. The dealer then picks a random $t \in T$ and publishes $(A, B) = (b^t, x^{c^t})$. Consequently, the participant will find that his secret share is $x = B^{(A^s)^{-1}}$. For verification, the dealer must prove that the pair (A, B) encrypts the element with which a public element N and n are conjugate. The dealer chooses a random $y, w \in G$ and publishes $N = n^x$, $t_h = b^w$ and $t_g = b^{y^w}$. The verifier publishes $r \in \{0, 1\}$. If $r = 0$, then the dealer sends $c = wt$. If $r = 1$, then the dealer sends $c = w$. Then the verifier can check that $t_h = A^c$.

2.3. VSS using non-abelian groups.
Suppose there are n participants and each is given a secret share so that at least $t = n - 1$ of them have to be together to obtain the secret s. Let G be a nonabelian group where the search conjugacy problem is hard and F be an abelian subset with n elements. The dealer secretly sends f_i to each participant P_i. Next, for every $i \leq n$, the following are published

$$S = (\prod_{i=1}^{n} f_i)^{-1} s \prod_{i=1}^{n} f_i \quad \text{and} \quad h_i = (\prod_{i \neq j} f_j)^{-1} s \prod_{i \neq j} f_j.$$

Any t participants can recover the secret by conjugating h_i by the inverse of the product of their shares, where i is the missing participant. In order for P_i to verify that his/her share is valid, s/he can check that $f_i^{-1} h_i f_i = S$. Lastly, if P_i and P_j want to verify that each other's shares are valid, then they can check that $f_i^{-1} h_j f_i = f_j^{-1} h_i f_i$ without making their secret shares known to the other participant.

Clearly, the platform group cannot be abelian as conjugation is heavily used. If the group is given by a presentation, then the elements in the subset F can be any elements that have their (pairwise) commutators in the presentation of the group. If there are not enough of these elements, then powers of any one of these elements can serve as another secret share; the only problem with this is that the scheme becomes less secure in this case. Examples of non-abelian groups that can be used are polycyclic and metabelian groups. Metabelian groups would be particularly convenient as a platform group because it would be easy to find commuting elements.

Alternatively, defining $h_i = (\prod_{j \in H_i} f_j)^{-1} s \prod_{j \in H_i} f_j$ where H_i is a subset of F with t elements allows for any threshold t. Similarly, any t participants can recover the secret by conjugating the appropriate h_i by the inverse of the product of their shares. However, the dealer has not published enough information for a participant to verfify that his share is vaild.

The requirement that the search conjugacy problem be hard in the platform group is necesarry for the security of the scheme. If the search conjugacy problem were efficiently solvable in the group, then an adversary could determine f_i from S and h_i and therefore recover the secret.

References

1. I. Anshel, M. Anshel and D. Goldfeld. *An Algebraic Method for Public-Key Cryptography.* Mathematical Research Letters, 6 (1999), pp. 287-291 MR1713130 (2000e:94034)
2. Johannes Buchmann. Introduction to Cryptography. Springer 2004. MR2075209 (2005f:94084)
3. B. Fine, M. Habeeb, D. Kahrobaei, G. Rosenberger, *Aspects of Non-Abelian Group Based cryptography: A Survey and Open Problems*, JP Journal of Algebra, Number Theory and Applications, Vol. 21(1) 1-41 (2011) MR2840945 (2012f:94107)
4. M.Habeeb, D.Kahrobaei, V.Shpilrain, *Secret sharing using non-abelian groups*, preprint, 2012.
5. D.Kahrobaei, B.Khan, *A Non-commutative generalization of the ElGamal key exchange using polycyclic groups* Proceeding of IEEE, Pages: 1-5 (2006)
6. K.H. Ko, S.J. Lee, J.H. Cheon, J.W. Han, J. Kang, C. Park. *New Public-Key Cryptosystem Using Braid Groups* CRYPTO 2000, LNCS 1880, pp.166-183, Springer 2000 MR1850042 (2002i:94057)
7. A. Myasnikov, V. Shpilrain, A. Ushakov. Non-commutative Cryptography and Complexity of Group-theoretic Problems. Mathematical Surveys and Monographs, American Mathematical Society, 2011. MR2850384
8. Dimitrios Panagopoulos, *A secret sharing scheme using groups* Computing Research Repository, 2010
9. Markus Stadler, *Publically verifiable secret sharing* Proceeding EUROCRYPT'96 Proceedings of the 15th annual international conference on Theory and application of cryptographic techniques. (1996)
10. Adi Shamir, *How to share a secret* Communivations of the ACM, Pages 612-613 (1979) MR549252 (80g:94070)
11. Silvio Micali, *Fair public-key cryptosystems* Technical Report 579, MIT Lab. for Computer Science, September 1993.

CUNY GRADUATE CENTER, AND CITY TECH, CITY UNIVERSITY OF NEW YORK
E-mail address: DKahrobaei@GC.Cuny.edu

CUNY GRADUATE CENTER, CITY UNIVERSITY OF NEW YORK
E-mail address: EVidaurre@GC.Cuny.edu

A Note on the Hyperbolicity of Strict Pride Groups

Matthias Neumann-Brosig

ABSTRACT. We show that finitely presented groups with sufficiently high-powered relators are hyperbolic, if the set of relators fulfills a certain condition. We then give another proof of this statement for the class of strict Pride groups, which includes braid groups, generalized triangle and tetrahedron groups, and Coxeter groups. We explicitly calculate the desired exponents in this special case.

1. Introduction and basic definitions

In his 1987 paper (cf. [7]), Pride investigated properties of groups that are known as Pride groups nowadays. These groups have the defining property that every relator can be expressed by using at most two of the generating elements. A few years later, Pride obtained results about the cohomology of such groups [8].

We are going to be concerned with so-called strict Pride groups in this paper. These groups have the additional property that they can be finitely presented in a way such that for every pair of generators, there exists at most one relator involving exactly these two generating elements.

DEFINITION 1. Let G be a group. If G admits a presentation $\langle S|R \rangle$ such that
 (i) $|S| + |R| < \infty$,
 (ii) every element of R involves at most two elements of S and
 (iii) for every (unordered) pair (s_1, s_2), there is at most one element of $r \in R$ such that r involves exactly the elements s_1 and s_2,
then G is called a strict Pride group.

Note that in condition (iii) of Definition 1, the elements s_1 and s_2 can be identical. Omitting condition (iii) leads to finitely presented Pride groups. Many interesting groups are strict Pride groups, for example all generalized triangle and tetrahedron groups, which appear in contexts as diverse as abstract group theory, topology, function theory and geometry. Other examples include Coxeter groups (including dihedral groups) and braid groups.

2010 *Mathematics Subject Classification.* Primary 20F67, 20F06.

The author wishes to thank Bettina Eick for helpful comments and reading the manuscript in depth. He is also very much indebted to Gerhard Rosenberger for fruitful discussions and comments as well as pointing him to corollaries 10 and 11. Last, but not least, he wishes to express his gratitude to the referee for insightful remarks.

Hyperbolic groups, on the other hand, are groups fulfilling a set of metric conditions on their Cayley graph, which leads to strong theorems about their structure. For example, hyperbolic groups admit a Dehn algorithm and are finitely presentable (cf. [1], [5], [3]). Before we can define them, we need the following definition.

DEFINITION 2. *Let Γ be a group. A set $S \subset \Gamma$ of generators of Γ with $1 \notin S$ is called symmetric if $x \in S \Rightarrow x^{-1} \in S$ holds for all $x \in S$.*

There are many different ways of defining hyperbolic groups. We choose a geometric variant here.

DEFINITION 3. *A group Γ having a symmetric, finite set of generators S is called δ-hyperbolic if the Cayley-graph for (Γ, S) is δ-hyperbolic (as a metric space). Γ is called hyperbolic (or word-hyperbolic) if there is a real number $\delta \in \mathbb{R}_0^+$ and a set of generators S of Γ such that (Γ, S) is δ-hyperbolic. (cf. [2], [1], [6])*

Even though much is known about the structure of Pride groups, the question about their hyperbolicity has not yet been answered.

2. Small Cancellation Theory

In this section, we recall the metric condition of Small Cancellation Theory. These basic definitions of Small Cancellation Theory can be found in many textbooks on group theory (cf. [6],[1]), but we list the ones we need for ease of reference.

DEFINITION 4. *Let X be a set, $F(X)$ the free group on X and $R \subset F(X)$ such that each element is cyclically reduced and the following condition holds:*

$r \in R \Rightarrow r^{-1} \in R$, *every cyclically reduced conjugate r' of r is also $\in R$.*

In this case, we say that R is symmetric.

Let X be a set, $G = F(X)$ free on X and $w, a, b \in F(X)$. By denoting $w \equiv ab$ we will mean $w = ab$ and $length(w) = length(a) + length(b)$.

Let $R \subset F(X)$ be symmetric. A piece (of a word in $F(X)$) is a word $b \in F(X)$ s.th. there exist $r_1, r_2 \in R$ with $r_1 \equiv bc_1$ and $r_2 \equiv bc_2$, where c_1 and c_2 are elements of $F(X)$ and $r_1 \neq r_2$. This brings us to the metric small cancellation condition.

DEFINITION 5. *Let $G = \langle S|R \rangle$ be a group, with symmetric set R of relators. We say that the condition $C'(\lambda)$ is fulfilled if for all $r \in R$ and $r \equiv bc$, where b is a piece, the inequality $|b| < \lambda|r|$ holds.*

The goal is to use the following well-known theorem:

THEOREM 6. *Let $G \cong \langle S|R \rangle$ be a group, s.th. R fulfills the condition $C'(\lambda)$ for $\lambda \leq \frac{1}{6}$. Then G is hyperbolic (cf. [6], [1])*

3. Strict Pride groups with sufficiently high-powered relators are hyperbolic

Before we move on to state the theorems, we need another definition.

DEFINITION 7. *Let $G \cong \langle S \mid R \rangle$ be a finitely presented group. Then the pair (S, R) is called admissible if S and R are symmetric, and if r_1, r_2 are two elements of R that are not inverse to each other, then neither r_1 nor r_2 is completely cancelled in the product $r_1 r_2$.*

Note that every finitely presented group has an admissible presentation. Given a finite presentation $\langle S \mid R \rangle$ with symmetric sets S and R, we can construct one by using Tietze transformations to replace relators (as well as their inverses and conjugates) if they are completely cancelled in a product. This makes the total length of the presentation smaller while not changing the group. Iterating the process a finite number of steps, we get a symmetric presentation that is as short as possible with respect to this process, and thus admissible.

We will now prove the main theorems. The first one will be the general case for finitely presented groups. The second one will deal with strict Pride groups. We give two theorems because in the first one, we will not give explicit expressions for the exponents, thus making the proof much easier and shorter. In the proof of the second theorem, we will explicitly state the equations defining the desired exponents, which requires much more work.

THEOREM 8. *Let $G \cong \langle S \mid R \rangle$ such that the pair (S, R) is admissible. Then there is a natural number $n \in \mathbb{N}$ such that $\langle S \mid \{r^n | r \in R\} \rangle$ is hyperbolic.*[1]

PROOF. We want to apply theorem 6. To this end, we look at products of the form $r_1^m r_2^m$, where both r_1 and r_2 are elements of R and $m \in \mathbb{N}$. Because the pair (S, R) is admissible and each element of R is cyclically reduced, no copy of r_1 or r_2 is cancelled out completely in this product. It follows that a piece of r_1^m in the presentation $\langle S \mid \{r^m | r \in R\} \rangle$ cannot be longer then a piece of r_1 in $\langle S \mid R \rangle$.

Choosing $m \geq 6$, we get a presentation that satisfies the metric small cancellation condition $C'(\frac{1}{6})$. By applying theorem 6, we deduce that $\langle S \mid \{r^m | r \in R\} \rangle$ is hyperbolic. □

Now we are going to restate the theorem for the case of strict Pride groups in another form. As we want to calculate these exponents, and make them as small as possible, we allow for different exponents on different (non conjugate) defining relations. This is important in many special cases, where the explicit form of the relators has great impact on the proofs, such as deciding the Tits alternative (or the existence of non-abelian free subgroups) for generalized triangle and tetrahedron groups.

THEOREM 9. *Let $G \cong \langle S|R \rangle$ be a strict Pride group and R as in definition 1. Then there are natural numbers $m_r (r \in R)$ such that $\langle S|r^{m_r}, r \in R \rangle$ is hyperbolic.*

PROOF. Instead of working with R directly, we take the smallest symmetric subset R' of $F(S)$ containing R. It is easy to see that this does not change the statement, as hyperbolicity does only depend on the isomorphism type of a group, and $\langle S|R \rangle$ is clearly isomorphic to $\langle S|R' \rangle$. The elements of R' can contain either exactly one or exactly two of the generating elements. We define

$$C_i := \{r \in R' | r \text{ involvex exactly i generators}\},$$

where $i \in \{1, 2\}$. Each of the subsets C_i is symmetric, and $C_1 \cup C_2 = R'$. For each $x \in S$ and $r \in C_2$ let $k(x, r)$ equal the maximal length of a word of the form x^i or x^{-i} in r or any cyclic conjugate of r. We get the connection $k(x, r) = k(x, r')$ for any cyclic conjugate r' of r.

---

[1] The author wishes to thank the referee for pointing him to this level of generality.

Now we consider the product $r_1^{m_1} r_2^{m_2}$ of powers of two elements r_1, r_2 of R'. In each of the four important cases, we want to show that the length of the longest cancellation in this product is bound from above by the length of a piece of r_1 or r_2.

(i) Both r_1 and r_2 are elements of C_1, i.e. they contain exactly one generating element. If the two generators are distinct, there is no cancellation. If they are not, either $r_1 = r_2$ holds, and there is still no cancellation, or $r_1 = r_2^{-1}$, and there is nothing to show, as long as we have $m_1 = m_2$ in the end.

(ii) Exactly one of r_1 and r_2 is in C_1. We can assume this to be r_1, the other case is similar. In that case, the length of the longest possible cancellation in the word $r_1^{m_1} r_2^{m_2}$ for any $m_1, m_2 \geq 1$ is bounded by $k(x, r_2)$, where x is the uniquely determined generating element of r_1. It follows that for any piece b of r_2 that is cancelled in such a product, we get $|b| \leq k(x, r)$.

(iii) Both r_1 and r_2 are in C_2, but they have not more than one common generator. If there is none, then there is no cancellation. Therefore, we assume the existence of a unique common generator $x \in S$. The longest possible cancellation would be if r_1 ended in a subword x^i and r_2 started with a subword x^{-i}. So the length of the cancellation is at most $min\{k(x, r_1), k(x, r_2)\}$.

(iv) The relators r_1 and r_2 are both in C_2 and consist of the same generating elements. In this case, r_2 is conjugate to r_1 or r_1^{-1}, as $\langle S|R \rangle$ was a strict Pride group. If $r_1 r_2 = 1$ in $F(S)$ holds, this is also true for $r_1^i r_2^i$, and there is nothing to show (as long as we have $m_{r_1} = m_{r_2}$ in the end). Therefore we assume $r_1 r_2 \neq 1$ in $F(S)$. As this product necessarily starts with a nontrivial, initial subword of r_1 and ends with a nontrivial, terminal subword of r_2, we don't get any further cancellation in $r_1^i r_2^j$ for any $i, j \geq 1$.

For any $r \in C_2$ we define $k'(r)$ by

$$k'(r) := max\{|r| - |rr'|/2 \mid r' \text{ is a cyclically reduced conjugate of } r \text{ or } r^{-1}, rr' \neq 1\}.$$

That means that $k'(r)$ is the maximal length of a piece that can be cancelled as in (iv). To summarize the above: a piece of any $r^n (n \in \mathbb{N})$, where $r \in C_2$, cannot be longer than $l_r := max\{\{k(x,r) | x \in S\}, k'(r)\}$. For the metric condition $C'(\frac{1}{6})$ to hold, we need to bring r to a power m such that $m > 6\frac{l_r}{|r|}$.

For r in C_1 we get (with $p_r := max\{k(x, r')|r' \in R\}$) the condition $6\frac{p_r}{|r|} < m$ for the power m in r^m. Thus the metric small cancellation condition holds whenever the m_r's are collectively big enough for these equations to hold, and therefore the group

$$\langle S|r^{m_r}, r \in R' \rangle \cong \langle S|r^{m_r}, r \in R' \rangle$$

is word-hyperbolic. □

As generalized triangle and tetrahedron groups are strict Pride groups, we get the following corollaries.

COROLLARY 10. *Let* $G = \langle x, y | x^p = y^q = R(x, y)^r = 1 \rangle$, *where* $R(x, y) = x^{n_1} y^{m_1} ... x^{n_k} y^{m_k}$, $k > 0, 0 < n_i < p, 0 < m_i < q$ *for all* i, $1 < p, q, r$, *be a generalized triangle group. If* $p > 6 \, max\{n_i \mid i = 1, ..., k\}$, $q > 6 \, max\{m_i \mid i = 1, ..., k\}$ *and* $r > 6$ *holds*, G *is hyperbolic.*

If we choose the n_i in the intervall $-\frac{p}{2}, ..., \frac{p}{2}$ and the exponents m_i in $-\frac{q}{2}, ..., \frac{q}{2}$, we can extend this result a bit. The same is true for the next corollary.

COROLLARY 11. *Let*
$$G = \langle x, y, z \mid x^p = y^q = z^r = R_1(x,y)^n = R_2(y,z)^m = R_3(z,x)^s = 1 \rangle$$
be a generelized tetrahedron group, where
$$R_1(x,y) = x^{n_1} y^{m_1} ... x^{n_k} y^{m_k},$$
$$R_2(y,z) = y^{p_1} z^{q_1} ... y^{p_k} z^{q_k},$$
$$R_3(z,x) = z^{r_1} x^{s_1} ... z^{r_k} x^{s_k}.$$
If $n, m, s > 6$ and
$$p > 6\,max\{\ n_i \mid i = 1, ..., k\} \cup \{\ s_i \mid i = 1, ..., k\},$$
$$q > 6\,max\{\ m_i \mid i = 1, ..., k\} \cup \{\ p_i \mid i = 1, ..., k\},$$
$$r > 6\,max\{\ q_i \mid i = 1, ..., k\} \cup \{\ r_i \mid i = 1, ..., k\},$$
then G is hyperbolic.

Note that a very similar proof can be applied to groups which are not strict Pride groups, but in which for every two relators r_1, r_2 there is a generating element $x_1 \in r_1$ not contained in r_2 and vice versa, or in Pride groups which are not strict, but where any two words in the same generators are of the same length.

References

[1] Camps, T., Rosenberger, G and V. Große Rebel, *Einführung in die kombinatorische und die geometrische Gruppentheorie*, Heldermann Verlag 2008. - (Berliner Studienreihe zur Mathematik ; 19). - ISBN 978-3-88538-119-8 MR2378619 (2009g:20094)
[2] de la Harpe, Pierre, *Topics in geometric group theory*, University of Chicago Press 2000 . - (Chicago Lectures in Mathematics) MR1786869 (2001i:20081)
[3] Epstein, D.B.A. and Holt, D. *Efficient computation in word-hyperbolic groups*, London Math. Soc. Lecture Note Series 275 (2000), S. 66-77 MR1776767 (2001f:20085)
[4] Lyndon, R. C. and Schupp, P. E., *Combinatorial group theory*, Springer-Verlag Berlin 2001 MR1812024 (2001i:20064)
[5] Neumann-Brosig, M. and Rosenberger, G., *A note on the homology of hyperbolic groups*, deGruyter Verlag 2010. - (Groups, Complexity, Cryptography). - MR2747149 (2012c:20118)
[6] Ohshika, K., *Discrete Groups*, American Mathematical Society 2001 . - (Translations Of Mathematical Monographs) . - ISBN 978-0-82182-080-3
[7] Pride, S. J., *Groups with presentations in which each defining relator involves exactly two generators*, Journal of the London Mathematical Society 36 (1987), S. 245-256 MR906146 (88m:20074)
[8] Pride, S. J., *The (CO)homology of groups given by presentations in which each defining relator involves at most two types of generators*, Journal of the Australian Mathematical Society 52 (1992), S. 205-218 MR1143189 (92m:20024)

INSTITUT COMPUTATIONAL MATHEMATICS, TU BRAUNSCHWEIG, POCKELSSTR. 14, 38106 BRAUNSCHWEIG, GERMANY
E-mail address: m.neumann-brosig@tu-bs.de

An Algorithm to Express Words as a Product of Conjugates of Relators

Ellen Ziliak

ABSTRACT. In this article we will look at an algorithm that uses properties of the Cayley graph to rewrite a trivial word in a finitely presented group as a product of conjugates of relators.

1. Introduction

The goal of this paper will be to give a more efficient algorithm to rewrite a word in a finitely presented group as a product of conjugates of relators. This algorithm can be used assuming the 2-cocycles are given to do arithmetic in group extensions. Current algorithms for arithmetic in group extensions use an augmented coset table as described in [2] and [5]. Using an augmented coset table one quickly runs into difficulty in storing the table for large groups. This new algorithm will decrease the storage requirement by using properties of the Cayley Graph for the factor group at run time.

2. Generating Set S

We assume that we are given a finitely presented group $Q \cong F/N$, where F is a free group and N is the normal closure of the subgroup generated by the relators. Since Q is finitely presented $Q = <X|R>$ where X is the set of generators for Q, and R is the set of relators for Q. From this definition, $F = <X>$ and $N = <R>^F$. For the relators R, we need to make two assumptions. First there is no relator that is a subword of another relator and second we do not have a generator that occurs only once in a relator. However, if we are given a presentation for Q where these assumptions are not satisfied we can apply a series of Tietze Transformations to get a new presentation where these assumptions are satisfied. We will also consider a second larger set of relators $\tilde{R}$. We define the set $\tilde{R}$ to be the multiplicative closure of the set of all cyclic conjugates of the elements in R and their inverses.

2010 *Mathematics Subject Classification.* Primary 20F05, 20F10; Secondary 68Q70.

I wanted to give a special thanks to my thesis advisor Alexander Hulpke, for his time and help with this project.

We assume further that we have an algorithm, which for every element $a \in Q$, written as a word in the generators, produces a *unique* word in the generators (depending only on a, not on the original word) representing a. We will call this word the *normal form* of a. We will require that a subword of a word in normal form is itself in normal form.

One way to obtain this normal form would be if we assume we also have a faithful permutation representation for Q, and we define the normal form word to be the smallest length-lexicographic representative. These representatives can be computed using enumeration of the elements of Q, e.g. following [1].

We begin the process of expressing a word in N as product of conjugates of relators by defining a free generating set S for N, fulfilling certain conditions. We then show existence of this set S for a given set of relators R. Next we shall describe how to rewrite elements of N as words in S. Finally we will show that this rewriting process can be done without computing the full generating set S.

3. The Cayley graph of Q

A principal tool for our algorithm will be the Cayley Graph $\mathcal{C}$ for Q. This is the graph with vertex set Q and edges labeled by generators of Q. For each edge e of the Cayley Graph $\mathcal{C}$ we denote the origin of e by $o(e)$ and the terminus of e by $t(e)$. Two vertices $a, b \in Q$ are connected by an edge labeled by the generator x, where $a = o(x)$ and $b = t(x)$ if and only if $a \cdot x = b$ in Q. We call the identity vertex $1 \in Q$ the *root* of $\mathcal{C}$. We shall identify paths in $\mathcal{C}$ with elements of Q (or of F) by taking the corresponding word in the generators.

The normal form for Q defines for every vertex $a \in \mathcal{C}$ a path in $\mathcal{C}$ from the root to a. Since prefixes of normal forms are normal forms themselves, these paths form a spanning tree $\mathcal{T}$ for $\mathcal{C}$, called the *geodesic tree*. We shall call edges in $\mathcal{C}$ which do not lie in $\mathcal{T}$ *nontree* edges.

Our next step will be to show that a generating set for N must contain exactly as many generators as there are positive nontree edges in $\mathcal{C}$:

DEFINITION 3.1. Suppose that for $b, c \in Q$ where $b = o(g)$, $c = t(g)$, the Cayley Graph $\mathcal{C}$ has the edge $g = b \to^g c$, then this edge is called a **positive edge** for b and a **negative edge** for c.

THEOREM 3.2. *The number of positive nontree edges in $\mathcal{C}$ equals* $\mathrm{rank}(N)$.

Proof: By the Nielsen-Schreier theorem, [4, Prop 3.9] we have

$$\mathrm{rank}(N) = [F:N](\mathrm{rank}(F) - 1) + 1 = [F:N](\mathrm{rank}(F)) - ([F:N] - 1).$$

On the other hand, $\mathcal{C}$ has exactly one positive edge for each generator and each element of Q, i.e. it has $|Q| \cdot \mathrm{rank}(F) = [F:N] \cdot \mathrm{rank}(F)$ positive edges in total. Of these there are $[F:N] - 1$ edges in $\mathcal{T}$, which leaves

$$[F:N] \cdot \mathrm{rank}(F) - ([F:N] - 1) = \mathrm{rank}(N)$$

nontree edges. $\square$

We will abbreviate the set of positive nontree edges by PNE. For each edge $p = o(p) \to^p t(p) \in PNE$ in the Cayley Graph $\mathcal{C}$, there exists a path in $\mathcal{T}$, denoted by b, which goes from the root to the origin of p. For any edge $p \in PNE$, there exists a closed path s_p that begins and ends at the root of $\mathcal{C}$, that contains p as an edge. Further we will require that $b \cdot p$ is the prefix of s_p if we think of s_p as a word in F. Now choose a relator $r \in \tilde{R}$ whose prefix contains some suffix d of $b \cdot p$, so that $b \cdot p = c \cdot d$ in F. We know that such a relator r exists, since we are guaranteed to have a relator which begins with p in $\tilde{R}$ by our assumptions on the relators. Then the edges of s_p are labeled by $c \cdot r \cdot c^{-1}$ which is a conjugate of a relator in $\tilde{R}$. Since the path s_p goes from the root of $\mathcal{C}$ back to the root, it represents an element of N. Note the construction of this generating set is modeled after the generating set presented in [3], however we make an additional requirement that our generators are conjugates of relators.

Let $S = \{s_p \mid p \in PNE\}$ be the set of all these paths. To summarize S consists of a set of words which are a product of conjugates of relators. We pick a unique word s_p for every edge $p \in PNE$ with the property that the prefix of that word is $b \cdot p$. Then $|S| = \text{rank}(N)$, so once we show that S generates N we can conclude (because of [4, Prop 2.2]) that S is a free generating set for N.

4. Generating N

Assuming we are given a set S with the conditions described above then we want to show that this set in fact generates N. Since the generators are elements of N we know that this set must generate a subgroup of N. To show that the set S generates all of N we will construct a graph $\mathcal{B}$ from S, which is commonly called a Stallings Graph of N.

DEFINITION 4.1. The **Stallings Graph of** N for S is the directed graph $\mathcal{B}$ obtained from the disjoint union of the paths in S by identifying the root vertices.

The following example will help us to visualize this definition. We use the notation $^c r$ to represent $r^{c^{-1}}$. This notation is natural since c is a subword of the prefix b. Later we will use b to identify s_p. The necessity of this notation will be seen when we give an algorithm to construct this generating set. Let $Q = S_3 \cong F/N$ then a set S which generates N is $S = \{a^2, b^2, {}^a b^2, {}^b a^2, {}^b (ababab), {}^{ab} a^2, ababab\}$. The Stallings Graph $\mathcal{B}$ looks like the following figure.

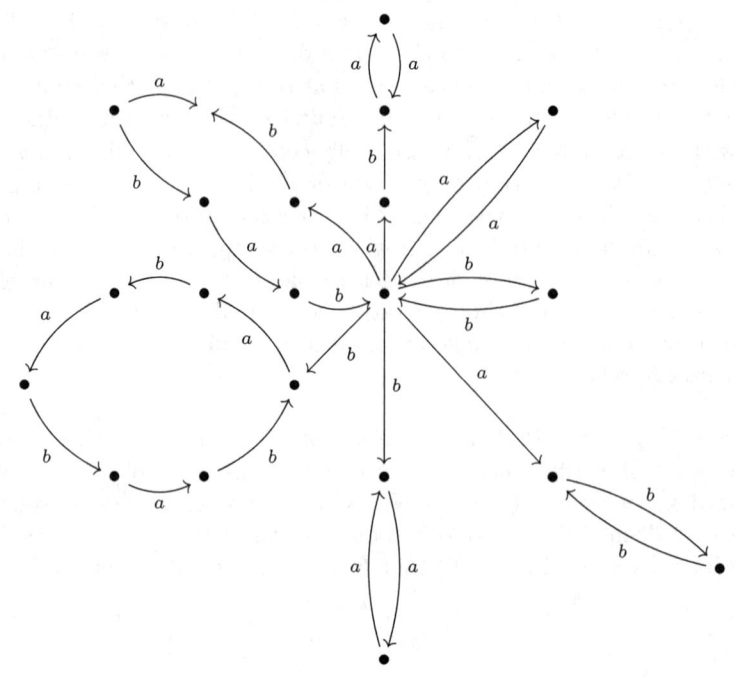

This graph was introduced by John R. Stallings [6, Algorithm 5.4]. Stallings proved as an immediate consequence of [6, Proposition 5.3] that if $\mathcal{B}$ contains a spanning tree of the Cayley Graph $\mathcal{C}$ then the generating set S used to construct the graph $\mathcal{B}$ is a free basis of N. In this argument Stallings shows that we can obtain the Cayley graph $\mathcal{C}$ by applying a series of folds to the bouquet $\mathcal{B}$. Since $\mathcal{T}$ the geodesic tree is a spanning tree of $\mathcal{C}$, if we show $\mathcal{T} \subseteq \mathcal{B}$ then by Stallings argument we can conclude that S generates N.

THEOREM 4.2. $\mathcal{T} \subseteq \mathcal{B}$

Proof: Let b be a path in $\mathcal{T}$ from the root to the vertex $o(p)$, where there exists an edge $p = o(p) \to^p t(p) \in PNE$ in $\mathcal{C}$. Hence there exists a path $s_p \in S$ such that the prefix of s_p is $b \cdot p$. We have thus shown that b is a path in $\mathcal{B}$. □

5. Rewriting

In this section we will show how a generating set S can be used to rewrite a word $w \in N$ as a product of generators in S. Recall S consists of a set of generators s_p for each positive nontree edge p. Each element of S has the property that the prefix is b, where b is the label of the path taken in $\mathcal{T}$ from the root to $o(p)$. We will use the prefix of b to identify the generator s_p used to rewrite $w \in N$.

Our first step will be to identify the longest prefix of $w \in N$ which is a path in $\mathcal{T}$. We will call this prefix r, then p is the first letter of $r^{-1}w$ which is the label of a nontree edge. We have two cases, either p is a positive edge or p is a negative edge.

If p is a positive edge, then there exists a generator $g_1 \in S$ which has rp as a prefix. We will choose g_1 as the first generator in the product and construct a new word $g_1^{-1}w \in N$ which we will now rewrite. This process will be repeated until we reach the empty word ϵ.

Next we will show that this process will terminate. Recall that S is a free generating set for N, therefore every word $w \in N$ can be expressed uniquely as a finite product of these generators $w = g_1 g_2 \cdots g_n$ where $g_i \in S$. We will prove by induction on the length of this product that our algorithm correctly identifies the generators g_i, for $1 \leq i \leq n$. If $w = g_1$ then g_1 must have rp as a prefix. Since we have a unique generator in S for each edge $p \in PNE$ we will correctly identify the generator g_1.

We now assume for any word that can be rewritten as a product of n generators in S the algorithm will correctly identify the generators. Consider a word $w \in N$, where w can be expressed uniquely as a product of $n+1$ generators, $w = g_1 g_2 \cdots g_n g_{n+1}$, and each $g_i \in S$ and $1 \leq i \leq n+1$. If w has rp as a prefix, then rp must be the prefix of the first generator in this product. If not, then some prefix of $g_2 \cdots g_{n+1}$ must cancel all of g_1 in F. In this case $g_1 = (g_2 \cdots g_k p_{k+1})^{-1}$ where p_{k+1} is the prefix of g_{k+1}. However, in Q, $(g_2 \cdots g_k p_{k+1})^{-1} = p_{k+1}^{-1} \neq g_1$ so therefore this cannot be the case in F since if $p_{k+1} = \epsilon$ this would contradict the rank of N. Hence the algorithm will correctly find the first generator in this product, after dividing off our new word $g_1^{-1}w$ can be expressed as a product of n generators in S, so by inductive hypothesis we will correctly rewrite $g_1 w^{-1}$. Therefore we conclude the process will terminate.

Next if p is a negative edge. Since p is a nontree edge and $\mathcal{T}$ is a spanning tree we know that p has a terminal vertex $t(p)$ which is a vertex in $\mathcal{T}$. Define s to be the label of the path in $\mathcal{T}$ from the root to $t(p)$. We can compute s by computing the normal form in Q of the word rp. We have the following picture to describe this situation, where s and r are both paths in $\mathcal{T}$.

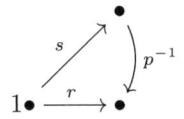

$$1 \bullet \xrightarrow{r} \bullet \xrightarrow{s} \bullet \xleftarrow{p^{-1}} \bullet$$

Notice that $sp^{-1}r^{-1}$ is an element of N since it is a closed path in $\mathcal{C}$. We will be able to rewrite $\tilde{w} = sp^{-1}r^{-1}$ as a product of generators since p^{-1} is a positive edge. After rewriting we have a new word $\tilde{w}w \in N$ where this product divides off (rps^{-1}) since $\tilde{w} = (rps^{-1})^{-1}$. We now continue to rewrite this new word $\tilde{w}w$ until we reach the empty word ϵ. Since the generators are from a free generating set for N this product is unique and we are computing the correct generators in this product. Therefore the rewriting process will terminate.

We end this section with an example of this rewriting. Consider again $Q = S_3 \cong F/N$ where N has the generating set $S = \{a^2, b^2, {}^a b^2, {}^b a^2, {}^b (ababab), {}^{ab} a^2, ababab\}$ and the geodesic tree using a length-lexicographic ordering looks like the following figure.

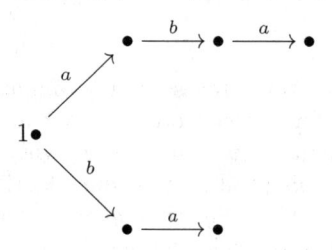

Suppose the word w that we wish to rewrite is $bab^{-1}aba \in N$. This process begins by finding the longest prefix r of w in $\mathcal{T}$. This gives $r = ba$ and $p = b^{-1}$, a negative edge so we have the following picture.

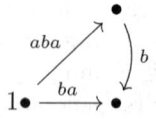

We begin by rewriting $\tilde{w} = ababa^{-1}b^{-1}$, and then update w. The generator which has $abab$ as a prefix is $g_1 = (ab)^3$, so we can rewrite $g_1^{-1}\tilde{w} = b^{-1}a^{-1}b^{-1}$. Continuing in this manner we will rewrite

$$w = bab^{-1}aba =^b (a^2) \cdot b^2 \cdot (ab)^{-3} \cdot^{ab} (a^2) \cdot^a (b^2) \cdot a^2.$$

Thus w is expressed as a product of elements in S.

6. Construction of S

In the previous section we showed that a set S as defined before can be used to rewrite a word as a product of conjugates of relators. However, it is not clear that for each positive edge $p \in PNE$ with $b = 1 \to^b o(p) \in Edge(\mathcal{T})$ we can find a conjugate of a relator s_p that meets these conditions. If the set $S = \{s_p | p \in PNE\}$ generates N then we must have $s_p \neq s_q$ for any other positive edge q. If this is not the case S would contain less than $\text{rank}(N)$ generators. What is still unclear is that a set S of size $\text{rank}(N)$ elements exists given a fixed set of relators R for Q. We will begin by showing that if two generators chosen for different positive edges are equal then the relators are cyclic conjugates.

LEMMA 6.1. *If $x, y \in PNE$ such that $s_x =^c (r) = s_y =^d (q)$, then r is a cyclic conjugate of q.*

Proof: Since $s_x = s_y$ we get $crc^{-1} = dqd^{-1}$, so $r =^{c^{-1}d} (q)$. We conclude that r is a conjugate of q. However, since we assumed that no relator is a subword of another we can conclude that r is a cyclic conjugate of q. □

Next if we assume we cannot find a conjugate of a relator with our prefix requirement for the edge $p \in PNE$ that has not already been chosen, then the conjugating element c where $s_p =^c (r)$ must be equal to the prefix $b = 1 \to^p o(p) \in Edge(\mathcal{T})$ which is the label of the entire path in $\mathcal{T}$.

LEMMA 6.2. *If we cannot find a conjugate of a relator s_p for $p \in PNE$ which has $b \cdot p$ as a prefix that is not equal to s_q for some other $q \in PNE$, then there is no prefix c of $b = 1 \to^b o(p) \in Edge(\mathcal{T})$, where $s_p =^c (r)$ and $s_p \neq s_q$ for any $q \in PNE$.*

Proof: Assume for every conjugate of a relator $s_p =^c (r)$ that has $b \cdot p$ as a prefix we cannot find a relator $r \in \tilde{R}$ that is not a cyclic conjugate of another relator chosen for another element s_q where $q \in PNE$. If $c \neq b$ then $b = ce_1e_2\cdots e_k$ where $e_1e_2\cdots e_k \neq \epsilon$ in F, then it could be possible to find a relator $\tilde{r} \in \tilde{R}$ so that $s_p =^{ce_1e_2\cdots e_j} (\tilde{r})$ with $j < k$ and $s_p \neq s_q$ for any other $q \in PNE$. However, if $b = c$ this possibility is removed. Since we cannot find a new conjugate of a relator s_p for $p \in PNE$ that has not already been used for some $q \in PNE$ we know this possibility does not exist. $\square$

Next we consider how many possible elements of N satisfy our conditions for s_p when the conjugating element is b.

LEMMA 6.3. *Within $\tilde{R}$ at least two relators begin with the letter p.*

Proof: Suppose not. Then p is a generator of F that occurs only once in a relator in R the original set of relators. This violates our original assumption on the set of relators. Thus at least two relators begin with p. $\square$

We will now show that a generating set S of this type exists, and therefore we can do the rewriting we seek.

THEOREM 6.4. *For each edge $p \in PNE$ with $b = 1 \to^b o(p) \in Edge(\mathcal{T})$ we can find a $s_p \neq s_q$ for any other $q \in PNE$ where the prefix of s_p is $b \cdot p$.*

Proof: Suppose there is an edge $p \in PNE$ where we cannot find a conjugate relator s_p with $b \cdot p$ as a prefix, and $s_p \neq s_x$ for any other $x \in PNE$. By lemma 6.1 we know that for each $x \in PNE$ where $s_x = s_p$ the relator for s_x is a cyclic conjugate of the relator choice for s_p. We also know that b is the only choice for our conjugating element by lemma 6.2. Finally we know that there exists at least two relators which begin with p by lemma 6.3.

Assume throughout the rest of this argument that all words are in F and are freely reduced as written. Suppose there are exactly two relators in $\tilde{R}$ which begin with p, namely pm and pn. Also suppose there are two other edges $x, y \in PNE$ such that a cyclic conjugate of pn and pm respectively was chosen for these edges. This means in particular that $n = n_1n_2xn_3$ and $m = m_1m_2ym_3$ where the n_i, m_i are subwords of n and m respectively so that $s_x =^d (n_2xn_3pn_1) =^b (pn)$ and $s_y =^f (m_2ym_3pm_1) =^b (pm)$. Now since all cyclic conjugates of relators are in $\tilde{R}$ we know $xn_3pn_1n_2, ym_3pm_1m_2 \in \tilde{R}$. Also by lemma 6.3 we know there exists at least two other nontrivial relators which begin with x and y, so $xe_x, ye_y \in \tilde{R}$ as well. Where $xe_x \neq xn_3pn_1n_2$ and $ye_y \neq ym_3pm_1m_2$ in F. However, since $\tilde{R}$ is multiplicatively closed we get $(xe_x)^{-1} \cdot (xn_3pn_1n_2) = e_x^{-1}n_3pn_1n_2 \in \tilde{R}$ and similarly $e_y^{-1}m_3pm_1m_2 \in \tilde{R}$.

This gives $pn_1n_2e_x^{-1}n_3$ and $pm_1m_2e_y^{-1}m_3$ as relators in $\tilde{R}$. If at least one of these is new we have found another relator that begins with p to choose from. The only way they would not be new is if these words are just a relabeling of our known relators that begin with p. If this is just a relabeling then, either $pn_1n_2e_x^{-1}n_3 = pn_1n_2xn_3$ and $pm_1m_2e_y^{-1}m_3 = pm_1m_2ym_3$, or $pn_1n_2e_x^{-1}n_3 = pm_1m_2ym_3$ and $pm_1m_2e_y^{-1}m_3 = pn_1n_2xn_3$. In the first case, if $pn_1n_2e_x^{-1}n_3 = pn_1n_2xn_3$ then $e_x^{-1} = x$ in F which implies $xe_x = \epsilon$ a contradiction to this relator being non trivial. Similarly $pm_1m_2e_y^{-1}m_3 \neq pm_1m_2ym_3$.

The second possibility is $pn_1n_2e_x^{-1}n_3 = pm_1m_2ym_3$ and $pm_1m_2e_y^{-1}m_3 = pn_1n_2xn_3$ in F. Manipulating the first equation we get $e_x^{-1}n_3m_3^{-1}y^{-1} = n_2^{-1}n_1^{-1}m_1m_2$. The second equation gives $n_2^{-1}n_1^{-1}m_1m_2 = xn_3m_3^{-1}e_y$. Putting these two equations together we get $e_x^{-1}n_3m_3^{-1}y^{-1} = xn_3m_3^{-1}e_y$. Therefore in F we have $x^{-1}e_x^{-1}n_3m_3^{-1}y^{-1}e_y^{-1} = n_3m_3^{-1}$, so we conclude that $x^{-1}e_x^{-1} = \epsilon$ and $y^{-1}e_y^{-1} = \epsilon$ in F since in a free group each word is equivalent to a unique reduced word. However that means $e_x^{-1} = x$ and $e_y^{-1} = y$, so $x \cdot e_x = (e_x^{-1} \cdot e_x) = \epsilon$ in F and similarly $ye_y = \epsilon$ in F. A contradiction to these being non trivial relators. Hence at least one of these relators must be new and this is not just a relabeling.

Suppose that this new relator was also chosen by another edge $z \in PNE$, then by a similar argument we can find another new relator which begins with p. Eventually since only rank(N)-1 generators can be assigned before this generator we will find a relator which begins with p that is not a cyclic conjugate of another chosen relator. Therefore we can compute a unique $s_p \in S$ for each edge $p \in PNE$. □

At this point we have shown how to construct a generating set S meeting our conditions and how the set can be used to do the rewriting we seek.

7. Rewriting Without Computing S

Previously we described how one can rewrite $w \in N$ using S, however $|S| = rank(N)$ is too large, so we want to be able to do this rewriting without computing the full generating set S. For the algorithm to be feasible we must reduce the quantity of generators we store. We begin by defining a graph which contains the elements of $\tilde{R}$.

DEFINITION 7.1. The **cyclically closed partial Cayley graph** (ccpcg) is a directed graph $\mathcal{P}$ obtained from folding the disjoint union of paths in $\tilde{R}$ by identifying the root vertices.

It should be noted this graph is the result of applying all possible foldings to a Stalling Graph defined for the set $\tilde{R}$. To give the reader a concrete example consider $Q = <a, b, c | a^2 = b^2 = abab = acac^{-1} = bcbc^{-1} = c^4>$ which is a transitive group of order 16. The ccpcg $\mathcal{P}$ looks like the following figure.

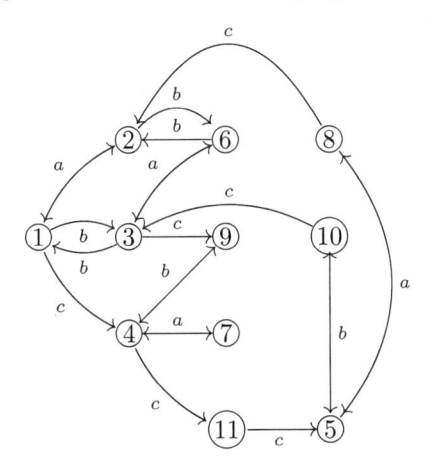

In general this graph is much smaller then the Cayley Graph, the number of vertices is bounded by the length of the elements of $\tilde{R}$. We will use this graph to compute the elements in S which generate N since all loops (closed paths) in $\mathcal{P}$ represent elements of $\tilde{R}$.

Now we describe how one can use the ccpcg $\mathcal{P}$ to compute the elements of S. We consider s_p for $p \in PNE$ with $b = 1 \to^b o(p) \in Edge(\mathcal{T})$ we use $\mathcal{P}$ to trace a prefix of $b \cdot p$ starting at the root vertex. Suppose $b = b_1 b_2 \cdots b_n$, if there is a path in $\mathcal{P}$ that begins at the root that is labeled by $b_1 b_2 \cdots b_n p$ we follow that path. If we are successful we reach a vertex $x = t(p)$. From x we can choose a series of edges which follow a path back to the root where we do not allow the first edge in this path to be p^{-1} (we do not backtrack). This loop would then be a relator r in $\tilde{R}$ with prefix $b \cdot p$ which we would choose as our generator $s_p = r$.

However, since $\mathcal{P}$ is not the full Cayley graph it is possible we will not be able to find a path in $\mathcal{P}$ that begins at the root that is labeled by $b_1 b_2 \cdots b_n p$ since some edges are missing. If we get stuck then we save b_1, by storing $c = b_1$ and we try to follow a path labeled by $b_2 b_3 \cdots b_n p$ the next longest suffix through $\mathcal{P}$. If we get stuck again, we then save the product $c = b_1 b_2$ and continue the process again. Eventually we have to find some suffix of $b \cdot p$ that is a path in $\mathcal{P}$, since at a minimum p is a path in $\mathcal{P}$ leaving the root. This is guaranteed by our assumptions on the relators and the inclusion of cyclic conjugates. After tracing this suffix, we will choose a path back to the root, the labels of this loop will represent an element $r \in \tilde{R}$. The element of S we will compute is $s_p =^c (r) = crc^{-1}$. Notice here it is more natural to conjugate by c^{-1}.

In this way we can use this object $\mathcal{P}$ to calculate the required generators during the rewriting process. Of course there is still the issue of the choice of path back to the root. Since we want S to be a free generating set for N, we must make the correct choice.

In lemma 6.1 we proved if two generators chosen for different positive edges are equal then the relators are cyclic conjugates. When we choose a path back to the root we need to be sure the relator defined by the labels of this closed path is not a cyclic conjugate of a relator that will be chosen for a different positive nontree edge.

Since we are not storing generators we need to have a way to determine which positive nontree edges get assigned the various relators. To break ties if a relator has a cyclic conjugate that could be assigned to more than one edge in PNE, we will assume that the edge $x \in PNE$ for which the algorithm computes the shortest conjugating element with respect to the well-ordering used for our normal form for Q will get assigned the appropriate cyclic conjugate of the relator.

Eventually we will either find a relator which works, or if possible make our conjugating element longer. If at some point it is not possible to make a longer conjugating element and we cannot find a relator that does not get assigned by our tie break to another element we must assign a generator. To make this choice we will choose the shortest path back to the root as long as this choice leaves more than one remaining choice for a relator for any other conflicting PNE edges. We will store for this word which we are rewriting a list of generators which cannot be chosen. At this point the rewriting process must be restarted with our original word. The list of conjugate relators we cannot use will be very small, since only words with the longest conjugating elements will be problematic. To give a bound on how many relators we need to store, consider a worst case suppose $w = {}^{c_1}(r_1)^{c_2}(r_2)\cdots^{c_n}(r_n)$. If each conjugating element c_i is the longest conjugating element for all other cyclic conjugates of r_i that are elements of S, then if l_i is the number of conflicting conjugates for r_i we would have to store $\sum_{i=1}^{n}(l_i - 1)$ words. This is significantly smaller than rank(N).

THEOREM 7.2. *We can choose a relator for s_p where $p \in PNE$ that will leave at least two other choices of relators for each conflicting edge in PNE.*

Proof: We showed in theorem 6.4 that for each edge $p \in PNE$ with $b = 1 \to^b o(p) \in Edge(\mathcal{T})$ we can find $s_p \neq s_q$ for any other $q \in PNE$ where the prefix of s_p is $b \cdot p$. We will now show that we can satisfy the additional criteria that there exists a relator $r \in \tilde{R}$ such that for any conflicting positive nontree edge q there are at least 2 other relators we can use to calculate s_q. Suppose not, that is suppose $q \in PNE$ is a conflicting edge where $r = px_1qx_2$ is a relator for s_p. Suppose further that qx_2px_1 and qe_q are the only possible choices for relators for s_q. Since we know by lemma 6.3 there exists another relator $pn \in \tilde{R}$ which begins with p, we can compute another element $qx_2n^{-1}x_1 \in \tilde{R}$. Since only two relators begin with q, and $pn \neq \epsilon$ in F we can conclude $qx_2n^{-1}x_1 = qe_q$. In this case we cannot choose r as our relator so then we must choose pn. Suppose $pn = pn_1tn_2$ also has a conflicting positive nontree edge t where only tn_2pn_1 and te_t are possible choices for s_t. Using a similar process we find another element $tn_2x_2^{-1}q^{-1}x_1^{-1}n_1 = te_t$. However at this point following an argument exactly the same as the one used in theorem 6.4 we find that either $tn_2x_2^{-1}q^{-1}x_1^{-1}n_1 \neq te_t$ or $qx_2n^{-1}x_1 \neq qe_q$ so we have at least one relator which satisfies our additional condition. □

8. Original Relators

At this point we have given an algorithm to rewrite words as a product of conjugates of an extended set of relators $\tilde{R}$. Recall $\tilde{R}$ is the multiplicative closure of the cyclic conjugates of the relators in R. Sometimes it is necessary, for example, in the group extension setting to rewrite as a product of conjugates of the given relators R. Using a similar algorithm we can rewrite all words in $\tilde{R}$ as a product of cyclic conjugates of relators in R. To do this second rewriting we must assume that our original set of relators R meets one further assumption that we cannot have two relators of the form abc and adc (or any cyclic conjugate of adc) where a, b, c, d are words in F, and no free cancellation takes place in forming abc. If our presentation does contain two relators of this form we can update our presentation using a series of Tietze Transformations.

To do this rewriting we consider a new smaller partial Cayley graph.

DEFINITION 8.1. A **partial Cayley graph** (pcg) is a directed graph $\mathcal{A}$ obtained from folding the disjoint union of paths in R by identifying the root vertices.

At this point we do the same rewriting algorithm as before using $\mathcal{A}$ however now our choice of path back is removed since we know the only type of conjugate we can have is a cyclic conjugate of the original relators. Our additional assumptions on R guarantee only one relator will be identified for each cyclic conjugate.

9. Conclusions

We have now given an algorithm to rewrite words in a finitely presented group as a product of conjugates of relators. Current algorithms require the storage and computation of an augmented coset table. To compare these algorithms let $|Q| = n$, and the number of generators for F be m, then in an augmented coset table requires we store $n \times m$ entries along with $n \times m$ augmented edges. Our algorithm requires storage of the cyclically closed partial Cayley graph $\mathcal{P}$ and the geodesic tree $\mathcal{T}$. The size of $\mathcal{P}$ is proportional to the length of the updated presentation which includes cyclic conjugates of the relators for Q. If the presentation has k relators of length $n_1, n_2, \ldots n_k$ respectively, then $\mathcal{P}$ will have at most $(n_1 - 1)n_1 + (n_2 - 1)n_2 + \cdots + (n_k - 1)n_k$ vertices and at most $n_1^2 + n_2^2 + \cdots + n_k^2$ edges. In practice however we generally have far less of each since from each vertex we can have at most rank(F) edges leaving. The geodesic tree stores the normal forms for Q, one for each element of Q, so it has size n. It is worth noting that both of these graphs are subgraphs of the Cayley graph which is used in the existing algorithm. Therefore our updated algorithm decreases the storage requirement at the expense of an increase in run time.

This algorithm has been implemented in GAP where we did several examples. In the implementation we do not build the entire geodesic tree, but rather we start with a subtree and only extend the tree if the word we are rewriting requires a longer branch. This will happen if our word requires two generators where an overlap is possible and by using a partial tree we had lost uniqueness in our rewriting by assigning a conjugate of a relator to an edge p which is not in PNE. Also, in the implementation we do not consider all possible products of cyclic conjugates of relators. Instead since this list is much longer than necessary, we only have a choice of two relators for each path in $\mathcal{P}$ which we compute as needed. We only extend this list if our condition in theorem 7.2 cannot be satisfied. If we increase the number of relators for a branch the rewriting for the current word must be restarted.

To look at some particular examples after the geodesic tree and the ccpcg were constructed for the group Q, the increase in run time to do the rewriting was not significant. To rewrite for example in $Q = PSp_6(2) =< a, b | a^2 = b^7 = (ab)^9 = (a^{-1}b^1a^{-1}b^{-1}a^{-1}b^{-1}ababab)^2 = (a^{-1}b^{-1}ab)^3 = (a^{-1}b^{-2}ab^2)^2 >$ which is a group of order $1,451,520$ the ccpcg $\mathcal{P}$ had 145 vertices and rewriting took less than half a second at run time. To get a better idea of the complexity of the algorithm, let us consider the rewriting of a few particular examples. For the word

$$word = a^{-3}ba^{-2} \cdot ba^{-3}b^{-1} \cdot a^{-1}b^2a^{-1} \cdot ba^{-2}b^{-1} \cdot ab^{-2}a^{-1} \cdot ba^{-1}b^{-2}a^{-1}$$

the original algorithm returned

$$(a^{-2}) \cdot (a^{-2}) \cdot (^{(ab)}(a^{-2})) \cdot (^{(ab^2)}(a^{-2})) \cdot (^{(ab^2)}(a^{-2})) \cdot$$
$$(^{(ab^2ab^{-1})}(a^{-2})) \cdot (^{(ab^2ab^{-1}ab^2)}(a^{-2})) \cdot (^{(ab^2ab^{-1}ab^2ab)}(a^{-2})) \cdot (^{(ab^2ab^{-1}ab^2)}(a^2))$$

which required the computation of 9 elements of S for a word of length 28. In rewriting this word there also were no updates made to the partial geodesic tree stored and no additional elements of $\tilde{R}$ from our initial set were required. Also we rewrote several other words and saw on average for a word of length n, where n has about 30 letters, we were required to compute about $\frac{n}{4}$ generators to do the rewriting. However these numbers are only relevant to this presentation and are highly dependent on the random words that were rewritten. The next example stems from a question asked by V. Mazurov in the GAP-Forum email list. As a specific example he sent the group $G =< x, y, z | x^2, y^2, z^2, (xy)^4, (xz)^4, (yz)^4, ((xy)^2z)^4, ((xz)^2y)^4, ((yz)^2x)^4, ((xy)^2(xz)^2)^4, ((xy)^2(yz)^2)^4, ((xz)^2(yz)^2)^4, (xyzy)^4, (xzyz)^4, (zxyx)^4 >$, and the word to be rewritten $(xyzxzyxyzxz)^2$. Using the algorithm the word rewrites to a product of 40 conjugate relators. Ignoring conjugates of the relators x^2, y^2, z^2 this product is

$$A :=^{(xyzxzyxy^2)}(((xz)^2y)^{-4}) \cdot^{(xyz)}((xzyz)^4) \cdot^{(xzxzyz)}((xz)^4) \cdot (xz)^4$$

This means that the word/A=B lies in $< x^2, y^2, z^2 >^G$. But since these relators form a confluent rewriting system, we easily rewrite

$$B = (z^{-1}x^{-1}z^{-1}x^{-1}x^{-1}y^{-1}x^{-1}y^{-1}x^{-1}y^{-1}x^{-1}x^{-1}z^{-1}x^{-1}z^{-1}x^{-1}z^{-1}$$
$$\cdot x^{-1}z^{-1}y^{-1}z^{-1}x^{-1}z^{-1}z^{-1}x^{-1}z^{-1}y^{-1})$$
$$\cdot (z^{-1}x^{1-}z^{-1}y^{-1}z^{-1}x^{-1}z^{-1}xyyxzxzyxzxzyzxzyxzxzzxzxyzxzxyxyzxz)$$

as a word in conjugates of these relators x^2, y^2, z^2. This expression has 19 conjugates of relators and can be expressed as follows:

$$(z^{-1}x^{-1}z^{-1})(x^{-2}) \cdot (z^{-1}x^{-1}z^{-1}y^{-1}x^{-1}y^{-1}x^{-1}y^{-1})(x^{-2})$$
$$\cdot(z^{-1}x^{-1}z^{-1}y^{-1}x^{-1}y^{-1}x^{-1}y^{-1}z^{-1}x^{-1}z^{-1}x^{-1}z^{-1}x^{-1}z^{-1}y^{-1}z^{-1}x^{-1})(z^{-2})$$
$$\cdot(z^{-1}x^{-1}z^{-1}y^{-1}x^{-1}y^{-1}x^{-1}y^{-1}z^{-1}x^{-1}z^{-1}x^{-1}z^{-1}x^{-1}z^{-1}y^{-1}z^{-1})(x^{-2})$$
$$\cdot(z^{-1}x^{-1}z^{-1}y^{-1}x^{-1}y^{-1}x^{-1}y^{-1}z^{-1}x^{-1}z^{-1}x^{-1}z^{-1}x^{-1}z^{-1}y^{-1})(z^{-2})$$
$$\cdot(z^{-1}x^{-1}z^{-1}y^{-1}x^{-1}y^{-1}x^{-1}y^{-1}z^{-1}x^{-1}z^{-1}x^{-1}z^{-1}x^{-1}z^{-1})(y^{-2})$$
$$\cdot(z^{-1}x^{-1}z^{-1}y^{-1}x^{-1}y^{-1}x^{-1}y^{-1}z^{-1}x^{-1}z^{-1}x^{-1}z^{-1}x^{-1})(z^{-2})$$
$$\cdot(z^{-1}x^{-1}z^{-1}y^{-1}x^{-1}y^{-1}x^{-1}y^{-1}z^{-1}x^{-1}z^{-1}x^{-1}z^{-1})(x^{-2})$$
$$\cdot(z^{-1}x^{-1}z^{-1}y^{-1}x^{-1}y^{-1}x^{-1}y^{-1}z^{-1}x^{-1}z^{-1}x^{-1})(z^{-2})$$
$$\cdot(z^{-1}x^{-1}z^{-1}y^{-1}x^{-1}y^{-1}x^{-1}y^{-1}z^{-1}x^{-1}z^{-1}x^{-1}y^{-1}z^{-1}x^{-1}z^{-1}x)(y^2)$$
$$\cdot(z^{-1}x^{-1}z^{-1}y^{-1}x^{-1}y^{-1}x^{-1}y^{-1}z^{-1}x^{-1}z^{-1}x^{-1}y^{-1}z^{-1}x^{-1}z^{-1})(x^2)$$
$$\cdot(z^{-1}x^{-1}z^{-1}y^{-1}x^{-1}y^{-1}x^{-1}zxzyxzx)(z^2) \cdot (z^{-1}x^{-1}z^{-1}y^{-1}x^{-1}y^{-1}x^{-1}zxzyxz)(x^2)$$
$$\cdot(z^{-1}x^{-1}z^{-1}y^{-1}x^{-1}y^{-1}x^{-1}zxzyx)(z^2)$$
$$\cdot(z^{-1}x^{-1}z^{-1}y^{-1}x^{-1}y^{-1}x^{-1}zxzy)(x^2) \cdot (z^{-1}x^{-1}z^{-1}y^{-1}x^{-1}y^{-1}x^{-1}zxz)(y^2)$$
$$\cdot(z^{-1}x^{-1}z^{-1}y^{-1}x^{-1}y^{-1}x^{-1}zx)(z^2) \cdot (z^{-1}x^{-1}z^{-1}y^{-1}x^{-1}y^{-1}x^{-1}z)(x^2)$$
$$\cdot(z^{-1}x^{-1}z^{-1}y^{-1}x^{-1}y^{-1}x^{-1})(z^2)$$

Thus word=(Expression for B)·A is the product we seek. This special treatment of relators of the form x^2 merits further consideration and will be the subject of further study.

This work is based on the author's PhD thesis at Colorado State University.

References

[1] G. Cooperman, L. Finkelstein, and N. Sarawagi. *Applications of cayley graphs.* Proceedings of the 8th International Symposium on Applied Algebra, Algebraic Algorithms and Error-Correcting Codes (1990), 367-378. MR1123964 (92f:68134)

[2] D F. Holt, B. Eick, and E A. O'Brien. *Handbook of computational group theory.* Chapman and Hall/ CRC. 2005. MR2129747 (2006f:20001)

[3] I. Kapovich and A. Myasnikov, *Stallings foldings and subgroups of free groups.* Journal of Algebra (2001), 608-668. MR1882114 (2003a:20039)

[4] R C. Lyndon and P E. Schupp, *Combinatorial group theory.* Springer-Verlag. 1977. MR0577064 (58:28182)

[5] C C. Sims. *Computation with finitely presented groups.* Cambridge University Press. 1994. MR1267733 (95f:20053)

[6] J R. Stallings, *Topology of Finite Graphs.* Invent. Math. 71 (1983), 551-565. MR695906 (85m:05037a)

DEPARTMENT OF MATHEMATICS, BENEDICTINE UNIVERSITY, LISLE, ILLINOIS 60532
E-mail address: `eziliak@ben.edu`